Advances in Process Systems Engineering: Selected Papers from China PSE Annual Meeting

Advances in Process Systems Engineering: Selected Papers from China PSE Annual Meeting

Guest Editors

Yufei Wang

Lei Zhang

MDPI

Basel • Beijing • Wuhan • Barcelona • Belgrade • Novi Sad • Cluj • Manchester

Guest Editors

Yufei Wang
State Key Laboratory of Heavy Oil Processing
China University of Petroleum
Beijing
China

Lei Zhang
School of Chemical Engineering
Dalian University of Technology
Dalian
China

Editorial Office
MDPI AG
Grosspeteranlage 5
4052 Basel, Switzerland

This is a reprint of the Special Issue, published open access by the journal *Processes* (ISSN 2227-9717), freely accessible at: https://www.mdpi.com/journal/processes/special_issues/4QEQ87F2J2.

For citation purposes, cite each article independently as indicated on the article page online and as indicated below:

Lastname, A.A.; Lastname, B.B. Article Title. *Journal Name* **Year**, *Volume Number*, Page Range.

ISBN 978-3-7258-5935-1 (Hbk)
ISBN 978-3-7258-5936-8 (PDF)
https://doi.org/10.3390/books978-3-7258-5936-8

Contents

About the Editors

Yufei Wang

Yufei Wang is a professor at the School of Chemical Engineering and Environment, China University of Petroleum (Beijing). Yufei Wang is an expert in Energy System and Process Integration. His research interests focus on energy system synthesis and renewable energy chemical engineering. For his excellent performance in academics, he received the 2nd Prize of the Nature & Science Award from the Chinese Ministry of Education. He has also authored or co-authored more than 100 papers in the leading refereed journals. He has edited three books focusing on energy saving and its application in the real industry and contributed 5 book chapters.

Lei Zhang

Lei Zhang is a professor at the School of Chemical Engineering, Dalian University of Technology. His current research interests include the development of methods and tools for chemical product and process design integrated with information and computing technologies, machine learning models, computational chemistry and molecular simulation. He received the National Science Fund for Excellent Young Scholars, the 2017 best paper award from the *Computers & Chemical Engineering* journal, the 2019 PSE young researcher award by PSE—Systems Engineering Society of China, and the 2023 RINENG Young Investigator Award. He was a postdoctoral research fellow at the Hong Kong University of Science and Technology and Technical University of Denmark, as well as a visiting scholar at Carnegie Mellon University. He received BS and PhD degrees from Tsinghua University in China.

processes

Editorial

Advances in Process Systems Engineering: Selected Papers from China PSE Annual Meeting

Yufei Wang [1,*] and Lei Zhang [2]

1 State Key Laboratory of Heavy Oil Processing, China University of Petroleum, Beijing 102249, China
2 School of Chemical Engineering, Dalian University of Technology, Dalian 116024, China; keleiz@dlut.edu.cn
* Correspondence: wangyufei@cup.edu.cn

The China Process Systems Engineering (PSE) Annual Conference is a series of academic annual conferences hosted by the Process Systems Engineering Professional Committee of the Chinese Society for Systems Engineering. This conference was held in 2024 in Dalian, China, where scholars from various universities in China's PSE field conducted in-depth exchanges and discussions on the theme of "PSE Assisting the Development of New Productivity". In the past year, Chinese PSE scholars have achieved significant results in the fields of "process simulation, analysis and optimization", "process control, scheduling and planning", and "process monitoring, risk analysis and safety". During the exchanges at the conference, it was observed that AI technology is currently being integrated more effectively with process systems, and by combining it with traditional algorithms, the modeling and computation processes have become more efficient [1,2]. Additionally, greater emphasis was placed on the integration of renewable energy with traditional chemical engineering, with the proposal of carbon reduction pathways for chemical processes from a systematic perspective [3,4]. This Special Issue, "Advances in Process Systems Engineering: Selected Papers from China PSE Annual Meeting", aims to show some recent progress on process systems engineering in the Chinese PSE community. This Special Issue is available online through the following link: https://www.mdpi.com/journal/processes/special_issues/4QEQ87F2J2 (accessed on 7 November 2025).

This Special Issue contains the latest research in key research direction of PSE, with a total of 11 research articles and 2 review articles. Among these articles, two articles focus on the recent advances in process simulation and analysis of different processes, including electrodialysis and gas turbines. Six articles focus on the recent developments in process optimization of different systems, such as solar PV deployment in manufacturing, diesel filtration in nuclear emergency systems, organic Rankine cycles, fresh agricultural product distribution, fluid catalytic cracking processes, and cogeneration systems. There are three articles that study process control and monitoring, such as distillation process control, fault judgment of catalyst loss, and corrosion state monitoring. The methods used in these works include traditional algorithms such as genetic algorithms and fuzzy algorithms, and many works have also adopted artificial intelligence algorithms to solve PSE problems. Since artificial intelligence methods are currently a popular research direction, one review article also delves into large language models. Additionally, another review article focuses on lubricant base oils and additives. Summaries of these studies follow.

In the direction of process simulation, Chen et al. (1) focused on the heat-stable salt removal process using electrodialysis technology and systematically examined the effects of different operating conditions. An industrial case is used to validate the process conditions. Meng et al. (2) proposed a method which combines random forest feature selection and the chaos game optimization algorithm to predict NOx in gas turbines; the results show that the random forest algorithm can select appropriate input variables, and the proposed method can extract the intrinsic links of the data and build a more accurate NOx prediction model. In the direction of process optimization, Zhang et al. (3) proposed a multi-objective evaluation method for working fluids selection in an organic Rankine cycle based on data envelopment analysis. This method takes into account the thermodynamic performance of the working fluid in the organic Rankine cycle, economic aspects, and environmental considerations. A total of 62 different working fluids were evaluated in the integrated technology. Briceño (4) introduces an integrated decision-making method combining a morphological matrix and fuzzy TOPSIS to systematically select and rank optimal PV system configurations. The results demonstrate better robustness compared to existing fuzzy TOPSIS-based methods for solar PV applications. Zhang et al. (5) proposes an efficient optimization framework that integrates genetic algorithm optimization, state-space networks, and computational fluid dynamics to enhance the reliability of emergency diesel generator systems and support nuclear safety operations. Zhu et al. (6) establishes an optimization model of cold chain logistics distribution routes for fresh agricultural products. In their work, a two-objective optimization model is proposed, considering customer satisfaction maximization and comprehensive cost minimization, and then a hybrid ant colony algorithm is designed to solve the model. The results show that the method can effectively optimize the cold chain logistics distribution routes of fresh agricultural products and reduce carbon emission and distribution cost. Yang et al. [5] proposed a novel solid oxide fuel cell-based cogeneration system integrating with an organic Rankine cycle for waste heat recovery. In this work, technical–economic and parametric analyses are conducted, and a multi-objective optimization is carried out. This work provides practical reference and pragmatic guidance for the integration of SOFC-based cogeneration systems. Li et al. [6] propose an integrated framework that combines hybrid modeling and surrogate model-based optimization for the operation optimization of a fluid catalytic cracking process. This approach combines plant and simulation data to train a multi-task learning prediction model, which then serves as a surrogate for operational optimization. The results show that the model predicts product yields with an error margin of under 4.84% and the optimization can achieve a 3.67% increase in product revenues. In the direction of process control and monitoring, Wang et al. (7) propose a neural network model based on process mechanism knowledge to enhance the prediction accuracy and interpretability of the distillation processes predictive control model. The proposed model is used to conduct the dynamic modeling of a benzene–toluene distillation column. Li et al. [7] proposed a fault judgment method of catalyst loss failures with quantitative criteria via a fault tree analysis method. In this method, the relationship model between flow field signals and faults in the FCC disengager is investigated by computational fluid dynamics. The method has been applied in cases of catalyst loss in two industrial disengagers and can accurately pinpoint the main factors leading to catalyst loss faults. Wang et al. (8) proposed a new deep learning framework to synergistically integrate acoustic emission and electrochemical noise signals at the algorithmic level for corrosion monitoring. In their work, a new feature alignment loss function and a probability distribution consistency loss function are designed to facilitate more effective feature learning to improve classification performance. The experimental results demonstrate that the method achieves superior accuracy in corrosion stage classification compared to other state-of-the-art models. This approach combines plant and simulation data to train a multi-task learning prediction model, which then serves as a surrogate for operational optimization. There are two review articles in this Special Issue. Khoo et al. (9) provides a comprehensive overview of the evolution and application of artificial intelligence and large language models in engineering, with a specific focus on chemical engineering. This review addresses the challenges and future considerations of integrating large language models into engineering workflows, emphasizing the need for domain-specific adaptations, ethical guidelines, and robust validation frameworks. Zhou et al. (10) provides a review and summary of the entire workflow for molecular simulations of lubricating oils, ranging from molecular modeling, with particular emphasis on the molecular representation of base oils, to simulation calculation methods and result analysis.

The Guest Editors would like to express their gratitude to all the authors for their high-quality manuscripts. We hope that this Special Issue will show the recent developments in the Chinese PSE community.

Author Contributions: Writing—original draft preparation, Y.W.; writing—review and editing, L.Z. All authors have read and agreed to the published version of the manuscript.

Conflicts of Interest: The authors declare no conflict of interest.

List of Contributions:

1. Chen, G.; Liu, Q.; Liu, L.; Zhang, S.; Li, G.; Li, H.; Wang, D. Research on the Process for Removing Heat-Stable Salts from Organic Amine Absorbents via Electrodialysis. *Processes* **2025**, *13*, 2519.
2. Meng, X.; Li, X.; Chen, J.; Fu, Y.; Zhang, C.; Nazir, M.S.; Peng, T. An Innovative NOx Emissions Prediction Model Based on Random Forest Feature Selection and Evolutionary Reformer. *Processes* **2025**, *13*, 107.
3. Zhang, L.; Wang, L.; Sun, X.; Xia, L.; Tao, S.; Xiang, S.; Jin, S. Multi-Objective Evaluation Strategy Based on Data Envelopment Analysis for Working Fluid Selection in the Organic Rankine Cycle. *Processes* **2025**, *13*, 1013.
4. Briceño, C.P.; Ponce, P.; Fayek, A.R.; Anthony, B.; Bradley, R.; Peffer, T.; Meier, A.; Mei, Q. Optimizing Solar PV Deployment in Manufacturing: A Morphological Matrix and Fuzzy TOPSIS Approach. *Processes* **2025**, *13*, 1120.
5. Zhang, L.; He, Y.; Zhou, Y.; Jiang, G.; Chu, X.; Ma, Q.; Liu, F.; Ye, H. A Comprehensive Optimization Framework for Diesel Filtration in Nuclear Emergency Systems: Integrating Genetic Algorithms, State-Space Networks, and Computational Fluid Dynamics. *Processes* **2025**, *13*, 648.
6. Zhu, X.; Liang, Y.; Wu, C.; Xiao, Y. Vehicle Routing Perfection for Fresh Agricultural Products Distribution Under Carbon Emission Regulation and Customer Satisfaction. *Processes* **2025**, *13*, 605.
7. Wang, Z.; Wang, H.; Du, Z. Deep Neural Network Model Based on Process Mechanism Applied to Predictive Control of Distillation Processes. *Processes* **2025**, *13*, 811.
8. Wang, R.; Shan, G.; Qiu, F.; Zhu, L.; Wang, K.; Meng, X.; Li, R.; Song, K.; Chen, X. Corrosion State Monitoring Based on Multi-Granularity Synergistic Learning of Acoustic Emission and Electrochemical Noise Signals. *Processes* **2024**, *12*, 2935.
9. Khoo, T.L.; Lee, T.S.; Bee, S.-T.; Ma, C.; Zhang, Y.-Y. A Comparative Review of Large Language Models in Engineering with Emphasis on Chemical Engineering Applications. *Processes* **2025**, *13*, 2680.
10. Zhou, K.; Che, X.; Wei, C.; Tang, Z.; Yu, H.; Wang, D.; Wang, J.; Zhang, L. The Molecular Modeling, Simulation, and Design of Base Oils and Additives in Lubricating Oils: A Review. *Processes* **2024**, *12*, 2407.

References

1. Li, H.; Qiu, T.; You, F. AI-Based Optimal Control of Fed-Batch Biopharmaceutical Process Leveraging Deep Reinforcemen Learning. *Chem. Eng. Sci.* **2024**, *292*, 119990. [CrossRef]
2. Chen, Z.; Xie, Y.; Xu, C.; Zhang, L. Scale-up of complex molecular reaction system by hybrid mechanistic modeling and dee transfer learning. *Nat. Commun.* **2025**, *16*, 8905. [CrossRef] [PubMed]
3. Wu, Y.; Zhao, T.; Tang, S.; Wang, Y.; Ma, M. Research on design and multi-frequency scheduling optimization method for flexibl green ammonia system. *Energy Convers. Manag.* **2024**, *300*, 117976. [CrossRef]
4. Wen, L.; Huang, C.; Dai, Z.; Nie, L.; Ji, X.; Dai, Y. Dynamic Simulation and Optimization for Load Regulation Strategies of th Green Ammonia Synthesis Process. *Ind. Eng. Chem. Res.* **2024**, *63*, 19662–19675. [CrossRef]
5. Yang, S.; Liang, A.; Jin, Z.; Xie, N. Investigation of a Cogeneration System Combining a Solid Oxide Fuel Cell and the Organi Rankine Cycle: Parametric Analysis and Multi-Objective Optimization. *Processes* **2024**, *12*, 2873. [CrossRef]
6. Li, H.; Zhao, Q.; Wang, R.; Xu, W.; Qiu, T. Integrated Hybrid Modelling and Surrogate Model-Based Operation Optimization c Fluid Catalytic Cracking Process. *Processes* **2024**, *12*, 2474. [CrossRef]
7. Li, Y.; Zhao, Y.; Li, Z.; Liu, N.; Zhu, C.; Li, S.; Shi, X.; Wang, C.; Lan, X. A Fault Judgment Method of Catalyst Loss in FC Disengager Based on Fault Tree Analysis and CFD Simulation. *Processes* **2025**, *13*, 464. [CrossRef]

processes

MDPI

Article

Integrated Hybrid Modelling and Surrogate Model-Based Operation Optimization of Fluid Catalytic Cracking Process

Haoran Li [1,†], Qiming Zhao [1,†], Ruqiang Wang [2], Wenle Xu [1] and Tong Qiu [1,*]

1 Department of Chemical Engineering, Institute of Process Systems Engineering, Tsinghua University, Beijing 100084, China; li-hr20@mails.tsinghua.edu.cn (H.L.); zqm21@mails.tsinghua.edu.cn (Q.Z.); xwl23@mails.tsinghua.edu.cn (W.X.)
2 PetroChina Planning and Engineering Institute, Beijing 100083, China; wangruqiang@petrochina.com.cn
* Correspondence: qiutong@tsinghua.edu.cn
† These authors contributed equally to this work.

Abstract: Fluid Catalytic Cracking (FCC) is one of the most important conversion processes in oil refineries, widely used to convert high-boiling, high-molecular-weight hydrocarbon components from crude oil into more valuable products like gasoline and diesel. Advanced simulation and optimization technologies are critical for improving the operational efficiency and economic performance of the FCC process. First-principles-based simulators rely on parameter estimation and are computationally intensive, making them unsuitable for online optimization. In recent years, with the development of deep learning, data-driven models have made significant progress in FCC modeling. However, due to their black-box nature and difficulty with extrapolation, they are rarely used for optimization. To bridge this gap, we propose an integrated framework that combines hybrid modeling and surrogate model-based optimization. This approach combines plant and simulation data to train a multi-task learning prediction model, which then serves as a surrogate for operational optimization. Validated on a large-scale FCC unit in southern China, the model predicts product yields with an error margin of under 4.84% for all products. Following optimization, yields of LNG, gasoline, and diesel rose by an average of 0.10 wt%, 1.58 wt%, and 1.05 wt%, respectively, resulting in a 3.67% increase in product revenues. This highlights the substantial potential of this framework for industrial applications.

Keywords: deep learning; optimization; catalytic cracking; multi-task learning

1. Introduction

Fluid Catalytic Cracking (FCC) is one of the key processes in the refining industry, transforming heavy fractions of crude oil into lighter and more valuable products such as gasoline and diesel [1]. During this high-temperature process, catalysts break down long-chain hydrocarbons into shorter ones [2]. FCC enhances fuel yields for refineries, playing an essential role in fulfilling the market demand for light hydrocarbon fuels [3]. In the petrochemical industry, the efficiency of the FCC process and the quality of its products directly impact the economic benefits and market competitiveness of refining enterprises [4]. Due to the complex and variable composition of crude oil and constantly changing demands for various products, precise simulation and optimization of the FCC process are extremely important [5].

In the era of smart manufacturing, advanced modeling and optimization are essential for the FCC process [6]. Process modeling helps engineers understand complex chemical reactions and fluid dynamics [7], which is significant for implementing process improvements, enhancing product quality, conserving energy and raw materials, and reducing operational costs [8]. Moreover, optimizing the FCC process ensures the production of products that better meet fluctuating market demands and operate in a more environmentally friendly manner [9], thus bringing significant economic and social benefits [10].

Classic methods for modeling the FCC process include rigorous mechanistic models founded on first principles. A major group of models are lumped kinetics models, which simplify complex reaction networks by grouping similar chemical species into "lumps" or pseudo-components [11]. The earliest model is a 3-lump model proposed by Weekman et al. [12,13] to predict the feedstock conversion rate and gasoline yield. Later, Arbel et al. reported a 10-lump model that allows feed and catalyst characterization based on laboratory experiments [14]. More recently, Ebrahimi et al. constructed a 9-lump model containing 18 kinetic parameters and 2 parameters for the catalyst de-activation for the vacuum gas oil (VGO) catalytic cracking process [15]. Another group of models is molecular kinetic models. Yang et al. presented a structure-oriented lumping (SOL) combined with the Monte Carlo (MC) method to simulate the secondary reaction process of FCC [16].

Kinetic parameters are crucial for the prediction accuracy of such models. Therefore, lots of researchers have explored parameter estimation methods for constructing mechanistic models. For example, John et al. reported a 6-lump model in which frequency factors, activation energies, and heat of reactions for the catalytic cracking kinetics and a number of model parameters are estimated using a model-based parameter estimation technique along with data from an industrial FCC unit [17]. Lee et al. proposed an approximate approach based on transition state theory for kinetic modeling of the catalytic cracking of paraffinic naphtha and estimated the parameters by a genetic algorithm [18]. Yang and Wang proposed a P system-based hybrid optimization algorithm (PHOA) to estimate the parameters of the FCC reactor–regenerator model [19]. Du et al. proposed an integrated model based on the equivalent reactor network (ERN) involving 8-reactor and 10-reactor networks to characterize complex hydrodynamics within the regenerator and the riser reactor [20]. Palos et al. presented a 6-lumped kinetic model describing the FCC process of tire pyrolysis oil (TPO) [21]. Nazarova et al. reported a prediction scheme of FCC units under feedstock base expansion by using oil fractions with a higher boiling point [22]. Based on the SOL method, Qin et al. constructed a molecular-level FCC-Gasoline Hydrotreating (GH) process coupling model containing 96 FCC reaction rules, 24 GH reaction rules, and about 120,000 reactions [23]. Chen et al. presented a heavy petroleum FCC process model based on the use of a hybrid structural unit and bond–electron matrix (SU-BEM) framework on a molecular level [24].

Due to the accurate simulation of mechanism models requiring extensive parameter calibration, it is necessary to recalibrate the parameters when the properties of the raw materials or operating conditions change, which demands significant effort. As a result, the applicability of mechanism models is limited. Moreover, these models are often highly complex, and their online deployment involves substantial computational load.

In recent years, with the development of deep learning and big data techniques, data-driven models have gained attention for FCC process modeling due to their powerful fitting capabilities and independence from parameter calibration. Chen et al. employed an adaptive immune genetic algorithm (AIGA) for variable screening and established a random forest (RF) model for the FCC process [25]. Long et al. proposed a hybrid approach that integrates the least absolute shrinkage and selection operator (LASSO) method for variable selection and an output-focused back-propagation neural network (BPNN) method [26]. Tian et al. designed a data-driven and knowledge-based fusion approach to predict the future trend of the key variable [27]. Yang et al. developed a hybrid predictive framework for FCC by integrating a data-driven deep neural network with a physically meaningful lumped kinetic model [28].

Despite the successful application of data-driven models in FCC process modeling, their direct application in optimization practice has yet to be realized due to several challenges. The first reason is the difficulty of extrapolation. Although data-driven models can perform well on training and test datasets, the response surface and gradient information of input variables may be unreasonable, potentially causing vanishing and exploding gradients. Therefore, ensuring the rationality of optimization results is challenging. Secondly, due to their black-box nature, the predictive outcomes lack interpretability. This reduces engineers' confidence in using these models for optimization in scenarios such as chemical manufacturing, where safe and stable operation is highly required.

To fill this gap, this work aims to develop an integrated framework for hybrid modeling and optimization of the FCC process by efficiently constructing accurate surrogate models that combine data with mechanistic knowledge and embedding these models into the optimization process to achieve yield optimization.

The key novelties of this study include the following:

- Proposed a unified framework for multi-source data collection, modeling, and optimization by integrating mechanistic model data with actual plant data;
- Constructed a multi-task learning prediction model to balance the patterns contained in different datasets, enhancing prediction accuracy and generalization ability;
- Formulated a non-linear programming (NLP) optimization model embedded with a data-driven surrogate model to improve gasoline yield in the FCC process.

The rest of this paper is organized as follows: Section 2 outlines the methodology of the proposed integrated hybrid modeling and optimization framework. Section 3 presents the results and discussions of the case study. Finally, the conclusions are provided in Section 4.

2. Materials and Methods

An integrated hybrid modeling and optimization framework is presented in Figure 1, consisting of three key modules: the hybrid data collection module, the multi-task learning prediction module, and the surrogate model-based optimization module. First, data from the plant and its simulation program are collected. Next, a multi-task learning model is trained on this hybrid dataset to predict product yields. Finally, the trained model is incorporated as a surrogate within an optimization framework to optimize the process's operating conditions for higher economic performance. The following subsections provide detailed explanations of these modules.

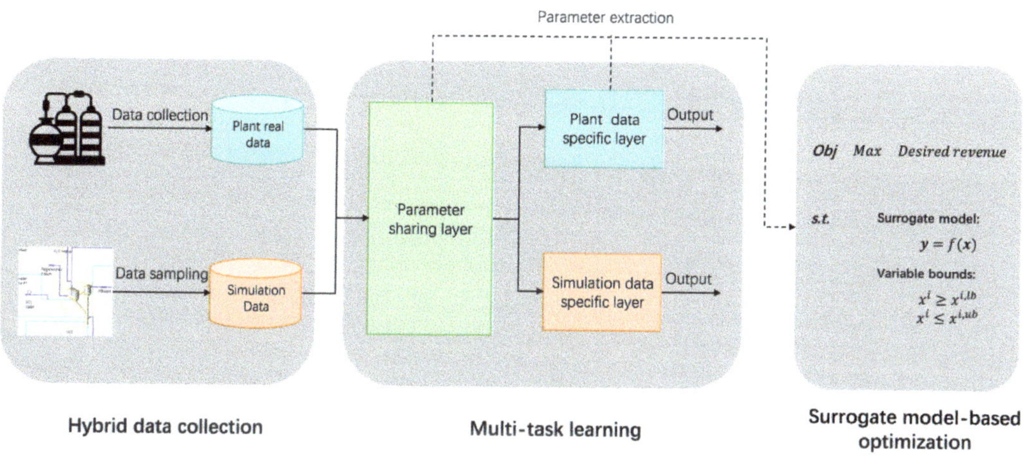

Figure 1. Scheme of the integrated hybrid modelling and optimization framework.

2.1. Hybrid Data Collection

The relative importance of process input and output variables was established through iterative communication with on-site engineers, leveraging their operational knowledge. From dozens of candidate process variables, twelve key process inputs and five key yields were identified from the reactor–regenerator section, the main fractionator, and the gas plant. Additionally, two laboratory test results, x_{13} and x_{14}, were included to capture fluctuations in feedstock properties. The sixth output variable y_6 represents the residue yield as a calculated mass balance term that represents coke formation and total substance loss.

The first step in gathering the dataset for modeling and optimization is selecting key process variables. The relative importance of process input and output variables was established through iterative communication with on-site engineers, leveraging their operational knowledge. From hundreds of candidate process variables, twelve key process inputs x_1 through x_{12} and six key yields y_1 through y_6 were identified from the reactor–regenerator unit and subsequent separation units. Additionally, two feedstock oil property-related variables, x_{13} and x_{14}, were included to capture fluctuations in feedstock properties. Density at 20 °C (x_{13}) is a fundamental property of petroleum fractions that indicates molecular composition. A higher density within a given boiling range correlates to lower alkane content in the feedstock, leading to higher proportions of cycloalkanes and aromatics in the products. Carbon residue content (x_{14}) quantifies the feedstock's propensity for non-catalytic coke formation. The residue, primarily composed of condensed polycyclic aromatic hydrocarbons, serves as a precursor to coke deposits. Conradson carbon tests are carried out to predict direct catalyst deposition from feed components. The variables selected are listed in Table 1.

Table 1. Key process input and output variables of interest after variable selection.

Notation	Variable	Unit	Notation	Variable	Unit
x_1	Reactor temperature	°C	y_1	Dry gas yield	wt%
x_2	Regenerator temperature	°C	y_2	Liquefied gas yield	wt%
x_3	Feed preheat temperature	°C	y_3	Gasoline yield	wt%
x_4	Feed flowrate	t/h	y_4	Diesel yield	wt%
x_5	Reflux flowrate	t/h	y_5	Slurry oil yield	wt%
x_6	Catalyst tank level	t	y_6	Residue yield	wt%
x_7	Fractionation tower cold reflux flow	t/h			
x_8	Light cycle oil extraction temperature	°C			
x_9	Fractionation tower bottom temperature	°C			
x_{10}	Supplemental absorbent flow rate	t/h			
x_{11}	Desorption tower top gas volume	Nm^3/h			
x_{12}	Desorption tower top temperature	°C			
x_{13}	Mixed feedstock density (20 °C)	kg/cm^3			
x_{14}	Mixed feedstock carbon residue content	wt%			

After choosing the variables, a plant dataset was collected from the plant's Distributed Control System (DCS), Manufacturing Execution System (MES), and Laboratory Information Management System (LIMS). The collected data span six months, containing two non-continuous three-month periods, March to May and July to September 2023. The variables of x_1 through x_{12} and y_1 through y_6 were gathered from the DCS and MES, while x_{13} and x_{14} were extracted from LIMS record sheets.

To augment the plant dataset mentioned above, a simulation program has been built using Petro-SIM v7.3 (KBC Advanced Technologies Ltd., Walton-on-Thames, UK) [29], a commercial software developed by KBC Advanced Technologies. Petro-SIM comprises several sub-modules, including FCC-SIM for catalytic cracking, HCR-SIM for hydrocracking, and NHTR-SIM for catalytic reforming. FCC-SIM incorporates comprehensive feedstock characterization and detailed kinetic models of risers, reactors, and regenerators. It is extensively utilized for process design, operational optimization, and production planning.

We first built an FCC-SIM model, with its process flow diagram (PFD) shown in Figure 2. We utilized the built-in 'refinery-large' component list, which includes 106 components such as hydrogen, carbon monoxide, and lighter hydrocarbons ($C < 6$). Additionally, hypothetical components were included based on the distillation curve at 10 °C intervals, ranging from 40 °C to 850 °C. The Peng–Robinson method was employed, as it is well-suited for property calculations in the FCC process. Then, operational data from the plant were used to calibrate this model. The overall feed masses and compositions were verified through mass balance criteria. After adjustments, these data were used in the calibration module to compute various parameters and calibration factors, such as reaction kinetics parameters.

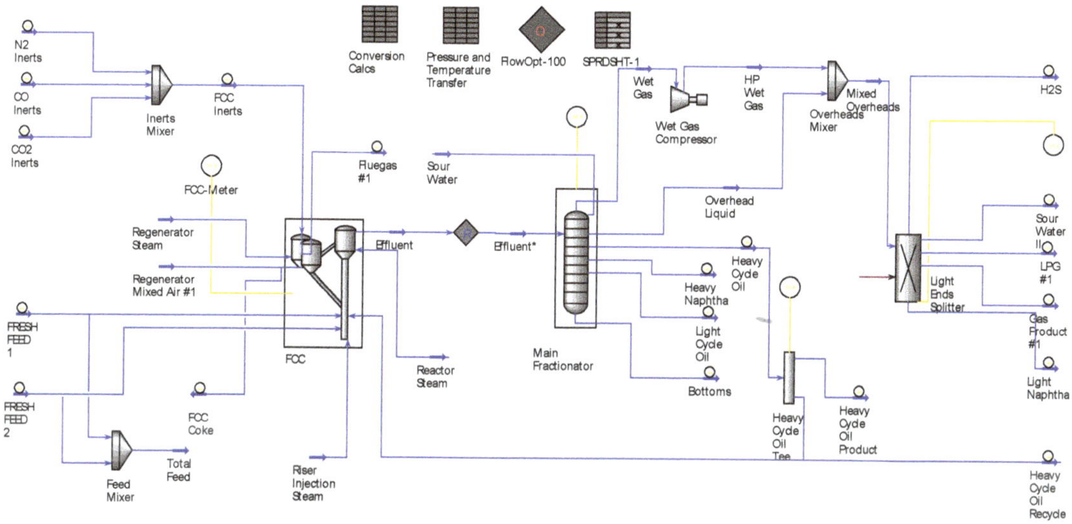

Figure 2. Process flow diagram of the FCC process simulation in Petro-Sim.

After obtaining the model calibrated to plant data, we generated the simulation dataset through sampling across different operation conditions. A Python v3.10 script was developed to commute with the Component Object Model (COM) in Petro-SIM and automatically changed the key variables of reactor temperature, regenerator temperature, and feed preheat temperature in ranges defined in Table 2. To explore the ranges and possible combinations of input variables more effectively, the Latin Hypercube Sampling (LHS) [30] method was adopted.

Table 2. The operational variables and their corresponding adjustment range.

Variable	Unit	Lower Bound	Upper Bound
Reactor temperature	°C	505	528
Regenerator temperature	°C	668	685
Feed preheat temperature	°C	220	240

2.2. *Multi-Task Learning Prediction Model*

The hybrid dataset covers a broader range of operating conditions compared to the pure plant dataset, thus providing a solid foundation for modeling. To make the best use of these data and precisely capture the mapping relationship between the feedstock properties and operation conditions (inputs) and product yields (output), considering the different distribution ranges of the two datasets, a multi-task learning model was proposed, as shown in Figure 3. The model is a deep neural network with two components: the parameter-sharing backbone and the task-specific heads.

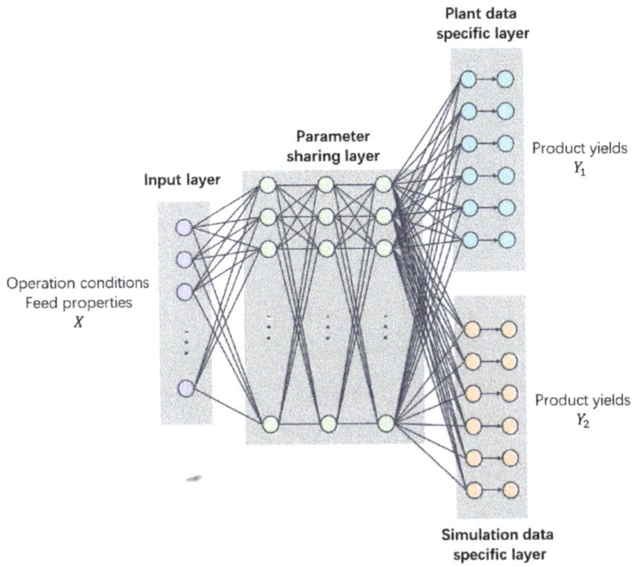

Figure 3. Structure of the multi-task learning model.

The parameter-sharing backbone has three layers of neurons that extract features common to hybrid inputs. The forward computation of these fully connected layers is shown in Equations (1) and (2).

$$a^0 = x \quad (1)$$

$$z^l = W^l a^{l-1} + b^l,\ l = 1,2,3 \quad (2)$$

where $W^l \in \mathbb{R}^{m_l \times m_{l-1}}$ is the weight matrix of the l-th layer, m_l is the number of neurons in the l-th layer, $a^{l-1} \in \mathbb{R}^{m_{l-1}}$ is the activation vector (i.e., output) of the l-1-th layer, $b^l \in \mathbb{R}^{m_l}$ is the bias vector of the l-th layer, and z^l is the linear combination result for the l-1-th layer neurons.

$$a^l = \sigma\left(z^l\right),\ l = 1,2,3 \quad (3)$$

where $\sigma(\cdot)$ is the activation function, applied element-wise to the vector z^l (Equation (3)). Here, SiLU is used as the activation function, which is continuous and differentiable, as shown in Equation (4).

$$SiLU\left(z^l\right) = z^l \cdot \frac{1}{1 + e^{-z^l}} \quad (4)$$

where after feature extraction, the task-specific part adopts specific output layers to predict the outputs for the plant data and simulation data, respectively, as expressed in Equations (5) and (6).

$$z_1^L = W^{L_1} a^{L-1} + b^{L_1} \quad (5)$$

$$z_2^L = W^{L_2} a^{L-1} + b^{L_2} \quad (6)$$

Finally, a SoftMax layer is added to normalize the output and compute the final prediction for the desired product yields, as shown in Equation (7).

$$y_1, y_2 = SoftMax\left(z_1^L, z_2^L\right) \quad (7)$$

During training, the loss function is designed as the weighted sum of mean squared error (MSE) loss for the prediction of y_1 and y_2, as shown in Equation (8).

$$Loss = \sum_i \frac{1}{N_1}(y_1 - \widehat{y_1})^2 + \lambda \sum_j \frac{1}{N_2}(y_2 - \widehat{y_2})^2 \tag{8}$$

where y_1, y_2 are the prediction values, $\widehat{y_1}, \widehat{y_2}$ are the true values, N_1, N_2 are the batch sizes for the two training datasets and λ is the weight to balance the learning performance of the prediction task for plant data and simulation data.

Before training, the plant and simulation datasets were divided into training, validation, and test sets in a 7:2:1 ratio. During the training phase, the batch size for the plant data and the simulation data was set proportionally according to the size of each dataset to ensure that the number of batches was the same.

2.3. Surrogate Model-Based Optimization Model

The trained multi-task prediction model is regarded as a surrogate model and represents the input–output mappings of the FCC process. With this model, we formulated a non-linear programming (NLP) optimization model for the FCC process.

The objective is to maximize the revenue from the desired product output (i.e., LNG, gasoline, diesel) per unit of feedstock oil, as expressed in Equation (9).

$$\boldsymbol{Max}\ obj = \frac{x_4(P_1y_1 + P_2y_2 + P_3y_3)}{100} \tag{9}$$

where $P_1,\ P_2,\ P_3$ are the prices of LNG, gasoline, and diesel, and are set at 5000, 9000, and 8000 CNY/t. The trained weight matrices of $\boldsymbol{W}^1, \boldsymbol{W}^2, \boldsymbol{W}^3, \boldsymbol{W}^{L_1}$ and biases of $\boldsymbol{b}^1, \boldsymbol{b}^2, \boldsymbol{b}^3, \boldsymbol{b}^{L_1}$ are extracted from the prediction model and the explicit form of the neural network is embedded as equality constraints in the optimization model, as shown in Equations (10)–(13).

$$h^1_{j_1} = \frac{\sum_i \left(W^1_{i,j_1} x_i + b^1_{j_1}\right)}{1 + e^{-\sum_i (W^1_{i,j_1} x_i + b^1_{j_1})}} \tag{10}$$

$$h^2_{j_2} = \frac{\sum_{j_1} \left(W^2_{j_1,j_2} h^1_{j_1} + b^2_{j_2}\right)}{1 + e^{-\sum_{j_1} (W^2_{j_1,j_2} h^1_{j_1} + b^2_{j_2})}} \tag{11}$$

$$h^3_{j_3} = \frac{\sum_{j_2} \left(W^2_{j_2,j_3} h^2_{j_2} + b^3_{j_3}\right)}{1 + e^{-\sum_{j_2} (W^2_{j_2,j_3} h^2_{j_2} + b^3_{j_3})}} \tag{12}$$

$$y_k = 100 * SoftMax\left(\sum_{j_3} W^{L_1}_{j_3,k} h^3_{j_3} + b^{L_1}_{j_4}\right) \tag{13}$$

where i is the index of input, $j_1,\ j_2,\ j_3$ are the indices of the neurons in the hidden layers (set to 256) and k is the index of the output.

The optimization variables are reactor temperature (x_1), regenerator temperature (x_2), and preheat temperature (x_3), all of which are constrained by respective upper bounds and lower bounds, as shown in Equations (14) and (15).

$$x_1,\ x_2, x_3 \geq x_1^{lo},\ x_2^{lo}, x_3^{lo} \tag{14}$$

$$x_1,\ x_2, x_3 \leq x_1^{up}, x_2^{up}, x_3^{up} \tag{15}$$

For on-site operational optimization, it is important to ensure operational continuity and consider integration with advanced process control (APC) systems by keeping the optimization space manageable. After discussions with plant operators, we selected an optimization range of ±5 °C for both reactor temperature and feed preheat temperature, and ±3 °C for regenerator temperature. The optimization program is implemented in

GAMS v38.2 on an Intel Core i7-9700CPU @ 3.00 GHz CPU with 16.0 GB RAM and optimized by the IPOPT [31] solver.

3. Results

3.1. Plant and Simulation Dataset Distribution

Following the methodology outlined in Section 2.1, a hybrid dataset was created, consisting of a plant dataset with 264,964 records spanning March to May and July to September 2023, along with a simulation dataset of 8760 records. Although the plant dataset is large, its coverage in the space spanned by key independent variables is limited. As shown in Figure 4, we present the distribution of the plant and simulation datasets across three variable dimensions. It can be observed that the plant dataset is concentrated in the central region of the cube. In contrast, the simulation dataset, sampled from the simulation program, nearly covers the entire area of the cube.

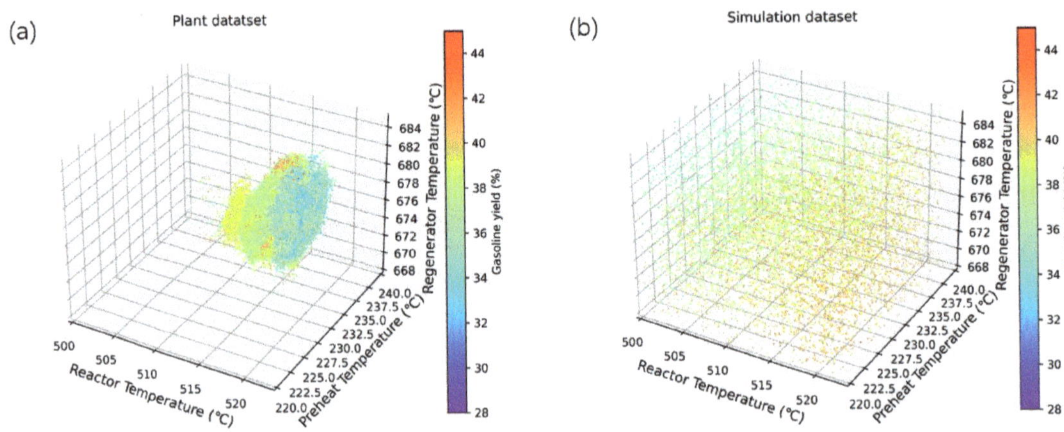

Figure 4. Data distribution of the (**a**) plant dataset and (**b**) simulation dataset.

3.2. Baseline Pure Data-Driven Model Prediction Results

As a baseline comparison, we adopt a pure data-driven model obtained using the plant dataset to train a deep neural network with three layers of hidden neurons. The input layer is x_1 to x_{14} and the output is y_1 to y_6 as defined in Table 1. The activation function is SiLU, and a SoftMax layer is added before the output. Figure 5 illustrates the trend of the loss function during the training process. As shown in the figure, although the loss continues to decrease, the training loss curve exhibits several sharp spikes, indicating instability in the training process and suggesting a risk of overfitting. The model is trained for 500 epochs.

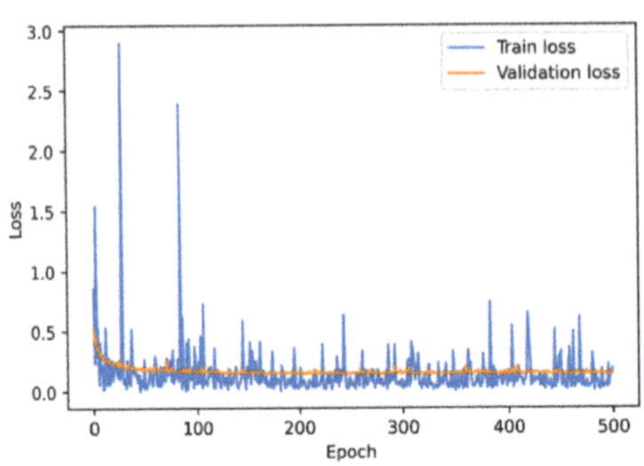

Figure 5. Train and validation loss of the pure data-driven model.

After training, the model is used to predict the product yields on both the test plant dataset and the simulation dataset, with the results shown in Figure 6. As illustrated, the model performs well on the plant dataset (cyan dots), with prediction points closely aligning with the diagonal line, and most predictions falling within ±10% of the actual data. However, the model's performance deteriorates significantly on the simulation dataset (pink dots), where the predictions are highly inaccurate.

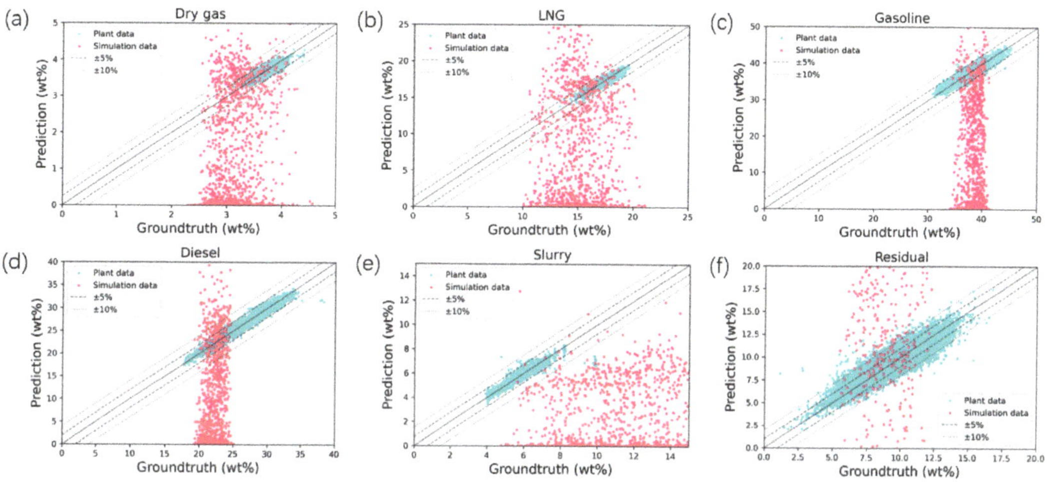

Figure 6. Prediction results of the pure data-driven model on the test datasets for (**a**) dry gas, (**b**) LNG, (**c**) gasoline, (**d**) diesel, (**e**) slurry, and (**f**) residual.

Table 3 demonstrates metrics comparisons of the prediction performance of this model on the plant dataset and simulation dataset. The purely data-driven model achieves high prediction accuracy on the plant dataset, with MSE ranging from 0.0029 to 0.3712 and mean absolute percentage error (MAPE) between 0.82% and 4.90%. However, on the simulation dataset, the predictions become highly inaccurate, with MSE reaching 6.3814 to 5413.0770 and MAPE soaring to 72.11% to 715.72%, indicating a failure in prediction on the simulation dataset.

Table 3. Prediction precision results of the pure data-driven model.

Yield (wt%)	Plant Dataset		Simulation Dataset	
	MSE	**MAPE**	**MSE**	**MAPE**
Dry gas	0.0029	1.15%	6.3814	72.11%
LNG	0.0337	0.82%	173.6118	78.71%
Gasoline	0.2757	1.08%	1073.6084	78.17%
Diesel	0.1372	1.02%	383.3285	78.65%
Slurry	0.0139	1.37%	128.6065	85.34%
Residual	0.3712	4.90%	5413.0770	715.72%

This phenomenon is because the purely data-driven model was trained solely on the plant dataset, which, as revealed earlier in Figure 4, has limited coverage of the independent variable space. Consequently, when this model is used to predict points in the simulation dataset that fall outside its data distribution range, it results in extrapolation failure.

The phenomenon of extrapolation failure can be further elucidated through trend surface analysis of the dependent variable with respect to the independent variables. Figure 7 displays the trend surface of gasoline yield as it varies with the three independent variables of reactor temperature, regenerator temperature, and feed preheat temperature at the first data point of the plant dataset. There are areas on the surface where the gasoline yield is 0, indicating regions not covered by the data distribution during the model training process. In these areas, extrapolation failure occurs, and gradient explosion phenomena can be observed in the positions adjacent to these regions. If this model is used for optimization, it would also present insurmountable challenges for the optimization solution.

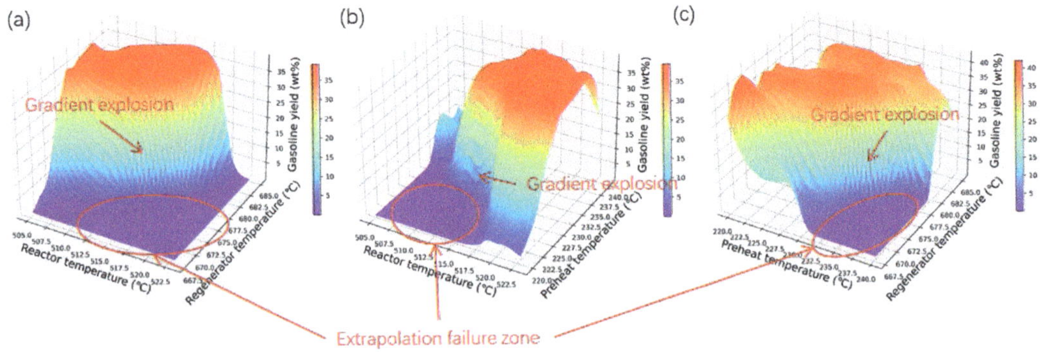

Figure 7. Trend surface of gasoline yield with respect to (**a**) reactor temperature and regenerator temperature, (**b**) reactor temperature and preheat temperature, and (**c**) regenerator temperature and preheat temperature in the pure data-driven model.

3.3. Multi-Task Model Prediction Results

A multi-task learning model proposed in Section 2.2 was trained using both the plant and the simulated dataset. The loss function curve during training is shown in Figure 8. As shown in the figure, unlike the pure data-driven model, the loss function decreases steadily as the training progresses. After 500 epochs, the loss functions for both the training and validation sets have converged.

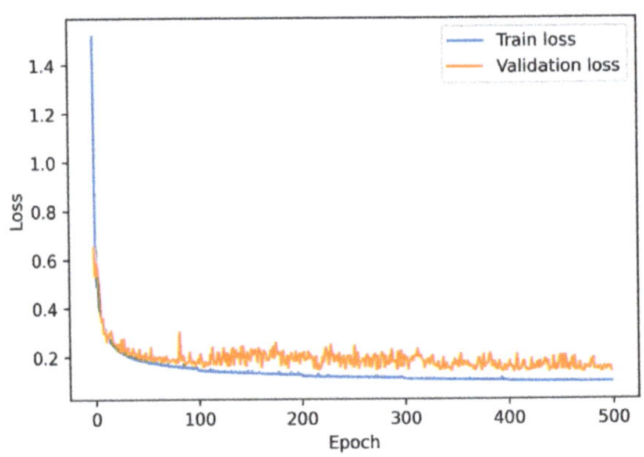

Figure 8. Train and validation loss of the multi-task learning model.

After training, the multi-task learning model is tested on the plant and simulation dataset, and the prediction results are shown in Figure 9. For all the product yields, the prediction results are around the diagonal lines for the plant and simulation datasets, suggesting satisfactory performances on both datasets.

Table 4 provides a detailed breakdown of the MSE and MAPE for the prediction results. For the plant dataset, the MSE for product yields ranges from 0.0032 to 0.3621, with MAPE between 0.98% and 4.84%. For the simulation dataset, the MSE ranges from 0.0002 to 0.0055, and MAPE from 0.09% to 0.47%. These results indicate that the multi-task learning model performs well on both the plant and simulation dataset prediction tasks.

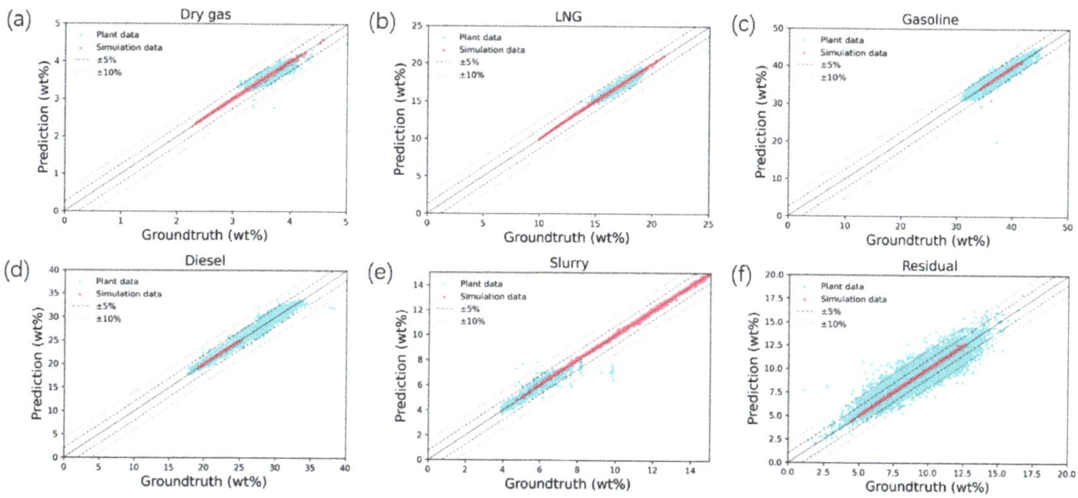

Figure 9. Prediction results of the multi-task learning model on the test datasets for (**a**) dry gas, (**b**) LNG, (**c**) gasoline, (**d**) diesel, (**e**) slurry, and (**f**) residual.

Table 4. Prediction precision results of the multi-task learning model.

Yield (wt%)	Plant Dataset		Simulation Dataset	
	MSE	MAPE	MSE	MAPE
Dry gas	0.0032	1.20%	0.0002	0.31%
LNG	0.0311	0.78%	0.0028	0.27%
Gasoline	0.2695	1.05%	0.0020	0.09%
Diesel	0.1297	0.98%	0.0015	0.13%
Slurry	0.0148	1.45%	0.0055	0.47%
Residual	0.3621	4.84%	0.0019	0.36%

Further analysis of the multi-task learning model revealed that the trend surface for gasoline yield concerning the three independent variables no longer exhibited regions of extrapolation failure, as shown in Figure 10. Additionally, the surface did not present characteristics detrimental to the optimization process, such as vanishing gradients. Therefore, this model is suitable for use as a surrogate model embedded within an optimization framework to enhance operational procedures.

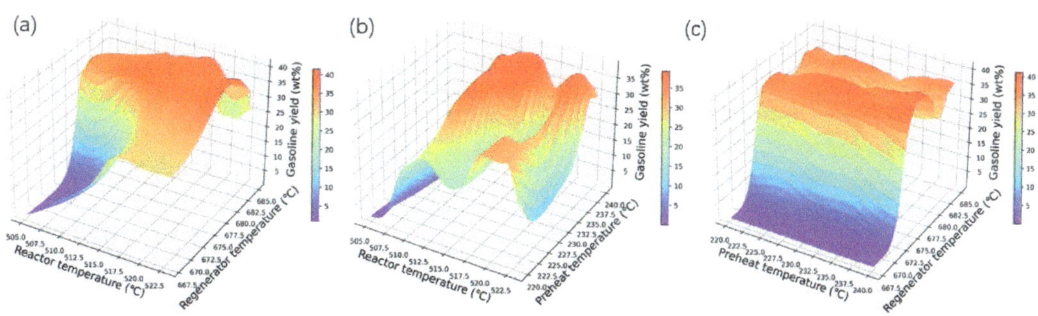

Figure 10. Trend surface of gasoline yield with respect to (**a**) reactor temperature and regenerator temperature, (**b**) reactor temperature and preheat temperature, and (**c**) regenerator temperature and preheat temperature in the multi-task learning model.

3.4. Optimization Results

Using the trained multi-task learning model parameters, an optimization model is constructed for the operational optimization of the FCC unit. A batch of 300 points from the dataset is sampled and run through the optimization model. Figure 11 presents the optimization results for four specific points, corresponding to the data for 26 March 2023, 17 April 2023, 6 May 2023, and 19 July 2023. As depicted in the figure, the optimization is centered on the plant's operational point, with the search space defined as a cube across three optimization variables: reaction temperature, regeneration temperature, and feed preheat temperature, each varying by ±5 °C, ±5 °C, and ±3 °C, respectively. The initial points are represented by circles, while the optimized points are indicated by stars. After optimization, the product revenues for the four points increased significantly.

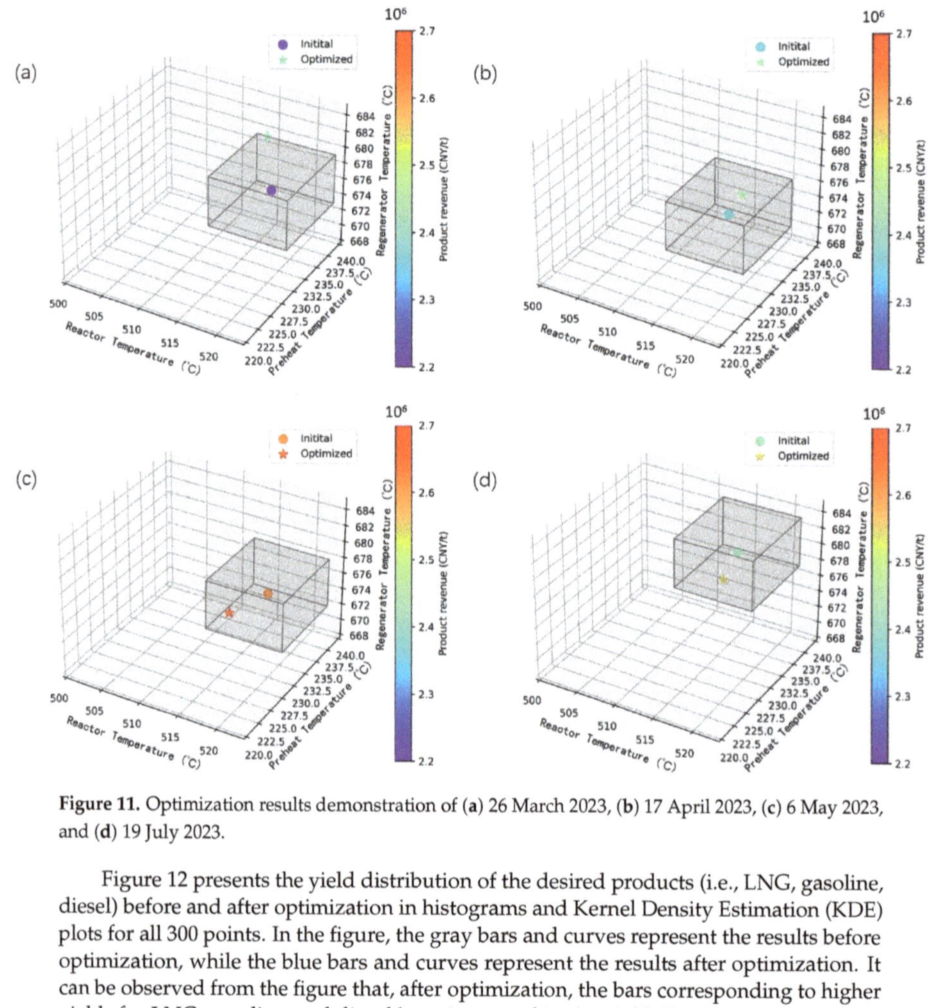

Figure 11. Optimization results demonstration of (**a**) 26 March 2023, (**b**) 17 April 2023, (**c**) 6 May 2023, and (**d**) 19 July 2023.

Figure 12 presents the yield distribution of the desired products (i.e., LNG, gasoline, diesel) before and after optimization in histograms and Kernel Density Estimation (KDE) plots for all 300 points. In the figure, the gray bars and curves represent the results before optimization, while the blue bars and curves represent the results after optimization. It can be observed from the figure that, after optimization, the bars corresponding to higher yields for LNG, gasoline, and diesel have increased in the yield distribution histograms. Additionally, the peaks of the probability density curves have shifted towards higher yields, indicating that the optimization has improved the yields of these desired products.

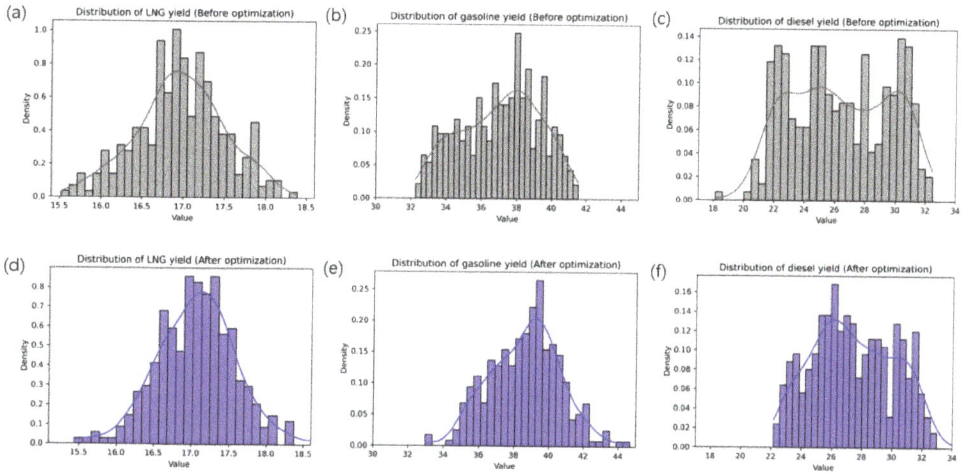

Figure 12. Product distribution of (**a**,**d**) LNG, (**b**,**e**) gasoline, and (**c**,**f**) diesel before and after optimization.

Table 5 provides a detailed overview of the optimization results for these 300 points. On average, the yield of LNG increased from 16.97 wt% to 17.07 wt%, the yield of gasoline from 37.06 wt% to 38.64 wt%, and the yield of diesel from 26.36 wt% to 27.41 wt%. Consequently, the product revenue, reflecting this improvement, increased from CNY 2,413,730 to CNY 2,502,356 per hour, representing a 3.7% increase. This illustrates the substantial potential of the optimization model for enhancing the economic performance of the FCC unit through adjustments in operational variables.

In practice, obtaining data from a plant that covers multiple operating conditions is often challenging. While there is an abundance of data available from the plant, most of the data are steady-state data with limited informational content. Therefore, relying solely on the available plant data for purely data-driven modeling is unreliable. This is why we introduce simulated data generated from mechanistic models for hybrid modeling. However, the accuracy of such hybrid models still largely depends on the quality of the dataset. When the plant uses new feedstock with properties significantly different from those in the dataset or operates under new conditions, the performance of the hybrid model may deteriorate significantly. In such cases, it is necessary to retrain the model using newly collected data to maintain its accuracy.

Table 5. Comparison of product yields and product revenues before and after optimization.

	Before Optimization	After Optimization	Relative Change
Average LNG yield	16.97 wt%	17.07 wt%	+0.59%
Average gasoline yield	37.06 wt%	38.64 wt%	+4.26%
Average diesel yield	26.36 wt%	27.41 wt%	+3.98%
Average product revenue	2,413,730 CNY/h	2,502,356 CNY/h	+3.67%

4. Conclusions

In this work, we address the operational optimization of the FCC process in an industrial setting by proposing an integrated framework that combines mechanistic and data-driven modeling with optimization. This framework involves using a simulation program to augment plant data samples collected from the field and applying a multi-task learning model to train the prediction model. This approach achieved product yield predictions with errors of less than 4.84% for the plant dataset and less than 0.47% for the simulation dataset. The trained multi-task learning model was then embedded as a surrogate model within the optimization framework to optimize operational variables.

Testing with 300 points demonstrated that this framework could improve LNG, gasoline, and diesel yields by an average of 0.10 wt%, 1.58 wt%, and 1.05 wt%, respectively, and increase final product revenue by 3.67%. This illustrates the substantial potential of this framework for enhancing the economic benefits of industrial FCC processes.

Author Contributions: Conceptualization, H.L. and Q.Z.; methodology, H.L.; software, W.X.; validation, H.L., R.W. and T.Q.; data curation, Q.Z.; writing—original draft preparation, H.L.; writing—review and editing, Q.Z.; visualization, H.L.; supervision, T.Q.; project administration, R.W.; funding acquisition, T.Q. All authors have read and agreed to the published version of the manuscript.

Funding: This research received no external funding.

Data Availability Statement: The data presented in this study are available on request from the corresponding author. Data sharing is restricted by the factory's confidentiality agreement.

Conflicts of Interest: Author Ruqiang Wang was employed by the PetroChina Planning and Engineering Institute. The remaining authors declare that the research was conducted in the absence of any commercial or financial relationships that could be construed as a potential conflict of interest.

References

1. Pinheiro, C.I.C.; Fernandes, J.L.; Domingues, L.; Chambel, A.J.S.; Graça, I.; Oliveira, N.M.C.; Cerqueira, H.S.; Ribeiro, F.R. Fluic Catalytic Cracking (FCC) Process Modeling, Simulation, and Control. *Ind. Eng. Chem. Res.* **2011**, *51*, 1–29. [CrossRef]
2. Xie, Y.; Zhang, Y.; He, L.; Jia, C.Q.; Yao, Q.; Sun, M.; Ma, X. Anti-deactivation of zeolite catalysts for residue fluid catalytic cracking *Appl. Catal. A Gen.* **2023**, *657*, 119159. [CrossRef]
3. Palos, R.; Gutiérrez, A.; Fernández, M.L.; Azkoiti, M.J.; Bilbao, J.; Arandes, J.M. Converting the Surplus of Low-Quality Naphtha into More Valuable Products by Feeding It to a Fluid Catalytic Cracking Unit. *Ind. Eng. Chem. Res.* **2020**, *59*, 16868–1687 [CrossRef]
4. Gholami, Z.; Gholami, F.; Tišler, Z.; Tomas, M.; Vakili, M. A Review on Production of Light Olefins via Fluid Catalytic Cracking *Energies* **2021**, *14*, 1089. [CrossRef]
5. Chen, Q.; Ding, J.; Chai, T.; Pan, Q. Evolutionary Optimization Under Uncertainty: The Strategies to Handle Varied Constraint for Fluid Catalytic Cracking Operation. *IEEE Trans. Cybern.* **2020**, *52*, 2249–2262. [CrossRef] [PubMed]
6. Sildir, H.; Arkun, Y.; Canan, U.; Celebi, S.; Karani, U.; Er, I. Dynamic modeling and optimization of an industrial fluid catalyti cracker. *J. Process Control* **2015**, *31*, 30–44. [CrossRef]
7. Lopes, G.; Rosa, L.; Mori, M.; Nunhez, J.; Martignoni, W. Three-dimensional modeling of fluid catalytic cracking industrial rise flow and reactions. *Comput. Chem. Eng.* **2011**, *35*, 2159–2168. [CrossRef]
8. Zhang, Y.; Liu, M.; Zhao, L.; Liu, S.; Gao, J.; Xu, C.; Ma, M.; Meng, Q. Modeling, simulation, and optimization for producin ultra-low sulfur and high-octane number gasoline by separation and conversion of fluid catalytic cracking naphtha. *Fuel* **202** *299*, 120740. [CrossRef]
9. Long, J.; Mao, M.S.; Zhao, G.Y. Model Optimization for an Industrial Fluid Catalytic Cracking Riser-regenerator Unit b Differential Evolution Algorithm. *Pet. Sci. Technol.* **2015**, *33*, 1380–1387. [CrossRef]
10. He, G.; Zhou, C.; Luo, T.; Zhou, L.; Dai, Y.; Dang, Y.; Ji, X. Online Optimization of Fluid Catalytic Cracking Process via a Hybri Model Based on Simplified Structure-Oriented Lumping and Case-Based Reasoning. *Ind. Eng. Chem. Res.* **2021**, *60*, 412–42 [CrossRef]
11. Zhao, X.; Sun, S. Lumped Kinetic Modeling Method for Fluid Catalytic Cracking. *Chem. Eng. Technol.* **2020**, *43*, 2493–250 [CrossRef]
12. Weekman, V.W.; Nace, D.M. Kinetics of catalytic cracking selectivity in fixed, moving, and fluid bed reactors. *AIChE J.* **1970**, 1 397–404. [CrossRef]
13. Weekman, V.W. Model of Catalytic Cracking Conversion in Fixed, Moving, and Fluid-Bed Reactors. *Ind. Eng. Chem. Process De Dev.* **1968**, *7*, 90–95. [CrossRef]
14. Arbel, A.; Huang, Z.; Rinard, I.H.; Shinnar, R.; Sapre, A.V. Dynamic and Control of Fluidized Catalytic Crackers. 1. Modeling the Current Generation of FCC's. *Ind. Eng. Chem. Res.* **1995**, *34*, 1228–1243. [CrossRef]
15. Ebrahimi, A.A.; Mousavi, H.; Bayesteh, H.; Towfighi, J. Nine-lumped kinetic model for VGO catalytic cracking; using cataly deactivation. *Fuel* **2018**, *231*, 118–125. [CrossRef]
16. Yang, B.; Zhou, X.; Chen, C.; Yuan, J.; Wang, L. Molecule Simulation for the Secondary Reactions of Fluid Catalytic Crackir Gasoline by the Method of Structure Oriented Lumping Combined with Monte Carlo. *Ind. Eng. Chem. Res.* **2008**, *47*, 4648–465 [CrossRef]
17. John, Y.M.; Mustafa, M.A.; Patel, R.; Mujtaba, I.M. Parameter estimation of a six-lump kinetic model of an industrial fluid cataly cracking unit. *Fuel* **2019**, *235*, 1436–1454. [CrossRef]
18. Lee, J.H.; Kang, S.; Kim, Y.; Park, S. New Approach for Kinetic Modeling of Catalytic Cracking of Paraffinic Naphtha. *Ind. E Chem. Res.* **2011**, *50*, 4264–4279. [CrossRef]

19. Yang, S.; Wang, N. A P systems based hybrid optimization algorithm for parameter estimation of FCCU reactor–regenerator model. *Chem. Eng. J.* **2012**, *211–212*, 508–518. [CrossRef]
20. Du, Y.; Sun, L.; Berrouk, A.S.; Zhang, C.; Chen, X.; Fang, D.; Ren, W. Novel Integrated Reactor-Regenerator Model for the Fluidized Catalytic Cracking Unit Based on an Equivalent Reactor Network. *Energy Fuels* **2019**, *33*, 7265–7275. [CrossRef]
21. Palos, R.; Rodríguez, E.; Gutiérrez, A.; Bilbao, J.; Arandes, J.M. Kinetic modeling for the catalytic cracking of tires pyrolysis oil. *Fuel* **2022**, *309*, 122055. [CrossRef]
22. Nazarova, G.; Ivashkina, E.; Ivanchina, E.; Oreshina, A.; Vymyatnin, E. A predictive model of catalytic cracking: Feedstock-induced changes in gasoline and gas composition. *Fuel Process. Technol.* **2021**, *217*, 106720. [CrossRef]
23. Qin, X.; Ye, L.; Liu, J.; Xu, Y.; Murad, A.; Ying, Q.; Shen, H.; Wang, X.; Hou, L.; Pu, X.; et al. A molecular-level coupling model of fluid catalytic cracking and hydrotreating processes to improve gasoline quality. *Chem. Eng. J.* **2023**, *451*, 138778. [CrossRef]
24. Chen, Z.; Feng, S.; Zhang, L.; Wang, G.; Shi, Q.; Xu, Z.; Zhao, S.; Xu, C. Molecular-level kinetic modeling of heavy oil fluid catalytic cracking process based on hybrid structural unit and bond-electron matrix. *AIChE J.* **2021**, *67*, e17027. [CrossRef]
25. Chen, C.; Zhou, L.; Ji, X.; He, G.; Dai, Y.; Dang, Y. Adaptive Modeling Strategy Integrating Feature Selection and Random Forest for Fluid Catalytic Cracking Processes. *Ind. Eng. Chem. Res.* **2020**, *59*, 11265–11274. [CrossRef]
26. Long, J.; Li, T.; Yang, M.-L.; Hu, G.; Zhong, W. Hybrid Strategy Integrating Variable Selection and a Neural Network for Fluid Catalytic Cracking Modeling. *Ind. Eng. Chem. Res.* **2019**, *58*, 247–258. [CrossRef]
27. Tian, W.; Wang, S.; Sun, S.; Li, C.; Lin, Y. Intelligent prediction and early warning of abnormal conditions for fluid catalytic cracking process. *Chem. Eng. Res. Des.* **2022**, *181*, 304–320. [CrossRef]
28. Yang, F.; Dai, C.; Tang, J.; Xuan, J.; Cao, J. A hybrid deep learning and mechanistic kinetics model for the prediction of fluid catalytic cracking performance. *Chem. Eng. Res. Des.* **2020**, *155*, 202–210. [CrossRef]
29. Li, X.; Zhang, L.; Sun, Y.; Jiang, B.; Zhang, R.; Li, X.; Yang, N. Investigation on steam injection condition in refining vacuum furnace. *Chem. Eng. Sci.* **2015**, *135*, 509–516. [CrossRef]
30. Loh, W.-L. On Latin hypercube sampling. *Ann. Stat.* **1996**, *24*, 2058–2080. [CrossRef]
31. Andrei, N. *Continuous Nonlinear Optimization for Engineering Applications in Gams Technology*; Springer Science+Business Media: New York, NY, USA, 2017.

processes

MDPI

Review

The Molecular Modeling, Simulation, and Design of Base Oils and Additives in Lubricating Oils: A Review

Kang Zhou [1,2], Xinhao Che [3], Chaoliang Wei [4], Zhongping Tang [2], Hai Yu [2], Dong Wang [4], Jianxin Wang [3] and Lei Zhang [3,*]

1 State Key Laboratory of Fluid Power and Mechatronic Systems, School of Mechanical Engineering, Zhejiang University, Hangzhou 310027, China
2 PetroChina Lanzhou Lubricating Oil R&D Institute, Lanzhou 730060, China
3 Institute of Chemical Process Systems Engineering, School of Chemical Engineering, Dalian University of Technology, Dalian 116024, China
4 PetroChina Dalian Lubricating Oil R&D Institute, Dalian 116024, China
* Correspondence: keleiz@dlut.edu.cn

Abstract: Lubricating oils play a crucial role in modern industrial production, mechanical manufacturing, aerospace, and other fields. This paper provides a review and summary of the entire workflow for molecular simulations of lubricating oils, from molecular modeling, especially the molecular representation of base oils, to simulation calculation methods and result analysis. The application prospects and values of the relevant simulation techniques are discussed in detail. The simulation methods, force fields, and software involved in the modeling and simulation process are also introduced, aiming to provide guidance and insights for more rigorous, rational, and accurate lubricant molecular simulations, so as to accelerate the modification and development of new high-quality lubricants.

Keywords: lubricating base oil; additive; molecular modeling; molecular simulation; molecular design

1. Introduction

A lubricating oil is a type of liquid or semi-solid lubricant. It has been widely used in modern industrial production, mechanical manufacturing, aerospace, and other fields [1], which is now indispensable. With the development of times, today's commercial lubricating oils generally perform multiple functions. In addition to the main function of reducing friction, they also play an important role in assisting cooling, cushioning, rust prevention, cleaning, sealing, etc. These functions are typically achieved through the addition of additives (also called oil modifiers) to oils. Finished lubricating oils consist of base oils and additives. Base oils are the main component of lubricating oils and serve as the functional carrier of additives, which make up more than 90% of the product by content and provide the basis for the lubrication effect [2]. The American Petroleum Institute (API) and Technical Association of the European Lubricants Industry (ATIEL) collaboratively developed the base oil classification standard API-1509 [3]. This standard categorizes lubricating base oils into five classes based on their saturated hydrocarbon content, sulfur content, and viscosity index. Classes I, II, and III are all mineral base oils, primarily obtained through solvent refining, hydrotreating, and hydrocracking processes, respectively. Class IV comprises synthetic polyalphaolefin (PAO) base oils. Other types of base oils are classified as class V. Mineral base oils are widely sourced and applied. Because of the lower costs, they have been considered the preferred raw materials for the processing and manufacturing of most commercially available lubricating oils [4]. Their production technology is also relatively mature. PAO base oils, known for their superior stability, safety, low evaporation loss, and low toxicity, are more suitable for high-end applications [5]. However, they have higher production costs and lower polarity, which results in poorer friction performance and solubility for polar additives. Other types of base oils, such as synthetic esters, have the advantages of a high viscosity index, low pour point, high flash point, and low volatility [6], but their production cost is also significantly higher than that of mineral base oils. Additives have a decisive influence on various physical and chemical properties of finished lubricating oils, although their content is relatively low. The function of these additives is either to enhance the already-existing properties of base oils or to impart new properties that they lack. Compared to base oils, additives are very diverse. Even when additives perform the same function, their structures may be different due to various mechanisms of action. Typical additives include antioxidants, viscosity modifiers, friction modifiers, antifoam agents, rust inhibitors, detergents, dispersants, and so on [7]. With the aim of saving energy and protecting environments [8], the standards of lubricating oils have been getting higher and higher in recent years. The development of lubricating oils is becoming more advanced, with a rapid evolution in the types and requirement quantities of base oils and additives.

In order to continuously improve the performance of lubricant products, as part of the past industrial production process, the composition and formulation of base oils and additives were usually optimized by experimental trial-and-error methods [9,10]. It often took several years, even over a decade, to scale from small-scale tests to pilot trials and finally full production, leading to long development periods, high costs, and low success rates. Nowadays, thanks to the continuous development of computer science, it has become one of the research hotspots to calculate and predict the macroscopic properties of oils through molecular simulation methods, which allows for the calculation or prediction of the macroscopic properties of oils by understanding their chemical composition from a molecular perspective. By revealing the intrinsic relationship between molecular composition and physical or chemical properties, researchers can achieve efficient and accurate modifications of lubricating oil products [11]. He et al. [12] focused on the research progress of molecular dynamics methods in the tribological behavior of lubricating oils in their review. The parameters used to evaluate friction and wear, including friction force, wear rate, and surface deformation, are easy to obtain by MD simulation. It is convenient to reveal how these parameters change with experimental factors (temperature, pressure, friction velocity, etc.). In addition, Martini et al. [13] further discussed the reactive molecular dynamics simulation method in tribochemistry. Srivastava et al. [14] discussed the molecular dynamics simulations in nanotribology. Although the modeling and interaction analysis of lubricating oil systems mainly now rely on molecular dynamics-based methods, there are also some other popular simulation methods, such as Monte Carlo simulation [15,16] for gas adsorption and desorption, molecular mechanics-based simulation [17,18] for conformation sampling, quantum chemistry-based simulation [19,20] for chemical bond breaking and formation, etc. Their strengths and potential should be further explored.

In this review, we introduced the entire workflow for the molecular simulations of lubricating oils, from molecular modeling, especially the molecular representation of base oils, to simulation calculation methods and result analysis. What is more, we discuss the application prospects and values of relevant simulation techniques, such as utilizing formulation optimization and molecular design methods to assist experimental trial-and-error. Besides the friction and lubrication properties, this paper involved multifaceted characteristics for the practical applications of lubricating oils from the perspective of the interaction between base oils and additives. The simulation methods, force fields, and software involved in the modeling and simulation process were also introduced, aiming at providing a reference and guidance for more reasonable and accurate molecular simulations of lubricating oils, so as to accelerate the modification and development of novel and high-quality lubricating oil products.

The molecular simulation of lubricating oil systems typically involves the following steps [12,21]: (1) molecular modeling of lubricating oils, such as the molecular representation of structures and composition of base oils, the determination of structures and the number of additives, the construction of simulation systems, etc.; (2) interaction simulations of base oils and additives, such as the selection of simulation methods and force fields, the determination of simulation protocols, the selection of analysis metrics and analysis methods, etc.; and (3) molecular design of novel lubricating oils, such as the establishment of prediction models for the physical or chemical properties of lubricating oils, molecular generation and design of base oils and additives, experimental verification, etc. The following sections will systematically present the research content involved in the above steps and provide an overview of the current research status.

2. Molecular Modeling of Lubricating Oils

In order to perform molecular simulations of a lubricating oil system, the first step is to establish a molecular model that can represent the real lubricating oil product. However, the composition of real oils is extremely complex, and their specific molecular structures and content cannot be accurately determined, even using the most advanced molecular characterization techniques. Experimental methods, such as elution chromatography, high-performance liquid chromatography, infrared spectroscopy, mass spectrometry, nuclear magnetic resonance, etc. [22], can only provide vague information on the structures and composition of oils from various aspects such as group composition, saturated/aromatic hydrocarbon content, carbon number distribution, and so on [23]. But molecular simulations must use explicit molecular structure and composition information as the initial input. Therefore, it is necessary to reasonably simplify and represent the real lubricating oil products.

A popular simplified method is to use a single representative component of lubricating oils. For example, Sun et al. [24] used two kinds of single molecular structures, pentaerythritol tetraheptanoate (PEC7) and pentaerythritol tetranonanoate (PEC9), to represent synthetic polyol ester base oil systems so that the density and viscosity of lubricating oils under extreme pressure could be calculated through simulations. Similarly, Wang et al. [25] also used neopentyl glycol dipentanoate (NPG-nC5), neopentyl glycol diheptanoate (NPG-nC7), and three other simplified molecules with similar structures to simulate the lubrication of pentaerythritol esters and Fe surfaces. For mineral base oil systems, considering that the ideal components of mineral base oils are hydrocarbons with few rings and long side chains, and the carbon number is distributed from C_{20} to C_{40} [1], four molecular structures of C_{20}, namely n-eicosane, 2,6,11,15-tetramethylhexadecane, tetradecylcyclohexane, and tetradecylbenzene, were selected in our work to represent different pure-component base oils [26], corresponding to the main components of mineral base oils, including linear paraffins, isoparaffins, naphthenes, and aromatic hydrocarbons. Although it is easier to obtain the molecular models of lubricating oils, and the costs of modeling and calculations are lower using pure components to represent real oil products, the models are far from real oil products. Therefore, Kondratyuk et al. [27] used a binary mixture of methylcyclohexane and 1-methylnaphthalene to study the viscosity property of the base oils with hydrocarbons as the main component under high pressure. For synthetic polyalphaolefin base oils, Kruse et al. [28] used a mixture of trimer, tetramer, and pentamer of C_{10} molecules with different mass fractions to represent real lubricating oil products. The above methods have contributed to the molecular representation of complex lubricant systems, but they still cannot reflect the properties of real oil products as much as possible. To further improve the accuracy of molecular representation, molecular reconstruction methods are increasingly recognized. Molecular reconstruction is a method through quantitative molecular structures and composition based on the vague experimental measurement and analysis data of real oil products, which can represent real oil products to the greatest extent [29]. Since molecular simulations need to use explicit molecular structures and composition information, this paper focuses on the molecular reconstruction methods using real molecular structure representation (rather than other representations such as the combinations of structure fragments [30]). At present, few studies have been reported on molecular reconstruction for specific lubricating oil products, and reconstruction methods are more commonly used for crude oils [31], petroleum [32], and diesel oils [29]. Lyu et al. [33] proposed a method to develop the average molecular model for hydrocarbons and petroleum fractions. Using ^{1}H-NMR, molecular weight, and elemental composition as the input parameters, they calculated a single representative molecule through the proposed reconstruction algorithm which could reflect the average chemical structures of real oil products. Wei et al. [34] coupled the molecular simulation method with the reconstruction process, and proposed a molecular reconstruction and property prediction method (MDs-NP) for naphtha. This method uses molecular dynamics simulations for the chemical property collections of molecules and uses Gamma distribution from real analysis data to calculate mole fractions of simulation mixtures. Compared with using a single molecule to represent the average chemical structure, the accuracy of this reconstruction method is greatly improved. Similar to the above research, we popularized and applied the reconstruction method to lubricating oil systems and then obtained reliable molecular models of mineral base oils for subsequent molecular simulations based on the reconstruction results, which makes it possible to predict relevant properties of base oils more accurately [35]. The algorithm workflow is shown in Figure 1.

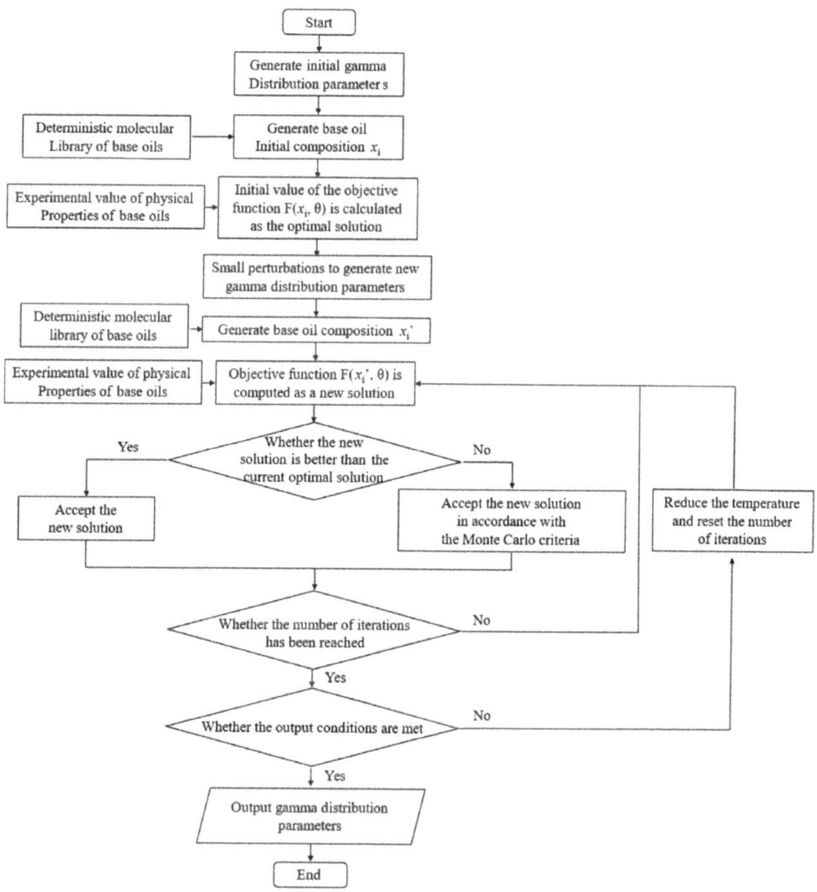

Figure 1. Molecular reconstruction algorithm workflow for mineral base oils.

First, the probability density function sampling and maximum information entropy optimization are combined to predict the molecular composition of mineral base oils by a simulated annealing algorithm according to the simulated distillation yield of real oil products. Then, the density of the reconstruction systems is calculated by molecular dynamics simulations, which is then compared with the experimental data to determine a final base oil model that can best represent the real oil product.

The above molecular representation methods of base oils are summarized in Figure 2. Compared to base oils, the representation method of additives is relatively simple. Since a small number of additive molecules can have a significant influence on the properties of specific aspects of the whole lubricating oil system, several representative additive molecules are added to analyze the possible effects on the base oil system [36,37].

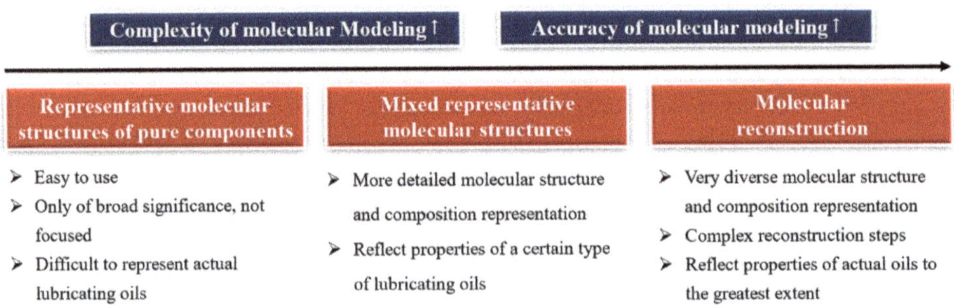

Figure 2. Molecular representation methods for base oil of lubricants.

After molecular composition and structures are reasonably simplified, a popular modeling approach is to use periodic 3D-structure units filled with base oil molecules and additive molecules to represent the real oil product. This approach uses a structure consisting of a certain number of molecules with 3D periodic boundary conditions to represent an actual macroscopic system. This way, the macroscopic properties of the actual system can be approximated by calculating the properties of a smaller structure unit. At the same time, the periodic modeling method also avoids boundary effects. Two kinds of widely used structure units are shown in Figure 3. In order to investigate the antioxidation mechanism of antioxidants in lubricating oils, Hu et al. [37] added trimethylolpropane trioleate base oil molecules, antioxidant molecules, and oxygen into a cubic periodic structure unit with a certain ratio, as shown in Figure 3a. Then, the effects of antioxidants on the diffusion and penetration of oxygen in the system were analyzed through molecular dynamics simulations. In our previous work [26], in order to investigate the effects of antifoam agents on the stability of oil-based foam, a typical "sandwich" model of additive-base oil additive was constructed to simulate the foam film structure composed of liquid–gas, as shown in Figure 3b. At the same time, a gas phase or vacuum layer of enough height is added on the top and bottom of the oil layer to avoid potential interference between the two adjacent periodic units.

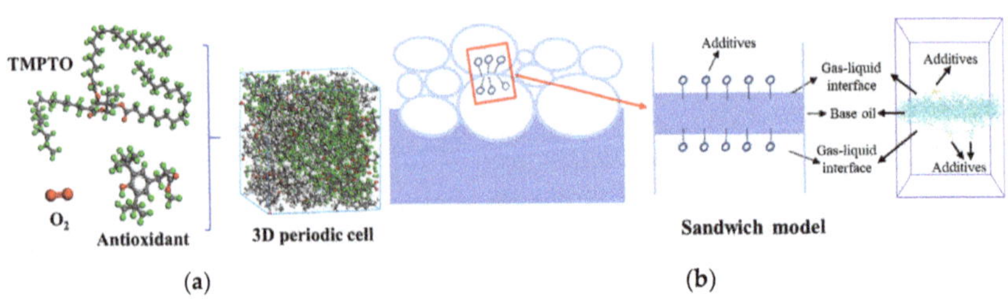

Figure 3. Examples of a molecular model of lubricating oil: (**a**) Hu et al. [37]; (**b**) Zhou et al. [26].

Next, the simulation scale should be determined according to the actual needs, as it will influence the refinement degree of the modeling. In general, a longer simulation time and finer simulation scale will result in fewer biases in the analysis results, making the simulation system closer to the real product. However, the costs of simulation for complex lubricating oil systems and large periodic units are still too high today. A single dynamics simulation task can take hours to days, depending on the computer's performance and simulation time. In addition, the comparison and analysis between different simulation systems, or more generally, multiple repeated experiments, still require considerable computational power. A feasible method is to use fuzzy atomic composition to represent the molecules in a periodic structure unit. With the representation of atoms in the all-atom model (usually used), the united-atom model [38], the coarse-grained model [39], and the dissipative particle dynamics (DPD) model [40] gradually becoming rougher, the simulation scale can also become larger and larger, which makes mesoscale simulation possible on a personal computer. The united-atom model simplifies molecular simulations by grouping certain atoms into larger pseudo-atoms. In this model, hydrogen atoms attached to carbon atoms are typically excluded, allowing for the simulation of larger molecules with fewer particles. This reduction in the number of degrees of freedom helps speed up computations while still retaining the essential structural and thermodynamic properties of the system. The coarse-grained model further reduces complexity by averaging over groups of atoms or molecules into single interaction sites. Each site represents several atoms, thus drastically decreasing the number of particles in the simulation. The model focuses on capturing the essential physics of the system over larger length and time scales, making it ideal for studying polymer systems in lubricating oils, such as PAOs and synthetic ester lubricants. DPD is a mesoscopic simulation method that combines molecular dynamics with fluid dynamics principles. It considers the whole molecule as a "particle", which is particularly effective for simulating complex fluids, such as colloidal suspensions and polymers.

3. Interaction Simulations of Base Oils and Additives in Lubricating Oils

Current molecular simulation methods can be broadly classified into four categories: Monte Carlo simulation, molecular mechanics-based simulation, molecular dynamics simulation, and quantum chemistry-based simulation. Their computational costs gradually increase. A Monte Carlo simulation is a statistical method used to understand the behavior of molecular systems. It employs random sampling to explore the configurational space of molecules and is widely used in calculating thermodynamic properties, phase transitions, etc. Because of its inability to capture dynamic processes, its computational costs are the lowest, but it does not simulate time evolution directly. A molecular mechanics simulation is based on classical physics principles, where atoms are treated as spheres and bonds as springs. The potential energy of the system is described by empirical force fields. It is extensively used for structural optimization, conformational analysis, etc. A molecular mechanics methods can struggle with accurately modeling electronic effects and show decreasing accuracy in the simulation of dynamic processes. A molecular dynamics simulation is a time-dependent simulation method that solves Newton's equations of motion for a system of atoms, which is the most widely used method in lubricant simulations. A quantum chemistry simulation leverages the principles of quantum mechanics to study molecular systems at an electronic level, which focuses on solving the Schrödinger equation to obtain the electronic structure of molecules. It is essential for understanding chemical reactivity, electronic properties, and the behavior of catalysts, but it is computationally demanding and often limited to small to medium-sized systems due to their high resource requirements. It is usually used to study a single base oil or additive molecule.

Except for quantum chemistry-based simulation methods, a suitable molecular force field needs to be selected for the preliminary optimization of the initial model, containing structural checks, energy minimization, etc. Popular molecular force fields for lubricant simulations include COMPASS II/III [41,42], DREIDING [43], and AMBER force fields [44]. In particular, MARTINI [45] is the most widely used coarse-grained molecular dynamics simulation force field, which can be used to simulate lubricant molecules at larger scales or for longer periods. For example, Sun et al. [39] developed a coarse-grained model of surfactant–water phase-oil phase using the MARTINI force field in order to investigate the foam bursting mechanism induced by the emulsified oil phase. They observed the dynamic process of bubble bursting during subsequent molecular simulations.

The interaction simulations of base oils and additives involve a wide range because there are many types of base oils and additives in real lubricating oil products. As shown in Figure 4, additives have various effects on base oils and the types of additives may be quite different even if they have the same effects. Therefore, it is necessary to formulate specific simulation protocols for each kind of additive. What is more, appropriate simulation parameters and analysis methods should be carefully determined in order to make the simulation results reflect the macroscopic properties of real oil products as much as possible so as to meet the actual needs. Here, some common types of additives in finished lubricating oils and their interaction simulation methods with base oils are introduced. Table 1 also shows the typical molecular structures of these additives.

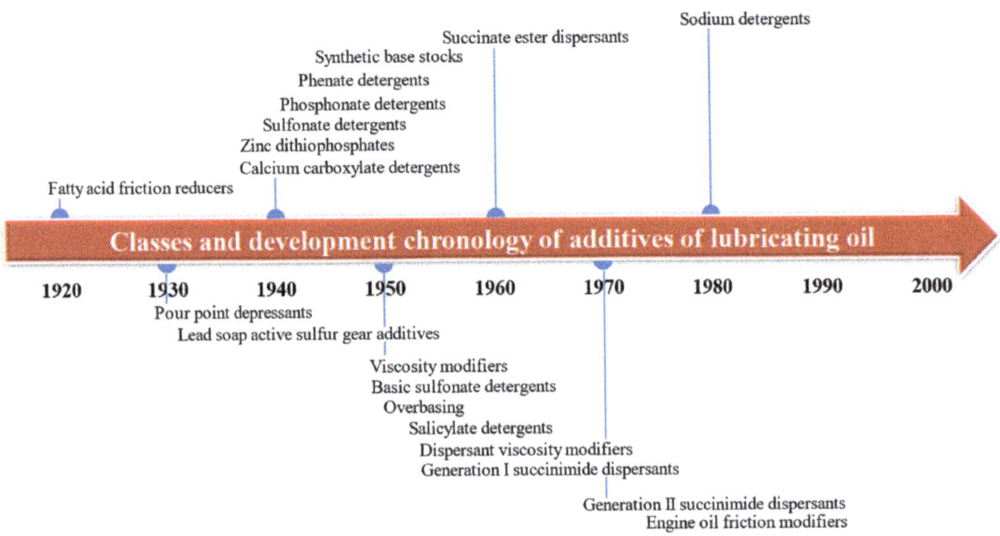

Figure 4. Classes and development chronology of additives of lubricating oil [1].

Antioxidants are used to prevent the deterioration of lubricating oils due to oxidation in storage or a high-temperature working environment. The oxidation of lubricating oils will result in a loss of performance and the formation of harmful by-products such as the deposits of sludge and varnish, which not only reduce the lubrication effectiveness but can also clog filters or increase wear of mechanical components. Hu et al. [37] analyzed the diffusion behavior of three kinds of phenolic antioxidants in the trimethylolpropane trioleate (TMPTO) base oil and the permeability of oxygen in this lubricating oil system by molecular dynamics simulations. The bond dissociation energies of active hydrogen atoms of different antioxidants were also calculated using quantum chemistry-based methods. According to the above simulation results, the effects of the antioxidants were quantitatively elucidated. Song et al. [46] combined molecular simulations and experimental methods to analyze the antioxidation performance of different alkylated diphenylamine antioxidants also in the TMPTO system, but this work is more systematic and complete. The permeability, binding energy, and solubility parameters of oxygen were analyzed in detail. The mean square displacement (MSD) and bond dissociation energy of the antioxidants are compared. The free volume of the simulation systems was calculated. The experimental results show that the antioxidant T558 with a larger molecular weight and longer amine para-alkyl chain shows better antioxidation performance, which is consistent with the simulation results.

Table 1. Examples of typical molecular structures of additives.

Additive Class	Typical Molecular Structure			
Antioxidants	Hindered Phenols 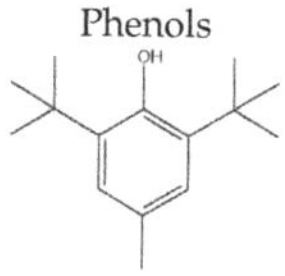	Hindered Amines 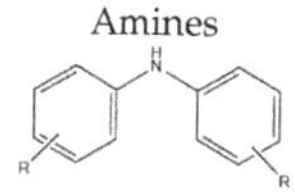	organic phosphorus/ sulfur compounds 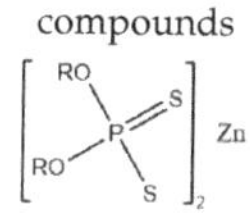	Metal Deactivators 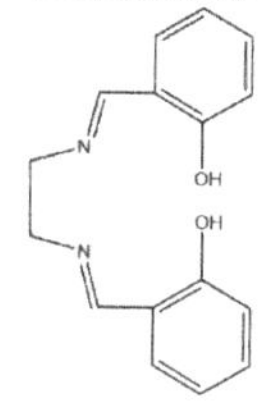
Viscosity modifiers	Polyisobutylene 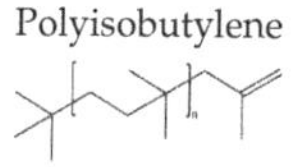	polymethacrylate 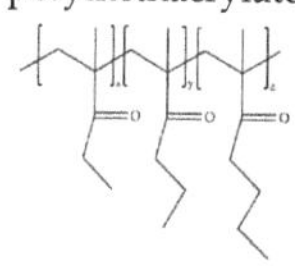	Olefin co-polymer 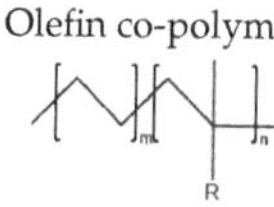	
Friction modifiers	Organic phosphorus compounds 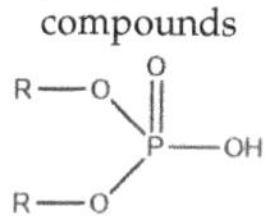	Organic molybdenum compounds 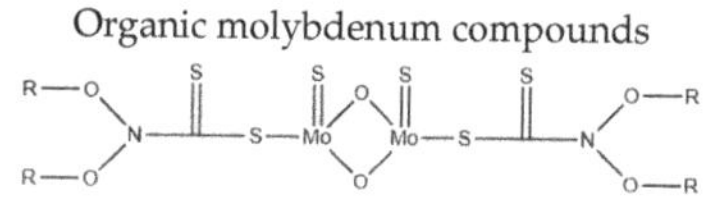		
Antifoam agents	Silicone-based 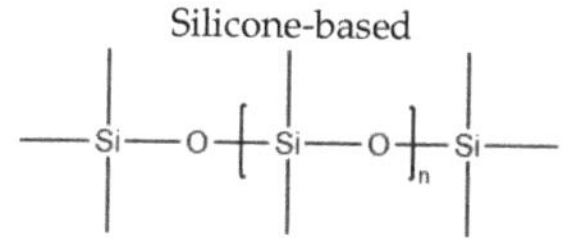	Non-silicone 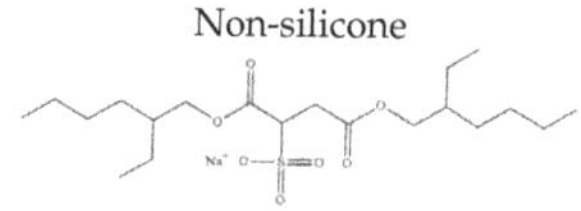		

Viscosity modifiers directly adjust the viscosity of lubricating oil products at a specific temperature or change the viscosity–temperature/pressure property. They can enable the lubricating oil to keep high viscosity at a high temperature while not being too viscous in cold environments, thus ensuring that the lubricating oil can maintain good fluidity over a wide temperature range. To investigate the effects of linear styrene–butadiene polymer structure on the viscosity–temperature behavior of the dodecane base oil system, Len et al. [47] analyzed the interactions between the viscosity modifiers with three different polymer configurations and the base oil by molecular dynamics simulations, which provides a theoretical basis for the design of novel viscosity modifiers with excellent performance.

Friction modifiers are used to improve the lubricating effects of base oils, increase anti-wear performance, or reduce friction between mechanical parts. Since the lubricating effects of lubricating oils are affected by many factors, including temperature, pressure, flow velocity, and the material and roughness of the contact surface, base oils alone may not provide a satisfactory lubricating effect in some cases. Therefore, it is necessary to add appropriate friction modifiers. In order to investigate the structure–activity relationship and mechanism of action of a series of non-ionic surfactant molecules derived from vegetable fat employed as friction modifiers, Davidson et al. [48] used mono- and di-alkanoyl glycerides as representative molecules of additives and added them into the pure-component base oil expressed by hexane. The friction effects between the systems with different friction modifiers and the surface layer of goethite were analyzed using molecular dynamics simulations, and the reliability of the simulation results was verified by experiments. Berro et al. [49] investigated the effects of added zinc dithiophosphate (ZDDP) on the anti-wear properties of the cetane base oil and anti-wear mechanism. The influence of the ZDDP additive and the surface–lubricant interaction on the mechanical and thermal interfacial response were studied in detail. The results show that mechanical and thermal slips were reduced by increasing the surface energy. Their simulations also demonstrate the migration of ZDDP molecules and their adsorption onto the solid surface resulting in a remarkable suppression of mechanical slip compared to pure hexadecane. Consequently, the effective shear rate is higher and so is the friction.

In addition to the common types of additives mentioned above, there are also some reported cases of other additives using molecular simulation methods to analyze the interactions between them and base oils, such as antifoam agents. At present, the research on the foam bursting mechanism focuses on the molecular simulation of water-based systems, with few reports on lubricating oil systems. Therefore, in our work, the effect of antifoam agents on the foam properties of lubricating oils was investigated by molecular dynamics simulation [26]. During the operation of mechanical equipment, air is inevitably involved, which results in the foam in lubricating oils. The foam phenomenon is more serious in real products because most of the added additives are surfactants. The function of antifoam agents is to reduce the foam generated during the use of lubricants. We took the bursting time of the liquid film composed of base oil in the simulation process as the evaluation index of foam stability. The effects of different base oils, antifoam agents, and other additives on the liquid film stability were analyzed according to the simulation results, which provided a new perspective on the molecular simulation of oil-based foam.

Although it is an effective way to understand the microscopic mechanism of base oils and additives through molecular simulation methods, it should also be pointed out that current simulation methods use a simplified representation of lubricating oil systems to calculate their microscopic properties to reflect, analyze, or predict the macroscopic properties of real oil products. The simulation results inevitably deviate from experimental data, but the simulation results can also be corrected by the previous experimental results or analyze the performance differences between various base oil and additive systems from the trend, that is, there are two kinds of analysis methods for the simulation results: quantitative and qualitative. Xiong et al. [36] investigated the antioxidation activities of three natural phenolic acid-based antioxidants by combining molecular simulations with experiments. Their solubility, compatibility, diffusivity, and ability to block oxygen in the pentaerythritol oleate (PETO) base oil were analyzed and compared. They also further explored the changes in the internal molecular structures and the viscosity of the whole system after oxidation. The antioxidation activities of the ferulate antioxidants were quantitatively elucidated with concrete microscopic property metrics and real measurement data. As the effectiveness of additives in improving friction performance is difficult to evaluate by a recognized index, Davidson et al. [48] gave the qualitative analysis results of the performance of mono-, di-, and tri-alkanoyl glyceride as friction modifiers by molecular simulations. The analysis results show that mono-alkanoyl glyceride molecules are able to pack more efficiently, forming significantly more intermolecular hydrogen bonds and occupying approximately half the volume needed by di-alkanoyl glyceride molecules. Density profiles were presented which show significant mixing of the hydrophobic tail groups and a non-polar solvent. The measured friction coefficients of equimolar solutions of the glycerides also show that the efficacy as friction modifiers varies in the order mono-, di- and the tri-oleyl glyceride, which is consistent with the efficacy of film formation predicted by the molecular dynamic calculations.

As shown in Figure 5, the combination of molecular simulation methods, computational theories of molecular interactions, and experimental verification is a trend in the development of novel lubricating oil products. Simulations and experiments complement each other, guiding the way for the modification of lubricating oils.

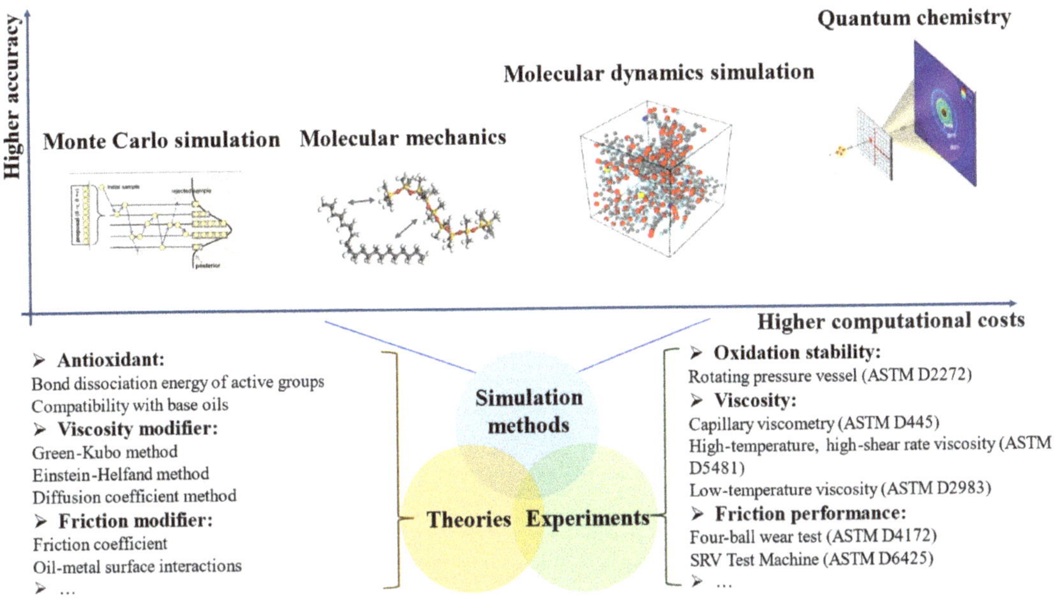

Figure 5. The combination of molecular simulation methods, theories, and experiments is a development trend for lubricating oil modification.

4. Novel Lubricating Oil Design

Along with the simulations and analysis based on known lubricating oil products, the physical or chemical properties of virtual lubricant systems can also be predicted by combining previous experimental data and simulation results. As a result, it is possible to screen or design novel lubricating oil products. In other words, the fundamental purpose of understanding lubricants is to modify them. Figure 6 shows the general design process of novel lubricating oils, which can be broadly classified into two categories: formula optimization and molecular design. Formulation optimization aims to answer the "how many" question, while molecular design aims to answer the "what is" question. They both aim to seek novel lubricant products with better performance. To design novel lubricating oils, the first step is to identify product needs and then, translate them to molecular properties. For example, to improve the antioxidation, the permeability, binding energy, and solubility parameters of oxygen in the whole lubricating oil system are the essential properties as mentioned in Section 3. However, these properties are calculated based on the known systems, and at the same time, the costs of molecular simulation limit the generalization of these properties to a large number of virtual systems. Therefore, developing accurate and quick quantitative structure–property relationship (QSPR) models for property prediction is necessary. The combination of the Group Contribution (GC) method [50] and artificial intelligence (AI) models has been widely used for QSPR model development. The GC method predicts the properties of molecules by describing a molecule as a combination of different groups [51–53]. It can be applied to both single-molecule and mixture systems, which makes the molecular design and formula optimization of lubricating oils the same kind of problem and thus makes it convenient for subsequent solutions. Furthermore, different neural network architectures make it possible to train models based on industrial, experimental, or simulation data of different quality. After the QSPR models are developed, various algorithms, such as the genetic algorithm [54], particle swarm algorithm [55], and so on, can be used to solve the optimal composition ratio or molecular structure. Although the global optimal solution is difficult to solve, the local optimal solutions, in other words better products than known lubricants, can be obtained.

Finally, new products that were designed scientifically and carefully need to be verified many times through simulations and experiments until they are put into production.

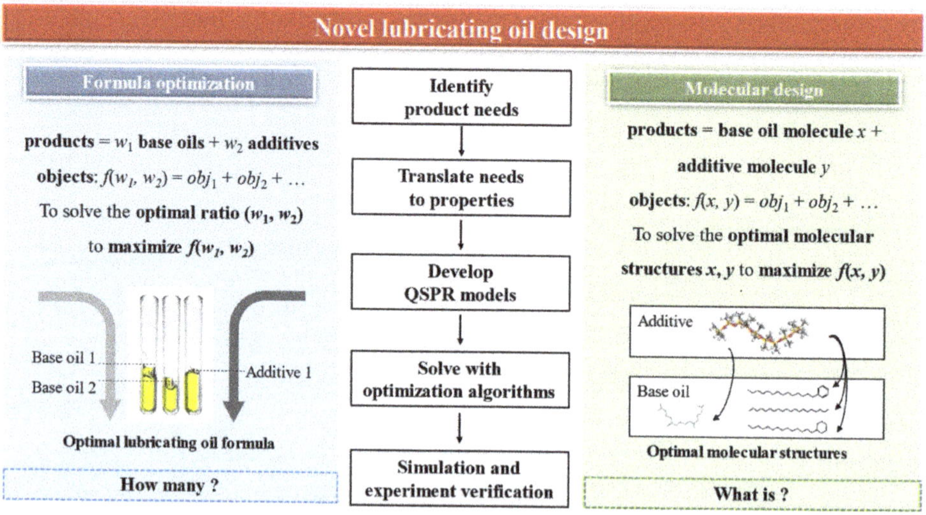

Figure 6. The general design process of novel lubricating oil products.

To optimize the lubricating oil formula, Bhaumik et al. [56] developed a molecular database based on the published literature on vegetable oil-based biodegradable lubricants (coconut oil, castor oil, and palm oil) with nano-friction modifiers (multi-walled carbon nanotubes and graphene) as additives. An artificial neural network (ANN) model was developed to predict the anti-friction performance of lubricating oils. According to the friction coefficients of lubricating oil systems with different ratios of additives and base oil predicted by the model, the product formula was optimized by a genetic algorithm. Two new lubricating oil products with better friction performance were designed. Similarly, Wu et al. [57] systematically investigated the intermolecular interactions between mixed natural antioxidants and the effects on the antioxidation of olive oil by combining multi-scale molecular simulations, experimental analysis, and ANN prediction methods. Based on the simulation results and experimental verification, two kinds of natural antioxidants, tert-butylhydroquinone and flavanol, were selected for further studies on the addition quantity. The optimal ratio and addition amount of the two antioxidants were predicted by ANN, and the prediction results were verified by infrared spectrum experiments, which further confirmed that the mixed antioxidants can effectively inhibit carbonyl formation in olive oil.

In addition to optimizing the composition formulas of lubricating oils, the molecular design of base oils and additives has become a hot topic. Computer-Aided Molecular Design (CAMD) [58] has demonstrated its strong advantages in the fields of material optimization [59] and drug discovery [60], and it is a feasible idea to apply it to the molecular design of lubricating oils. The designed new molecules can also be tested by chemical synthesis and subsequent characterization experiments. Although it is still unrealistic to directly produce the molecules designed by computers at present, the molecular design of base oils and additives provides clues and guidance for the discovery of novel lubricant products. In another word, lubricants can be modified in a variety of ways to get as close as possible to the "ideal" molecular structures and composition designed by computers so as to save lots of experimental costs. For example, to design novel ionic liquid-based lubricants used for automotive engines that meet industrial standards, Valencia et al. [61] solved the CAMD model as a Mixed-Integer Nonlinear Programming (MINLP) problem using classical optimization methods, and then, the expected properties and chemical feasibility of the designed lubricants were verified by molecular dynamics simulations. The simulation results agreed well with the experimental results. Abdulfatai et al. [62] concluded the structure–property relationship of 30 additives. Combined with molecular dynamics simulations, they determined the dynamic binding strength between the sliding interface of hydrogen-containing diamond-like carbon and additive molecules. Based on the discovery, two multifunctional additive molecules were designed, which showed better anti-friction performance than the predicted results of the commercially available additive ZDDP. In our previous work [63], a reinforcement learning framework was proposed to personalize the generated mineral base oil molecules according to the required objective properties. Diverse molecules meeting multiple properties can be obtained without any given sample data. We have illustrated the feasibility and effectiveness of the proposed method by two real cases: molecular generation of base oils for the purpose of exploring the ideal molecular composition and for the purpose of supporting molecular characterization in experiments, which is expected to further guide the molecular design of base oils.

There are various optimization algorithms and their derivatives to solve the optimal formulation or the optimal molecular structure. Each has its advantages and disadvantages, and it is difficult to assess which is better in a practical lubricant modification problem. Experimental verification is therefore essential. In order to improve the experimental success rate as much as possible and reduce the costs of trial and error, the accuracy of lubricant formulation evaluation and molecular property prediction in the optimization process is crucial, which directly determines the reliability of the optimization results. Poor data are the main reason currently limiting the accuracy of prediction models, especially the absence of publicly available and comprehensive databases on the composition and physical properties of oil products. Most of the oil measurement data are commercial secrets, making them inaccessible, which poses a common challenge for both academia and industry.

5. Open and Commercial Simulation Sources

Software that can be used for the molecular modeling and simulation of lubricating oil systems are widely available. Among the open-source tools, GROMACS [64] is widely recognized for its high performance in large-scale simulations. It benefits from an active community that offers extensive documentation and tutorials, although its command–line interface may present a steep learning curve for newcomers. Another available option is LAMMPS [65], which features a modular design allowing users to choose different force fields and techniques. It requires advanced knowledge for proper setup, and its documentation can be challenging for beginners.

On the commercial side, COMSOL Multiphysics is a powerful tool for molecular simulation in the field of lubricating oils [66], particularly valued for its capability to handle multiphysics problems. Its applications include tribological studies, where it simulates friction and wear characteristics, and heat transfer analysis to understand the impact of temperature on lubricant performance. Additionally, COMSOL facilitates the modeling of chemical reactions, aiding in the formulation of stable lubricants by analyzing degradation over time. Its user-friendly interface allows researchers to set up complex simulations without extensive programming knowledge, while its robust post-processing tools enhance data visualization and interpretation. Materials Studio [67] is another versatile software suite that plays a crucial role in molecular simulation across various scientific fields, including materials science, chemistry, and pharmaceuticals. Materials Studio supports a wide range of molecular simulation techniques, including molecular dynamics, Monte Carlo simulations, quantum mechanical calculations, and so on.

6. Application Potential of Artificial Intelligence in Lubricant Development

With the development of time, artificial intelligence technology has penetrated most stages of novel lubricant development, which can also maximize the application of the molecular simulation results. In addition to combining AI-based QSPR models with molecular simulation methods to optimize the formula or design the molecular structures of lubricating oils or accelerating the molecular dynamics process described above, AI-based methods, especially deep learning methods, are expected to solve the subsequent synthesis problem of the designed novel lubricant products. Although deep learning methods have played an important role in the single-molecule organic synthesis pathway search [68] and inverse synthesis analysis [69], how to apply similar methods to the synthesis planning of mixture systems such as PAO base oils and synthetic ester base oils is still a question worth considering. Taking a longer view, with the development of Internet of Things (IoT) technology, smarter systems may emerge in the future that can automatically collect real-time data during lubricating oil use, and combine the data and AI-based models and algorithms for real-time analysis to provide users with instant suggestions or automatically adjust lubrication strategies. One of the development trends is that large language models (LLM) have been widely used in the industry [70]. It is believed that the development of lubricants will also realize intelligent design and manufacturing based on AI in the near future.

7. Discussions

Thanks to the development of computer science, molecular simulation and artificial intelligence methods have also developed rapidly in recent years. Their applications in lubricating oil systems continue to deepen, becoming more and more rational, detailed, and accurate. Through molecular simulations, we can not only understand the microscopic mechanism of lubricating oils, but also design and modify lubricating oils, which is of great significance to industrial production and national development.

However, it should be noted that there is still a long way to go to improve the current research methods. For example, although molecular dynamics simulation plays a dominant role in lubricating oil systems due to its acceptable costs and accuracy, how to choose an appropriate simulation method and relevant simulation parameters according to the actual needs is still an essential issue to think about, which directly determines whether the simulation matches the true situation. Molecular dynamics simulation is not omnipotent. Different simulation methods have different computational costs, accuracy, and applicable scopes. Their respective advantages should be fully exploited or used in combination to verify the simulation results from multiple aspects and perspectives.

For the molecular dynamics method itself, accelerating the simulation process in parallel using GPUs has been an important advance in recent years [71]. Simulation time has been greatly reduced, and it is no longer difficult to achieve a millisecond all-atom molecular dynamics simulation on a personal computer. For another thing, mesoscale simulation is available by reasonably simplifying the molecular structures by coarse-grained dynamics simulation or dissipative particle dynamics method. However, compared to the general all-atom dynamic simulation method, the above methods sacrifice some accuracy. An emerging solution that can be considered in the future is the development of molecular force fields based on machine learning potentials [72,73], such as DeepMD [74,75], to achieve the development of better molecular dynamics simulations with the advantages of the speed and accuracy of artificial intelligence methods.

8. Conclusions

Understanding and transforming nature has been a long-standing pursuit for humanity. Studying the structure, composition, and mechanisms of lubricants at the molecular level is an effective approach to lubricant modification, and molecular simulation effectively makes up for the shortcomings of experimental characterization. This paper summarizes the current research status in molecular simulation of lubricant systems and shows its wide application prospect. Although there are still many challenging issues between simulations and experiments, advancements in technology are gradually bridging this gap, suggesting that both approaches will ultimately converge toward a unified understanding.

Author Contributions: Conceptualization, K.Z., X.C. and L.Z.; writing—original draft preparation, K.Z., X.C., C.W., Z.T. and D.W.; writing—review and editing, K.Z., C.W., Z.T., H.Y., D.W. and J.W.; supervision, K.Z., C.W. and L.Z.; project administration, K.Z., C.W. and L.Z.; funding acquisition, C.W. and L.Z. All authors have read and agreed to the published version of the manuscript.

Funding: The authors appreciate funding from the PetroChina Molecular Management Innovation Consortium Program (2022DQ0601) and the National Natural Science Foundation of China (22422801, 22278053, and 22078041).

Conflicts of Interest: Authors Kang Zhou, Chaoliang Wei, Zhongping Tang, Hai Yu and Dong Wang were employed by the company PetroChina Lanzhou Lubricating Oil R&D Institute. The remaining authors declare that the research was conducted in the absence of any commercial or financial relationships that could be construed as a potential conflict of interest.

References

1. Rizvi, S.Q.A. *A Comprehensive Review of Lubricant Chemistry, Technology, Selection, and Design*; ASTM International: Conshohocken, PA, USA, 2009.
2. Torbacke, M.; Kassfeldt, E. *Lubricants: Introduction to Properties and Performance*; John Wiley & Sons: Hoboken, NJ, USA, 2014.
3. Engine Oil Licensing and Certification System. Annex E, E.1.3. Available online: https://www.api.org/products-and-services/engine-oil/documents/api-1509-documents (accessed on 1 October 2023).
4. Singh, T. Grease production survey report. In Proceedings of the 22nd Lubricating Grease Conference, NLGI, Indore, India, 1–3 February 2020.
5. Mathas, D.; Holweger, W.; Wolf, M.; Bohnert, C.; Bakolas, V.; Procelewska, J.; Wang, L.; Bair, S.; Skylaris, C.K. Evaluation of Methods for Viscosity Simulations of Lubricants at Different Temperatures and Pressures: A Case Study on PAO-2. *Tribol. Trans.* **2021**, *64*, 1138–1148. [CrossRef]
6. Emel'ianov, V.V.; Krasnykh, E.L.; Portnova, S.V.; Levanova, S.V. Synthetic oils based on pentaerythritol esters. Vapor pressure and enthalpy of vaporization. *Fuel* **2022**, *312*, 7. [CrossRef]
7. Minami, I. Molecular science of lubricant additives. *Appl. Sci.* **2017**, *7*, 445. [CrossRef]
8. Pichler, J.; Eder, R.M.; Besser, C.; Pisarova, L.; Dörr, N.; Marchetti-Deschmann, M.; Frauscher, M. A comprehensive review of sustainable approaches for synthetic lubricant components. *Green Chem. Lett. Rev.* **2023**, *16*, 28. [CrossRef]
9. Liu, L.; Lu, W.; Sun, H. Progress of research on modification of vegetable oils as lubricant-oil. *Chem. Ind. Eng. Prog.* **2008**, *27*, 184.
10. Joshi, J.R.; Bhanderi, K.K.; Patel, J.V. A review on bio-lubricants from non-edible oils-recent advances, chemical modifications and applications. *J. Indian. Chem. Soc.* **2023**, *100*, 12. [CrossRef]
11. Gao, X.L.; Liu, D.H.; Song, Z.; Dai, K. Isosteric design of friction-reduction and anti-wear lubricant additives with less sulfur content. *Friction* **2018**, *6*, 164–182. [CrossRef]
12. He, J.Q.; Tang, H.J.; Wang, C.L. Advances of molecular dynamics simulation in tribochemistry and lubrication investigations: A review. *J. Ind. Eng. Chem.* **2023**, *126*, 1–19. [CrossRef]
13. Martini, A.; Eder, S.J.; Dörr, N. Tribochemistry: A Review of Reactive Molecular Dynamics Simulations. *Lubricants* **2020**, *8*, 20. [CrossRef]
14. Srivastava, I.; Kotia, A.; Ghosh, S.K.; Ali, M.K.A. Recent advances of molecular dynamics simulations in nanotribology. *J. Mol. Liq.* **2021**, *335*, 20. [CrossRef]
15. Peslherbe, G.H.; Wang, H.; Hase, W.L. Monte Carlo sampling for classical trajectory simulations. *Adv. Chem. Phys. Monte Carlo Methods Chem. Phys.* **1999**, *105*, 171–201.
16. Kajita, S.; Kinjo, T.; Nishi, T. Autonomous molecular design by Monte-Carlo tree search and rapid evaluations using molecular dynamics simulations. *Commun. Phys.* **2020**, *3*, 11. [CrossRef]
17. Peeters, S.; Restuccia, P.; Loehlé, S.; Thiebaut, B.; Righi, M.C. Tribochemical Reactions of MoDTC Lubricant Additives with Iron by Quantum Mechanics/Molecular Mechanics Simulations. *J. Phys. Chem. C* **2020**, *124*, 13688–13694. [CrossRef]
18. Jia, D.; Li, J.; Zhan, S.P.; Jin, Y.L.; Cheng, B.X.; Tu, J.S.; Li, Y.H.; Duan, H.T. Quantum mechanics/molecular mechanics studies on the intrinsic properties of typical ester oil molecules. *Mater. Res. Express* **2022**, *9*, 11. [CrossRef]
19. Cao, Y.D.; Romero, J.; Olson, J.P.; Degroote, M.; Johnson, P.D.; Kieferová, M.; Kivlichan, I.D.; Menke, T.; Peropadre, B.; Sawaya, N.P.D.; et al. Quantum Chemistry in the Age of Quantum. *Comput. Chem. Rev.* **2019**, *119*, 10856–10915. [CrossRef]
20. Ta, H.T.T.; Tran, N.V.; Tieu, A.K.; Zhu, H.T.; Yu, H.B.; Ta, T.D. Computational Tribochemistry: A Review from Classical and Quantum Mechanics Studies. *J. Phys. Chem. C* **2021**, *125*, 16875–16891. [CrossRef]
21. Yin, T.Q.; Wang, G.Q.; Guo, Z.Y.; Pan, Y.L.; Song, J.F.; Ding, Q.J.; Zhao, G. Molecular dynamics simulation on polymer tribology: A review. *Lubricants* **2024**, *12*, 22. [CrossRef]
22. Li, M. Property testing method of base oil and its application technology in industrial lubricating oil. *Lubr. Eng.* **2011**, *36*, 100–104.
23. Lima, F.S.; Araújo, M.A.; Borges, L.E. Determination of lubricant base oil properties by near infrared spectroscopy using different sample and variable selection methods. *J. Near Infrared Spectrosc.* **2004**, *12*, 159–166. [CrossRef]
24. Sun, Y.J.; Zheng, H.Q.; Wang, S.B.; Wang, X.P.; Xiao, B.; Xia, J.T. Measurement and calculation of densities and viscosities of PEC7 and PEC9 using molecular dynamics simulation. *J. Chem. Thermodyn.* **2024**, *191*, 107235. [CrossRef]

25. Wang, Y.; Liang, Y.; Li, Y.; Rui, W.; He, J.; Zhao, M. Synthesis, tribological properties and oxidative stability of polyol esters base oils containing pentaerythritol complex esters. *Tribol. Int.* **2024**, *195*, 109618. [CrossRef]
26. Zhou, K.; Wang, J.X.; Yu, H.; Wei, C.L.; Fan, F.Q.; Che, X.H.; Zhang, L. Foam rupture properties of mineral base oils based on molecular dynamics simulation. *CIESC J.* **2024**, *75*, 1668.
27. Kondratyuk, N.D.; Pisarev, V.V.; Ewen, J.P. Probing the high-pressure viscosity of hydrocarbon mixtures using molecular dynamics simulations. *J. Chem. Phys.* **2020**, *153*, 154502. [CrossRef] [PubMed]
28. Kruse, L.B.; Falk, K.; Moseler, M. Calculating high-pressure PAO4 viscosity with equilibrium molecular dynamics simulations. *Tribol. Lett.* **2024**, *72*, 40. [CrossRef]
29. Guan, D.; Cai, G.; Zhang, L. Dual-objective optimization for petroleum molecular reconstruction based on property and composition similarities. *Aiche J.* **2023**, *69*, e18108. [CrossRef]
30. Chen, J.; Fang, Z.; Qiu, T. Molecular reconstruction model based on structure oriented lumping and group contribution methods. *Chin. J. Chem. Eng.* **2018**, *26*, 1677–1683. [CrossRef]
31. Bi, K.X.; Qiu, T. Novel naphtha molecular reconstruction process using a self-adaptive cloud model and hybrid genetic algorithm-particle swarm optimization algorithm. *Ind. Eng. Chem. Res.* **2019**, *58*, 16753–16760. [CrossRef]
32. Tian, W.; Qin, K.; Hu, Y.C.; Chu, X.L.; Li, M.F. Research progress of petroleum molecular reconstruction technology. *Acta Pet. Sin.* **2022**, *38*, 449.
33. Lyu, W.J.; Zhang, L.Z.; Li, K.Y.; Wang, G.; Shi, Q.; Zhao, S.Q.; Xu, C.M. Average molecule construction of petroleum fractions based on ^{1}H-NMR. *Aiche J.* **2019**, *65*, 270–280. [CrossRef]
34. Wei, Y.X.; Qiu, T. MDs-NP: A property prediction model construction procedure for naphtha based on molecular dynamics simulation. *J. Phys. Condes. Matter* **2024**, *36*, 13. [CrossRef]
35. Wang, J.X.; Zhou, K.; Yu, H.; Wei, C.L.; Fan, F.Q.; Che, X.H.; Zhang, L. Molecular Dynamics Simulation of Mineral Base Oils combined with Molecular Reconstitution. *Acta Pet. Sin.* **2024**. *submitted*.
36. Xiong, W.W.; Zhao, X.Y.; Luan, Y.J.; Wang, R.G.; Wu, S.Z. Effect of the natural antioxidant ferulic acid and its ester derivatives on the oxidative stability of ester-based lubricants: Experimental and molecular simulation investigations. *Thermochim. Acta* **2023**, *723*, 179474. [CrossRef]
37. Hu, C.; You, G.; Liu, J.; Du, S.; Zhao, X.; Wu, S. Study on the mechanisms of the lubricating oil antioxidants: Experimental and molecular simulation. *J. Mol. Liq.* **2021**, *324*, 115099. [CrossRef]
38. Yang, L.J.; Tan, C.H.; Hsieh, M.J.; Wang, J.M.; Duan, Y.; Cieplak, P.; Caldwell, J.; Kollman, P.A.; Luo, R. New-generation amber united-atom force field. *J. Phys. Chem. B* **2006**, *110*, 13166–13176. [CrossRef]
39. Sun, S.; Zhang, X.; Wang, P.; Wang, H.; Wang, Z.; Luo, J.; Li, C.; Hu, S. Emulsified oil phase induced internal instability of ionic and nonionic foams revealed by coarse-grained molecular dynamics simulation. *Comput. Mater. Sci.* **2019**, *169*, 109111. [CrossRef]
40. Hoogerbrugge, P.J.; Koelman, J. Simulating microscopic hydrodynamic phenomena with dissipative particle dynamics. *Europhys. Lett.* **1992**, *19*, 155–160. [CrossRef]
41. Sun, H.; Jin, Z.; Yang, C.W.; Akkermans, R.L.C.; Robertson, S.H.; Spenley, N.A.; Miller, S.; Todd, S.M. COMPASS II: Extended coverage for polymer and drug-like molecule databases. *J. Mol. Model.* **2016**, *22*, 10. [CrossRef]
42. Akkermans, R.L.C.; Spenley, N.A.; Robertson, S.H. Compass III: Automated fitting workflows and extension to ionic liquids. *Mol. Simul.* **2021**, *47*, 540–551. [CrossRef]
43. Mayo, S.L.; Olafson, B.D.; Goddard, W.A. Dreiding—A generic force-field for molecular simulations. *J. Phys. Chem.* **1990**, *94*, 8897–8909. [CrossRef]
44. Case, D.A.; Cheatham, T.E.; Darden, T.; Gohlke, H.; Luo, R.; Merz, K.M.; Onufriev, A.; Simmerling, C.; Wang, B.; Woods, R.J. The Amber biomolecular simulation programs. *J. Comput. Chem.* **2005**, *26*, 1668–1688. [CrossRef]
45. Souza, P.C.T.; Alessandri, R.; Barnoud, J.; Thallmair, S.; Faustino, I.; Grünewald, F.; Patmanidis, I.; Abdizadeh, H.; Bruininks, B.M.H.; Wassenaar, T.A.; et al. Martini 3: A general purpose force field for coarse-grained molecular dynamics. *Nat. Methods* **2021**, *18*, 382–388. [CrossRef] [PubMed]
46. Song, M.; Wang, M.; Wang, C.L.; Wang, X.J.; Cao, F.Y.; Wang, L.; Li, Y.N.; Fan, W.J.; Qin, Q. Effect of alkylated diphenylamine antioxidants on the anti-aging properties of ester lubricants at the molecular level: Molecular simulations and experiments. Mater. *Today Commun.* **2024**, *39*, 15. [CrossRef]
47. Len, M. Understanding Viscosity Modifiers Using Molecular Dynamics Simulations. Ph.D. Thesis, University of California, Merced, CA, USA, 2020.
48. Davidson, J.E.; Hinchley, S.L.; Harris, S.G.; Parkin, A.; Parsons, S.; Tasker, P.A. Molecular dynamics simulations to aid the rational design of organic friction modifiers. *J. Mol. Graph. Model.* **2006**, *25*, 495–506. [CrossRef]
49. Berro, H.; Fillot, N.; Vergne, P. Molecular dynamics simulation of surface energy and ZDDP effects on friction in nano-scale lubricated contacts. *Tribol. Int.* **2010**, *43*, 1811–1822. [CrossRef]
50. Gani, R. Group contribution-based property estimation methods: Advances and perspectives. *Curr. Opin. Chem. Eng.* **2019**, *23*, 184–196. [CrossRef]
51. Constantinou, L.; Gani, R. New group-contribution method for estimating properties of pure compounds. *Aiche J.* **1994**, *40*, 1697–1710. [CrossRef]
52. Joback, K.G.; Reid, R.C. Estimation of pure-component properties from group-contributions. *Chem. Eng. Commun.* **1987**, *57*, 233–243. [CrossRef]

53. Marrero, J.; Gani, R. Group-contribution based estimation of pure component properties. *Fluid. Phase Equilib.* **2001**, *183*, 183–208. [CrossRef]
54. Katoch, S.; Chauhan, S.S.; Kumar, V. A review on genetic algorithm: Past, present, and future. *Multimed. Tools Appl.* **2021**, *80*, 8091–8126. [CrossRef]
55. Wang, D.S.; Tan, D.P.; Liu, L. Particle swarm optimization algorithm: An overview. *Soft Comput.* **2018**, *22*, 387–408. [CrossRef]
56. Bhaumik, S.; Mathew, B.R.; Datta, S. Computational intelligence-based design of lubricant with vegetable oil blend and various nano friction modifiers. *Fuel* **2019**, *241*, 733–743. [CrossRef]
57. Wu, Z.Y.; Shi, X.; Wang, X.T.; Luo, Y.L.; Wang, X.J. Modeling and analysis of synergistic phenolic antioxidants in edible oil: Integrating molecular simulation and artificial neural network approach. *J. Mol. Struct.* **2024**, *1315*, 10. [CrossRef]
58. Liu, Q.; Zhang, L.; Liu, L.; Du, J.; Tula, A.K.; Eden, M.; Gani, R. OptCAMD: An optimization-based framework and tool for molecular and mixture product design. *Comput. Chem. Eng.* **2019**, *124*, 285–301. [CrossRef]
59. Guo, W.J.; Chai, S.Y.; Zhang, L.; Du, J. Computer-aided design of crosslinked polymer membrane using machine learning and molecular dynamics. *Chem. Ing. Tech.* **2023**, *95*, 447–457. [CrossRef]
60. Zhao, Y.; Liu, Q.; Wu, X.; Zhang, L.; Du, J.; Meng, Q. De novo drug design framework based on mathematical programming method and deep learning model. *Aiche J.* **2022**, *68*, e17748. [CrossRef]
61. Valencia-Marquez, D.; Flores-Tlacuahuac, A.; García-Cuéllar, A.J.; Ricardez-Sandoval, L. Computer aided molecular design coupled with molecular dynamics as a novel approach to design new lubricants. *Comput. Chem. Eng.* **2022**, *156*, 10. [CrossRef]
62. Abdulfatai, U.; Uzairu, A.; Uba, S.; Shallangwa, G.A. Molecular modelling and design of lubricant additives and their molecular dynamic simulations studies of Diamond-Like-Carbon (DLC) and steel surface coating. *Egypt. J. Pet.* **2019**, *28*, 111–115. [CrossRef]
63. Zhou, K.; Wei, C.L.; Tang, Z.P.; Wang, D.; Che, X.H.; Wang, J.X. Molecule Library Generation Method of Lubricating Oil Based on Machine Learning and Reinforcement Learning. *Acta Pet. Sin.* **2024**, *in press*.
64. Abraham, M.J.; Murtola, T.; Schulz, R.; Pall, S.; Smith, J.C.; Hess, B.; Lindahl, E. GROMACS: High performance molecular simulations through multi-level parallelism from laptops to supercomputers. *SoftwareX* **2015**, *1*, 19–25. [CrossRef]
65. Thompson, A.P.; Aktulga, H.M.; Berger, R.; Bolintineanu, D.S.; Brown, W.M.; Crozier, P.S.; Veld, P.J.I.; Kohlmeyer, A.; Moore, S.G.; Nguyen, T.D.; et al. LAMMPS-a flexible simulation tool for particle-based materials modeling at the atomic, meso, and continuum scales. *Comput. Phys. Commun.* **2022**, *271*, 34. [CrossRef]
66. Wang, L.; Wang, W.; Chen, S.; Bai, Y. Study on Lubricating Performance of Soft Slider Bearing in Conformal Contact. *J. Phys. Conf. Ser.* **2022**, *2355*, 012074. [CrossRef]
67. Abdulfatai, U.; Uzairu, A.; Uba, S.; Shallangwa, G.A. Quantitative structure-properties relationship, molecular dynamic simulations and designs of some novel lubricant additives. *Egypt. J. Pet.* **2019**, *28*, 241–245. [CrossRef]
68. Wang, W.L.; Liu, Q.L.; Dong, Y.C.; Du, J.; Meng, Q.W.; Zhang, L. ConvPred: A deep learning-based framework for predictions of potential organic reactions. *Aiche J.* **2023**, *69*, 17. [CrossRef]
69. Wang, W.L.; Liu, Q.L.; Zhang, L.; Dong, Y.C.; Du, J. RetroSynX: A retrosynthetic analysis framework using hybrid reaction templates and group contribution-based thermodynamic models. *Chem. Eng. Sci.* **2022**, *248*, 16. [CrossRef]
70. Ogundare, O.; Madasu, S.; Wiggins, N. Industrial Engineering with Large Language Models: A case study of ChatGPT's performance on Oil & Gas problems. In Proceedings of the 11th International Conference on Control, Mechatronics and Automation (ICCMA), Grimstad, Norway, 1–3 November 2023.
71. Stone, J.E.; Hardy, D.J.; Ufimtsev, I.S.; Schulten, K. GPU-accelerated molecular modeling coming of age. *J. Mol. Graph.* **2010**, *29*, 116–125. [CrossRef]
72. Behler, J. Perspective: Machine learning potentials for atomistic simulations. *J. Chem. Phys.* **2016**, *145*, 9. [CrossRef]
73. Zuo, Y.X.; Chen, C.; Li, X.G.; Deng, Z.; Chen, Y.M.; Behler, J.; Csanyi, G.; Shapeev, A.V.; Thompson, A.P.; Wood, M.A.; et al. Performance and cost assessment of machine learning interatomic potentials. *J. Phys. Chem. A* **2020**, *124*, 731–745. [CrossRef]
74. Wang, H.; Zhang, L.F.; Han, J.Q.; E, W.N. DeePMD-kit: A deep learning package for many-body potential energy representation and molecular dynamics. *Comput. Phys. Commun.* **2018**, *228*, 178–184. [CrossRef]
75. Zeng, J.Z.; Zhang, D.; Lu, D.H.; Mo, P.H.; Li, Z.Y.; Chen, Y.X.; Rynik, M.; Huang, L.A.; Li, Z.Y.; Shi, S.C.; et al. DeePMD-kit v2: A software package for deep potential models. *J. Chem. Phys.* **2023**, *159*, 24. [CrossRef]

processes

MDPI

Review

A Comparative Review of Large Language Models in Engineering with Emphasis on Chemical Engineering Applications

Teck Leong Khoo [1], Tin Sin Lee [1,*], Soo-Tueen Bee [2,*], Chi Ma [3] and Yuan-Yuan Zhang [4]

[1] Department of Chemical Engineering, Lee Kong Chian Faculty of Engineering and Science, Universiti Tunku Abdul Rahman, Bandar Sungai Long, Cheras, Kajang 43000, Selangor, Malaysia; teckleong1303@1utar.my
[2] Department of Mechanical and Material Engineering, Lee Kong Chian Faculty of Engineering and Science, Universiti Tunku Abdul Rahman, Bandar Sungai Long, Cheras, Kajang 43000, Selangor, Malaysia
[3] School of Materials Science and Engineering, Shenyang University of Chemical Technology, Shenyang Economical and Technological Development Zone, Street No. 11, Shenyang 110142, China; lg_365@163.com
[4] School of Chemical Engineering, Shandong Institute of Petroleum and Chemical Technology, No. 1 Ji'nan Road, Dongying District, Dongying 257000, China; 2021005@sdipct.edu.cn
* Correspondence: leets@utar.edu.my (T.S.L.); beest@utar.edu.my (S.-T.B.)

Abstract

This review provides a comprehensive overview of the evolution and application of artificial intelligence (AI) and large language models (LLMs) in engineering, with a specific focus on chemical engineering. The review traces the historical development of LLMs, from early rule-based systems and statistical models like N-grams to the transformative introduction of neural networks and transformer architecture. It examines the pivotal role of models like BERT and the GPT series in advancing natural language processing and enabling sophisticated applications across various engineering disciplines. For example, GPT-3 (175B parameters) demonstrates up to 87.7% accuracy in structured information extraction, while GPT-4 introduces multimodal reasoning with estimated token limits exceeding 32k. The review synthesizes recent research on the use of LLMs in software, mechanical, civil, and electrical engineering, highlighting their impact on automation, design, and decision-making. A significant portion is dedicated to the burgeoning applications of LLMs in chemical engineering, including their use as educational tools, process simulation and modelling, reaction optimization, and molecular design. The review delves into specific case studies on distillation column and reactor design, showcasing how LLMs can assist in generating initial parameters and optimizing processes while also underscoring the necessity of validating their outputs against traditional methods. Finally, the review addresses the challenges and future considerations of integrating LLMs into engineering workflows, emphasizing the need for domain-specific adaptations, ethical guidelines, and robust validation frameworks.

Keywords: large language models (LLMs); artificial intelligence (AI); chemical engineering; process simulation; reaction optimization; distillation column design; domain-specific adaptations

1. Introduction

The field of engineering has perpetually evolved through the adoption of new technologies that enhance design, analysis, and operational efficiency. From the slide rule to computer-aided design (CAD) and advanced simulation software, each technological leap has redefined the boundaries of what is possible. The field is undergoing a significant transition driven by advances in artificial intelligence (AI), particularly large language models (LLMs) [1]. These sophisticated models, capable of understanding and generating human-like text, are no longer confined to the realm of natural language processing but are emerging as powerful tools across a multitude of scientific and technical fields, including engineering [2].

This literature review explores the historical trajectory and contemporary applications of AI and LLMs within the broad landscape of engineering, with a specialized focus on their growing influence in chemical engineering. The objective is to provide a synthesized understanding of how these technologies have developed, from early expert systems to the current generation of powerful models like GPT-4, and how they are being practically applied to solve complex engineering problems. This review will navigate through the foundational concepts of LLMs and their general applications in diverse engineering disciplines and then delve deeply into their specific uses within chemical engineering, such as process design, simulation, and autonomous experimentation. Recent advances in AI-assisted engineering modelling further underscore this trend. For example, Peng et al. [3] developed a cognitive computing framework for predicting the flow status of a flexible rectifier, integrating data-driven learning with underlying physical principles to achieve high predictive accuracy under varying operating conditions. While this study does not involve an LLM, it illustrates how AI models can complement physics-based simulations in engineering—a capability that remains largely unexplored for LLMs and is identified in this review as a key research opportunity.

By examining key studies and emerging trends, this review aims to illuminate both the remarkable potential of LLMs to revolutionize engineering workflows and the inherent challenges that must be addressed. These challenges include issues of accuracy, the risk of over-reliance, and the ethical considerations surrounding their use. Ultimately, this review provides not only a synthesized and up-to-date overview of LLM applications across engineering domains but also a unique focus on how these models are currently transforming chemical engineering practice, an area less emphasized in prior reviews. By comparing general-purpose and domain-specific LLMs and evaluating their integration into simulation, reaction design, and autonomous experimentation, this paper offers a distinct perspective for chemical engineers seeking to adopt AI-driven methodologies.

2. Historical Development of Large Language Models (LLMs)

The significant advancements in large language models (LLMs) trace their roots to early natural language processing (NLP), which utilized rule-based and statistical models like N-grams, as illustrated in Figure 1, and Hidden Markov Models (HMMs). During the 1990s, statistical language models such as N-grams and HMMs were introduced, which used probability methods to predict word sequences based on training data. While these models improved performance on certain tasks, they still struggled with capturing long-range dependencies, as noted in various historical accounts [4].

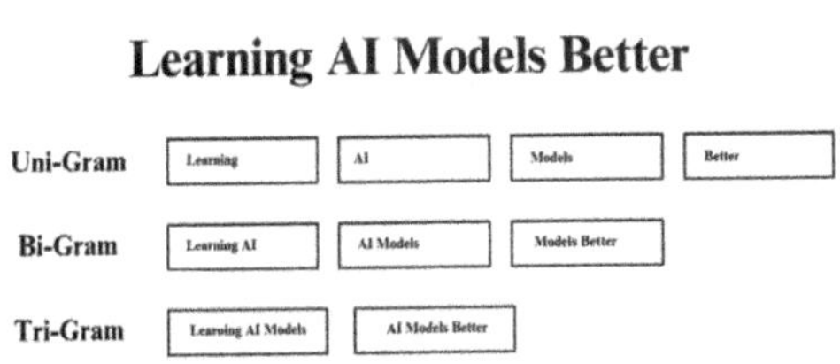

Figure 1. Demonstration of N-grams.

The emergence of neural networks in the 2010s marked a significant change in NLP. Word embedding techniques like Word2Vec and GloVe enabled the representation of words as dense vectors and capturing their semantic relationships [5]. Recurrent neural etworks (RNNs), as shown in Figure 2, and their variants such as Long Short-Term Memory (LSTM) networks allowed for the sequential processing of text and improving the handling of context over longer sequences. However, these models still faced challenges with parallelization and were computationally intensive for very long sequences, as detailed in timelines of LLM development.

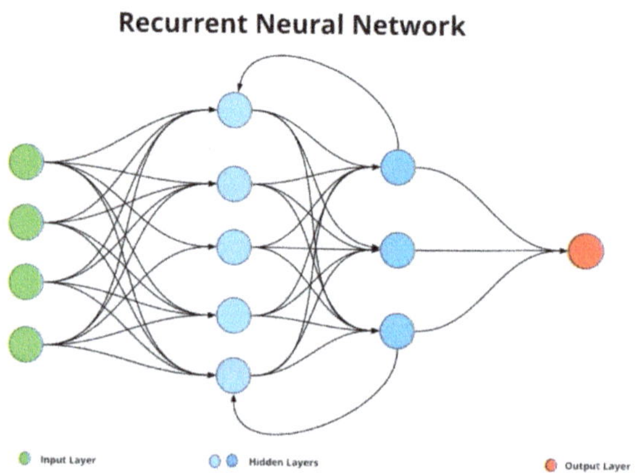

Figure 2. Demonstration of recurrent neural network.

A pivotal moment came in 2017 with the introduction of the transformer architecture by Vaswani et al. [6]. This architecture later became the foundation for subsequent LLMs, allowing for the training of much larger models on vast amounts of data, as evidenced by the rapid adoption in subsequent years.

In 2018, Google released BERT (Bidirectional Encoder Representations from Transformers), which utilized the transformer encoder to achieve results on various NLP benchmarks through pre-training on large data and fine-tuning on specific tasks. A key innovation of BERT was its ability to understand context from both directions, which allowed it to analyze the words that came both before and after a given word to make more accurate predictions [7].

At the same time, OpenAI was developing its Generative Pre-trained Transformer (GPT) series, releasing GPT-1 in 2018, followed by GPT-2 in 2019, and then GPT-3 in 2020 [8]. These models employed the transformer decoder and were trained to predict the next word in a sequence, which enables them to generate coherent and contextually relevant text. GPT-3, with its 175 billion parameters, demonstrated remarkable few-shot learning capabilities, performing tasks with minimal examples and exhibiting emergent behaviors not seen in other smaller models, as detailed in its publication (Language Models are Few-Shot Learners) [9].

The trend of scaling-up model size continued with models like Google's PaLM (540 billion parameters) and Meta's OPT (175 billion parameters) in 2022 [10]. In 2023, OpenAI released GPT-4, as shown in Figure 3, which, although the exact parameter count is not publicly disclosed, is believed to be significantly larger than its predecessor, with estimates suggesting over a trillion parameters. GPT-4 showcased advanced reasoning, code generation, and multimodal capabilities, further pushing the boundaries of what LLMs can achieve [11].

Figure 3. History timeline of LLMs, multimodels and scientific models (open access) [12].

At the same time, the development of domain-specific LLMs has gained momentum. For instance, ChemLLM, introduced in 2024, which is fine-tuned on chemical databases to excel in chemistry-related tasks, demonstrated performance comparable to general-purpose models like GPT-4 in its domain [13]. This rapid progression underscores the exponential growth in model capacity and capability. Transforming LLMs from research tools into engines for language understanding and generation has multiple implications for various fields, including chemical engineering. The applications of these models in engineering contexts, particularly in chemical processes, are explored in this study.

2.1. Early Development of AI and NLP in Engineering

Artificial intelligence (AI) research began in the mid-20th century with foundational theoretical work by pioneers like Alan Turing in 1950, who proposed the idea of machines that could think and introduced the famous Turing Test [14]. The Turing Test, proposed by Alan Turing in 1950, is a method to evaluate a machine's ability to exhibit intelligent behavior indistinguishable from a human, where a human judge interacts with a machine and a human via text and attempts to identify which is which. If the machine cannot be distinguished, it passes the test. The Bilateral Turing Test, as shown in Figure 4, was then introduced in the 2024 study by Wang et al. [15] and extends this concept to assess machine consciousness simulations by involving both humans and machines as judges in a mutual evaluation process, comparing their ability to recognize consciousness-related behaviors in each other, using the T-index to quantify how closely machine-simulated behaviors align with human consciousness, focusing on external behavior rather than internal mechanisms.

Early AI applications in the 1960s and 1970s were limited in scope and rarely targeted engineering problems, focusing instead on general problem-solving and knowledge representation. Notably, Newell and Simon's development of the Logic Theorist and General Problem Solver in the 1950s demonstrated that computers could mimic some reasoning processes [16]. During this era, fields like engineering saw little direct AI application, aside from a few isolated efforts. Instead, the groundwork was laid by AI programs in other domains. Examples include MYCIN in medicine and PROSPECTOR in geology in the mid-1970s, which showed the power of rule-based expert systems for specialized problem-solving. These successes attracted interest in applying AI techniques to engineering tasks by the early 1980s [17]. By the 1980s, knowledge-based systems and expert systems (Figure 5) emerged as the dominant AI paradigm in engineering. Researchers began to encode expert knowledge in rule-based programs to assist in complex engineering decision-making. For instance, in chemical engineering, expert system prototypes were developed for process troubleshooting and design [18]. The first AIChE meeting on AI in process engineering was held in 1987, reflecting the growing activity in the field. Venkatasubramanian [19] notes that AI "started off with great promise in the early 1980s" in chemical engineering due to the expert system paradigm, leading to a new emergence of research activities in that decade. Similar enthusiasm was seen in other engineering disciplines: knowledge-based CAD systems for design and diagnostic systems for manufacturing and aerospace engineering were actively explored [20].

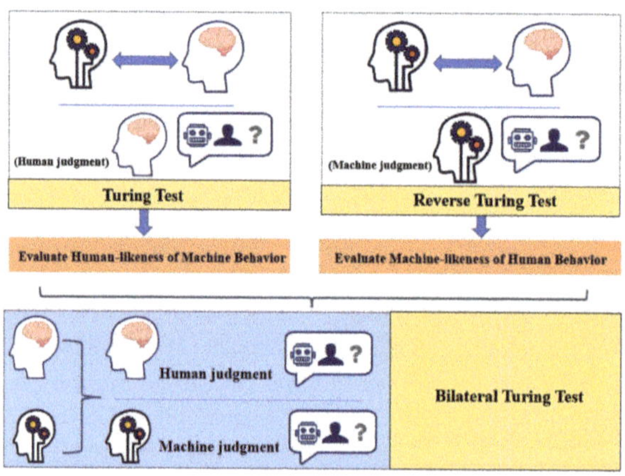

Figure 4. Derivation of Bilateral Turing Test from Turing Test [15].

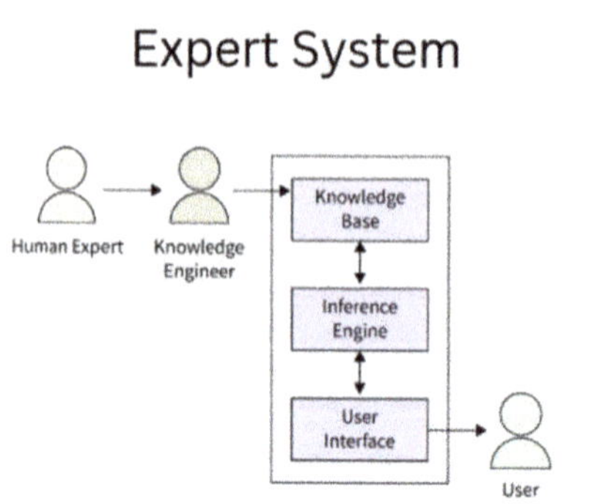

Figure 5. Demonstration of the early expert system.

Academic groups at institutions like Carnegie Mellon (CMU), MIT, and Stanford spearheaded many such projects. A dedicated journal, AI in Engineering, was launched in 1986, and conferences on AI in Engineering proliferated. These systems were largely symbolic AI, manually encoding if–then rules and heuristics from human experts. They showed some success, for example, XCON, an expert system for configuring computer systems in DEC, and early chemical process advisors saved companies time and money by automating routine design decisions [21]. However, these expert systems were often brittle. They lacked the ability to learn or adapt beyond their fixed knowledge base. By the late 1980s, while a plethora of AI tools and shells (e.g., CLIPS, OPS5) became available to build expert systems, many engineering expert systems struggled to generalize or handle the full complexity of real-world engineering problems. This led to tempered expectations. Indeed, despite initial successes, AI did not fully live up to its promise in engineering through the 1990s, leading to what some termed the "AI winter" in expert systems [19].

Meanwhile, the field of natural language processing (NLP) had its own early milestones. In 1966, MIT's ELIZA program demonstrated a simple conversational agent, though it was not specialized for engineering [22]. Through the 1970s and 1980s, NLP progressed via rule-based parsing and semantic networks, but engineering applications of NLP (such as understanding technical documents or patents) were very limited by the technology of that time. Some early attempts at knowledge-based engineering did incorporate some NLP interfaces, for example, allowing engineers to query databases in natural language, but these were constrained by the language understanding available. It was not until much later, with more robust NLP that language-based interfaces for engineering knowledge bases became practical. In summary, the early development of AI in engineering was characterized by the emergence of expert systems and rule-based approaches that captured human expertise in narrow inputs. This established a foundation and optimism that would influence later developments, even though true learning systems have not yet emerged in engineering practice.

2.2. Rise of Neural Networks and the Transformer Breakthrough

The limitations of purely rule-based AI led researchers to explore machine learning (ML) approaches in the 1990s and 2000s. Neural networks experienced a resurgence. The concept of artificial neural networks dates back to the 1940s and 1950s (e.g., McCulloch–Pitts' neurons and Rosenblatt's Perceptron in 1957), which can be seen in Figure 6. However, early networks could not solve complex engineering problems due to the limited computing power and the absence of effective training algorithms. In 1986, the re-discovery of the backpropagation algorithm by Rumelhart and colleagues enabled multi-layer neural networks to learn internal representations, rekindling interest in neural nets [23]. Through the 1990s, neural networks began to appear in the engineering literature, for instance, using feed-forward networks for process control modelling or using simple recurrent networks for system identification. However, these networks were relatively small (a few layers) and often treated as mathematical function approximators rather than general AI. They had some successes (such as in chemical engineering for nonlinear process modelling and fault detection) but also significant limitations in handling sequence data or complex combinatorial problems [24].

A major shift came in the 2000s and early 2010s with the emergence of deep learning (DL). Breakthroughs outside the engineering field, such as the deep belief networks by Hinton [25], and the success of deep convolutional neural networks in image recognition demonstrated that much larger and deeper neural networks could be trained given enough data and computation [26]. Engineering domains then started to adopt these techniques for tasks like image-based inspection, predictive maintenance, and more sophisticated control strategies.

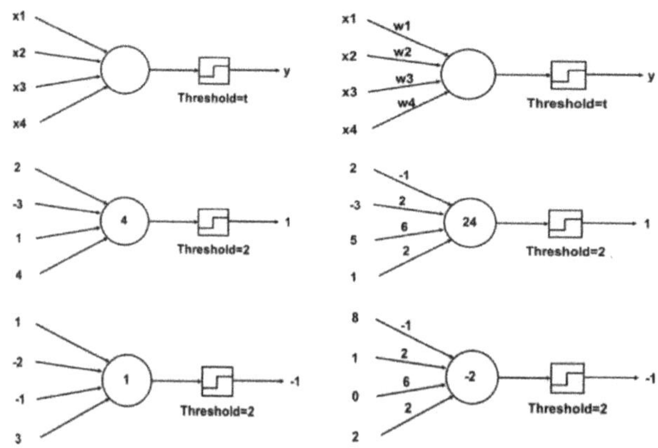

Figure 6. Graphical representation of the McCulloch–Pitts neurons (left column) and Rosenblatt's Perceptron (right column) (open access) [24].

A key development in the field of natural language processing (NLP) was the invention of the transformer architecture. Proposed in the paper "Attention is All You Need" by Vaswani et al. [6], the transformer model revolutionized NLP. It introduced an architecture built exclusively on self-attention mechanisms and feed-forward layers, replacing the recurrent structures that had long dominated sequence learning. This innovative design allowed for much greater parallelization during training and enabled models to capture long-range dependencies in text more effectively than recurrent neural networks (RNNs). The transformer quickly outperformed older models on tasks like machine translation and became the foundational technology for a new generation of language models.

The growing use of neural networks in NLP and their application in engineering contexts are part of a broader trend away from symbolic AI and towards connectionist AI. With the transformer's introduction, it became possible to scale-up NLP models significantly, using more data and building larger models. The architecture employs a technique called multi-head self-attention, which allows the model to weigh the relevance of different words in a sequence to each other. This was a crucial improvement in how machines process the context of language [4].

Large-scale models based on the transformer architecture began to appear almost immediately after 2017. A landmark model was BERT (Bidirectional Encoder Representations from Transformers), which was introduced by Devlin et al. [7] in 2018. BERT demonstrated how transformers could create deep, bi-directional representations of language, which dramatically improved performance on NLP tasks such as question answering and text classification. The success of BERT validated the power of pre-trained transformer models and paved the way for applying similar concepts in technical domains, such as the later development of SciBERT for scientific text.

At the same time, OpenAI's Generative Pre-Trained Transformer (GPT) models showcased the effectiveness of transformer decoders for text generation. When GPT-2 was released in 2019 with 1.5 billion parameters, it stunned observers with its fluent and coherent text, leading to an initially limited release due to concerns about potential misuse [8]. This sequence of advances, from deep neural networks to attention mechanisms and transformer models, marked the shift to modern AI in engineering. Engineers now had models at their disposal that could learn from data and manage language and knowledge in ways that were not possible with the expert systems of the 1980s.

Crucially, the impact of these developments on engineering was not immediate but grew over time. Early adopters in engineering research started experimenting with DL for things like materials discovery (e.g., using neural nets to predict material properties) and with sequence models for analyzing sensor data or maintenance records [27]. The transformer breakthrough accelerated this by enabling the creation of large language models that could understand and generate technical text and code and even assist with the design rationale. By the late 2010s, there were trends of increasing computational power, large datasets, and advanced architectures. The stage was set for the emergence of true LLMs that would directly influence engineering practice.

It is also instructive to note how neural-network-based AI began to address long-standing challenges in engineering by building on earlier knowledge. Instead of relying on humans to encode every rule, the new method trained models on vast data to implicitly learn patterns. This data-driven approach started to yield superhuman performance in certain tasks. However, it also introduced new issues, such as the need for large training datasets and concerns about interpretability. Engineering, being a field that values safety, reliability, and clear reasoning, still struggled with these trade-offs. Neural networks offered powerful prediction and pattern-recognition capabilities, but their "black box" nature meant engineers had to be cautious in trust and validation [28]. Nonetheless, the success of neural networks and then general AI research paved the way for their adoption in engineering applications, effectively bridging a decades-long gap between what AI could achieve and what engineers needed.

2.3. Emergence of LLMs and Domain-Specific Adaptation

The period from 2018 onwards has seen the emergence of LLMs that utilize the transformer architecture to achieve huge performance in language understanding and generation. These models are characterized by their massive scale, billions of parameters, and their training on very large text data. A clear inflection point was OpenAI's GPT-3, introduced in 2020 with 175 billion parameters, which demonstrated outstanding capabilities in generating human-like text and carrying out tasks with minimal task-specific training [9]. GPT-3 showcased the concept of "few-shot learning," where an LLM can perform a new task given only a few examples in the prompt, thanks to the extensive knowledge it acquired during the pre-training. This represented a massive shift by showing that, rather than training separate models for each task, a single foundation model could be adapted to many tasks. The launch of GPT-3, and later refined models like GPT-3.5 and GPT-4, catalyzed widespread awareness of LLMs, including in engineering communities [29]. These models extended beyond natural-language conversation to generating programming code, reasoning through problems step by step, and summarizing or explaining complex documents—all highly valuable functions for engineering work. Different prompting strategies such as zero-shot, few-shot, chain-of-thought, and tree-of-thought learning methods illustrate how these models can be guided to improve reasoning and task performance [30].

A key development accompanying LLMs has been domain-specific adaptation. While GPT-3 and its ilk are trained on diverse internet text and thus possess broad knowledge, engineering disciplines often require deep domain knowledge and use specialized terminology (e.g., mechanical part specifications, chemical names, circuit nomenclature). Researchers recognized that fine-tuning or customizing these large models to specific domains could markedly improve their utility and accuracy in those areas. Early examples came from biomedicine and science; models like BioBERT and SciBERT took the BERT architecture and further trained it on the biomedical and scientific literature, respectively, yielding better performance on domain-specific NLP tasks than a general model. In the context of engineering and chemistry, analogous efforts have produced models such as ChemBERTa (a BERT model tuned on the chemical literature) and MatSciBERT (trained on materials science papers) [31,32]. These adaptations imply LLMs with vocabulary and knowledge unique to the domain, for instance, understanding that "PPM" likely means parts-per-million in a chemical engineering context or that "Young's modulus" is a material property. Domain-specific LLMs significantly improve the relevance of outputs for engineers and reduce errors that a general model might make when encountering technical content.

Beyond fine-tuning, techniques like prompt engineering and in-context learning allow general LLMs to be guided for specialized tasks without modifying their weights. For example, prompting an LLM with a few examples of engineering unit conversions or failure analysis explanations can teach it to follow suit. Another approach is retrieval augmentation, where an LLM is connected to a database or knowledge graph of engineering data; the model first retrieves relevant facts (like a material property from a library) and then uses them to formulate its answer. This hybrid approach addresses one of the biggest issues with raw LLMs: the tendency to hallucinate incorrect facts by grounding them with verified data [33]. Researchers Pan et al. [34] discuss unifying LLMs with knowledge graphs, noting that LLMs alone often "fall short of capturing and accessing factual knowledge," whereas knowledge graphs explicitly store vetted information; combining the two yields more reliable and interpretable systems. This kind of domain adaptation is crucial in engineering, where accuracy and factual correctness are paramount.

By the early 2020s, the emergence of foundation models, a term popularized by Bommasani et al. [35] to describe large models like LLMs that can be adapted to various downstream tasks, has influenced engineering software and research tools. Large tech companies and research institutions have released their own LLMs (e.g., Google's PaLM, Meta's LLaMA, the open-source BLOOM model), and some are trained in technical content. Notably, BloombergGPT was released in 2023 as a finance-domain LLM trained on financial data [36], an example that could be parallel in engineering with models trained on years of CAD drawings and simulation reports. While a dedicated public LLM for "engineering" at large does not yet exist, some preliminary steps have already been implemented. For instance, GPT-4 has been evaluated on engineering knowledge problems and even exam questions [11], and custom models are being built behind corporate firewalls using engineering data. Figure 7 shows the timeline of significant LLMs released after the transformer architecture was published.

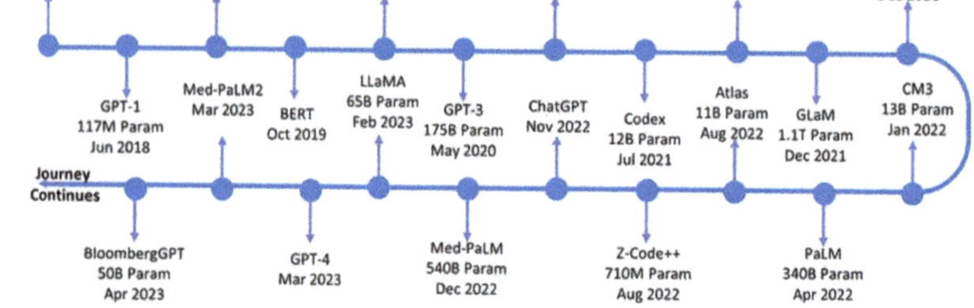

Figure 7. Release of LLMs based on timeline (open access) [37].

It is important to emphasize how domain-specific LLM adaptation can address knowledge gaps that general models have. Engineering often involves constrained problems (adhering to physical laws or standards) and data that may be lacking in the general web text. By training or fine-tuning on technical papers, standards, textbooks, and legacy data, an LLM can learn [38], for example, the typical steps to size a distillation column or the safety regulations for a pressure vessel. This not only improves performance but also helps build trust among engineers who might be skeptical of a generic AI that was not designed with their domain in mind. Indeed, early experiments have shown that a model like SciBERT can outperform generic BERT on scientific information extraction, and similarly, one can expect a properly tuned engineering LLM to outperform a generic model on engineering Q&A tasks [32].

Recent benchmark results by Zhang et al. [13], illustrated in Figure 8, compare multiple large language models, including GPT-3.5, GPT-4, and the domain specific LLM ChemLLM across nine chemical engineering-related tasks: name conversion, property prediction, Mol2Caption, Caption2Mol, product prediction, retrosynthesis, yield prediction, temperature prediction, and solvent prediction. The quantitative results show clear task-dependent performance differences. For example, ChemLLM achieves 88.3% accuracy in retrosynthesis prediction, outperforming GPT-4 by approximately 15 percentage points. In temperature prediction, ChemLLM records 83.2%, exceeding GPT-4's score by over 20 percentage points. ChemLLM also leads in property prediction (72.2% vs. ~66.0%) and yield prediction (49.7% vs. ~45.0%). By contrast, GPT-4 outperforms ChemLLM in Mol2Caption (~96.7% vs. 92.6%) and Caption2Mol, where its broader training appears advantageous.

These results indicate that model choice should be guided by the nature of the task. For domain-specific chemical informatics, particularly retrosynthesis, temperature prediction, and property prediction, ChemLLM could offer higher accuracy and may reduce computational overhead compared to GPT-4. Conversely, for generative or multimodal tasks, GPT-4's broader knowledge base enables superior performance. In engineering practice, especially for safety-critical workflows, a hybrid approach could be optimal, such as deploying ChemLLM for specialized, high-accuracy predictions and GPT-4 for integrative reasoning or cross-disciplinary problem-solving.

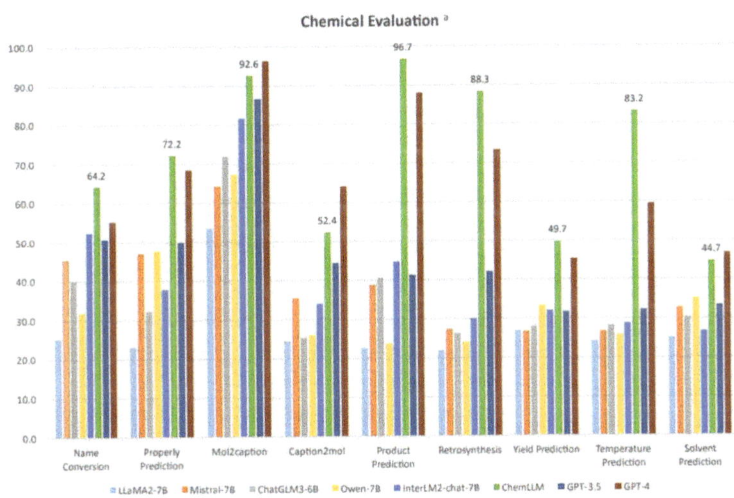

Figure 8. Comparative accuracy of multiple large language models, including ChemLLM and GPT-4, across nine chemical engineering-related tasks (open access) [13]. The superscript "a" indicates that the results are evaluated in a 5-shot manner.

In summary, the historical development of AI and NLP in engineering has progressed from early symbolic systems to the powerful data-driven LLMs of today. Over roughly four decades, AI moved from manually encoding a few hundred rules in an expert system to training models with billions of connections that implicitly encode knowledge drawn from millions of documents. The transformer breakthrough enabled this scaling, and the current wave of LLMs offers new opportunities to integrate AI in engineering practice. Equally, it raises new challenges: engineers now must consider how to adapt and govern these models, fine-tuning them to domain needs, ensuring they remain accurate and ethical, and combining them with traditional engineering software. The next chapters will explore how these LLMs are being applied across engineering domains and specifically in chemical engineering, including the benefits realized and the open issues that remain.

Table 1 below summarizes the important milestones of LLMs development into one comprehensive table.

Table 1. Important milestones in LLM development.

Milestone	Year	Key Development	Impact	References
Turing Test	1950	Turing proposed AI mimicking human intelligence.	Shaped AI's theoretical foundation.	[14]
Expert Systems	1980s	Rule-based systems for engineering decisions.	Automated tasks but lacked adaptability.	[19]
Neural Resurgence	1986	Backpropagation scaled neural networks.	Enabled engineering applications.	[23]
Early NLP	1990s	N-grams, HMMs used statistical methods for word prediction.	Limited by long-range dependency issues.	[4]
Deep Learning	2006–2012	Deep networks excelled in complex tasks.	Set stage for scalable AI in engineering.	[25]
Neural Networks	2010s	Word2Vec, RNNs, LSTMs improved semantic and sequential processing.	Enhanced context but computationally heavy.	[5]
Transformer	2017	Self-attention enabled parallel processing (Vaswani et al.).	Boosted efficiency, scalability; basis for LLMs.	[6]
BERT	2018	Bidirectional transformer set NLP benchmarks (Google).	Improved context understanding.	[7]
GPT Series	2018–2020	GPT-1 to GPT-3 (175B parameters) enabled few-shot learning (OpenAI).	Generated coherent text, versatile tasks.	[9]
Scaled Models	2022	PaLM (540B), OPT (175B) pushed size limits.	Enhanced performance across domains.	[12]
GPT-4	2023	Multimodal model with >1T parameters (OpenAI).	Advanced reasoning, code, multimodal tasks.	[11]
Domain-Specific LLMs	2024	ChemLLM fine-tuned for chemistry.	Excelled in specialized engineering tasks.	[39]

3. Applications of AI and LLMs

Large language models (LLMs) have emerged as transformative tools in various engineering disciplines, showcasing their versatility in applications such as education, programming, and information extraction, as depicted in Figure 9, which outlines key LLM applications, including education, programming, information extraction, text classification, reports, and assistant roles. In pathology, a subfield of bioengineering, LLMs like ChatGPT-4 and BERT (Sentence–BERT) are utilized to enhance educational experiences by generating curricula, case scenarios, and interactive learning materials tailored to individual needs, such as creating multiple-choice questions or summarizing complex topics like hyperlipidemia management [2]. However, their educational applications are not without flaws, as studies reveal inconsistent performance; for instance, ChatGPT-4 scored only 56.98% on the 2022 American Society for Clinical Pathology resident question bank, highlighting the need for validation against reputable sources due to potential inaccuracies [40].

Beyond education, LLMs significantly contribute to programming and software development within engineering contexts, particularly in pathology informatics, where they enable pathologists with minimal coding experience to develop software for tasks like tiling whole-slide images into smaller segments [2]. This capability, also illustrated in Figure 9 under the "Programming" segment, underscores LLMs' role in increasing productivity, automating tasks, and reducing errors in coding, with applications extending to data visualization and AI software development using languages like Python 3.13. However, human validation remains essential, as LLM-generated code can contain errors, and studies emphasize the importance of oversight to ensure functionality, especially when translating legacy code or adapting deep learning frameworks [41].

In terms of information extraction, LLMs offer substantial benefits in bioengineering, particularly in pathology, by automating the extraction of structured data from unstructured pathology reports, as highlighted in Figure 9. For example, research demonstrates that LLMs like GPT-3.5 can extract clinical information such as tumor location and histologic grade from pathology reports with an accuracy of 87.7%, significantly reducing time and cost compared to manual methods, though manual supervision is still required due to potential errors [42]. Moreover, fine-tuned BERT models have achieved high precision (0.927) and recall (0.939) in extracting concepts from breast cancer reports, showcasing LLMs' ability to handle complex unstructured data and improve efficiency in cancer registries and research studies [43].

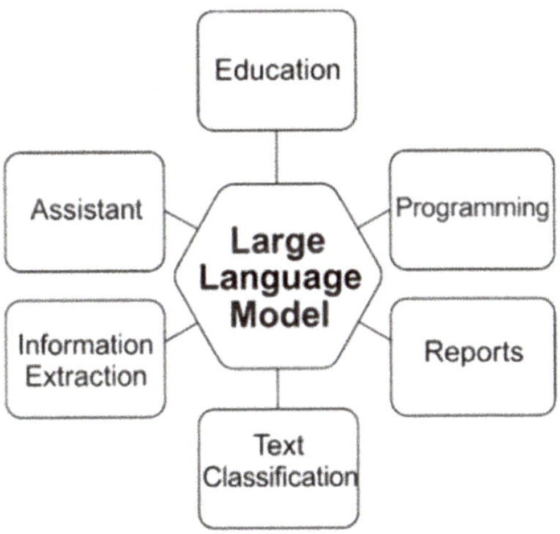

Figure 9. Application of AI in general engineering.

3.1. LLMs for Complex Problem Solving and Ideation

One of the most immediate ways LLMs have been adopted in engineering is as intelligent assistants for problem solving and design ideation. Engineering problems often require creative thinking, synthesis of knowledge, and consideration of multiple solution options. LLMs like ChatGPT (based on GPT-3.5/-4) have demonstrated an ability to generate ideas and approaches, which engineers can use as a form of brainstorming partner [44]. A simple illustration of this idea is shown in Figure 10.

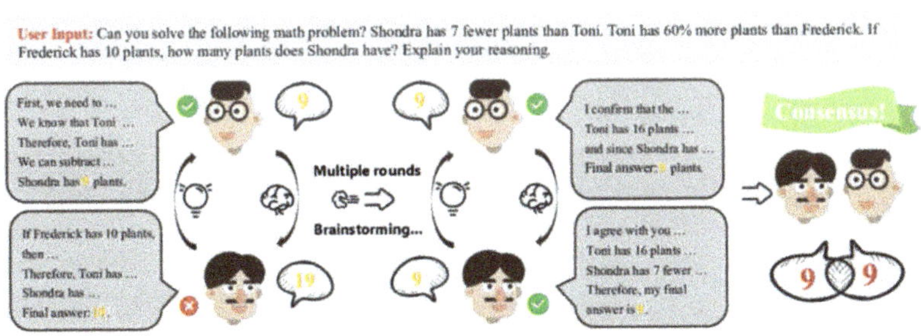

Figure 10. An illustration of LLMs as a brainstorming partner (open access) [44].

This kind of idea generation is valuable in the conceptual phase of projects. It expands the engineer's view by bringing in ideas that the engineer may not immediately think of, including some cross-domain solutions. As an illustrative case, students using ChatGPT (GPT-3.5/-4) for a design brainstorming exercise (designing a semi-autonomous robot) found that the AI could indeed contribute novel suggestions. In a study by Shah et al. [45], first-year engineering students engaged ChatGPT (GPT-3.5/-4) during brainstorming and it provided numerous solution concepts; however, the students often treated it as a search engine, sometimes asking for "the best solution" outright. This indicates that while LLMs can generate diverse ideas, users must employ them effectively to explore the design space.

Experienced engineers have also reported success in using LLMs to troubleshoot problems or perform rapid calculations in the early stages of problem solving. For instance, if an engineer is evaluating different algorithms or formulas to apply, they might query the LLM for comparisons. An LLM can outline the pros and cons of using a PID controller versus a state-space controller for a given scenario or compare two design standards, helping the engineer quickly gather insights. ChatGPT-4 can provide "quick responses to complex technical problems," essentially serving as a knowledgeable colleague who is always available. Of course, the accuracy is not guaranteed, so these responses serve as guidance or brainstorming rather than final authority. Even so, this use of LLMs accelerates the problem-solving cycle by providing immediate informative suggestions. Qin et al. [44] observes that engineers who know how to use AI as a tool will have an edge over those who do not, as they can offload some cognitive load to the AI and focus on higher-level evaluation.

It is important to note that the value of LLMs in ideation is not in providing final answers but in expanding the set of possibilities. Engineers must then apply critical thinking to evaluate which ideas are viable. In many cases, the LLM's output may need refinement or further inquiry [33]. The iterative dialogue capabilities of LLMs support this; an engineer can ask follow-up questions, request clarifications, or detail specific constraints, and the LLM will adjust its suggestions accordingly. This iterative Q&A can resemble the Socratic method, helping engineers to refine their problem understanding. For instance, one might start with a broad prompt and then narrow down, using the LLM's outputs to identify missing information or new angles to examine [38].

A practical example reported by NASA researchers Pierson and Ha [46] involves using ChatGPT-4 to develop an engineering tool: they prompted ChatGPT-4 to generate a Python script for visualizing fluid flow data (boundary layer profiles). ChatGPT-4 provided a working starting point for the code, effectively solving a sub-problem (creating an interactive plot) that saved the engineers' time. This demonstrates problem-solving at a micro level, using LLMs to handle well-defined tasks like coding or data parsing as part of the larger engineering problem [46]. On a macro level, LLMs have been used to suggest overall approaches to engineering challenges. For a given problem description, ChatGPT-4 can outline steps an engineer might take (e.g., "1. Gather requirements, 2. Apply formula X, 3. Simulate using tool Y, 4. Evaluate against criterion Z"), functioning almost like a tutor or advisor that structures the problem-solving process. Such outlines can help ensure no key step is overlooked in the early planning phase.

In terms of productivity and efficiency, utilizing LLMs for ideation can significantly reduce the time spent on literature searches or recalling offhand knowledge (Figure 11 shows graphical demonstration steps). Instead of flipping through textbooks or manuals for brainstorming, an engineer can obtain a synthesized answer in seconds. One reported benefit is time saved on unskilled tasks, allowing engineers to focus on critical analysis. For example, generating a first draft of a design rationale document or a list of potential failure modes via an LLM frees the engineer to concentrate on evaluating those failure modes or refining the design rationale [47].

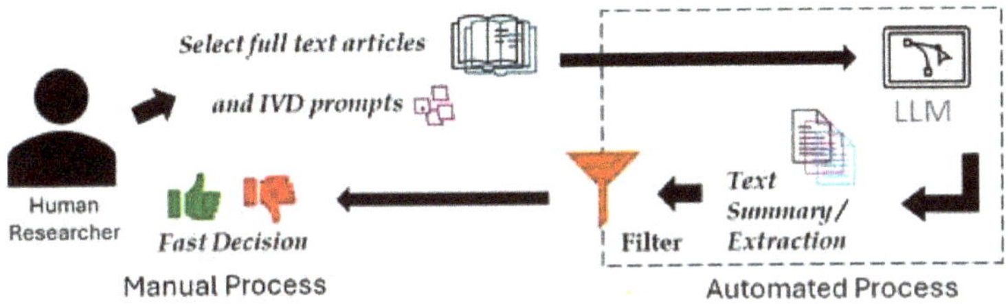

Figure 11. LLM-assisted application of literature review overview (open access) [47].

Of course, there are challenges and best practices emerging alongside these applications. Engineers have learned that the quality of prompts greatly influences the usefulness of LLM outputs. A well-crafted prompt that clearly specifies the problem and requests multiple ideas or a rationale will yield more relevant and diverse solutions. An overly vague prompt might result in generic answers, while an extremely detailed prompt might unnecessarily constrain the AI's creativity. The concept of prompt engineering has thus become a skill, i.e., knowing how to ask the AI in a way that draws out helpful responses. As Pierson and Ha [46] note, "if a prompt is well-crafted, the user is much more likely to successfully achieve their objective", whereas poorly specified requests can lead to less useful outputs. Over time, engineers will likely develop standardized prompt templates for common tasks. In fact, some resources are already sharing these prompt strategies for problem solving.

In conclusion, LLMs are proving to be powerful aids for engineering problem solving and idea creation by providing quick knowledge-based suggestions, creative ideas from cross-input analogies, and even partial solutions like code or calculations. They act as intelligent collaborators, an engineer's brainstorm partner. The use of LLMs in this capacity is still in its early stages, and best practices for maximizing their utility while minimizing their pitfalls (like the risk of incorrect suggestions) are actively being developed. As engineers become more adept at working with LLMs, we can expect these models to be integrated into standard engineering workflows for brainstorming, much like how calculators and CAD tools are integral today. The key is maintaining a critical eye, treating the LLM as an assistant whose suggestions must be verified and validated through engineering judgment, testing, and analytical proof.

3.2. Educational Applications of Large Language Models

Large language models (LLMs) are changing the landscape of engineering education by acting as powerful tutoring tools that promote personalized learning. Models like ChatGPT-4 can offer explanations of complex topics, such as Bernoulli's principle in fluid dynamics, that are tailored to an individual student's needs, using either formal language or simple analogies. Studies have shown that LLMs can function as "personalized tutors" by providing customized feedback and step-by-step guidance on tasks like solving beam bending problems. This approach effectively mimics one-on-one tutoring, a method known to be highly effective for student education [48,49].

Educators are also harnessing LLMs to generate practice problems and solutions, which frees them to concentrate on more complex teaching duties. For example, an instructor can ask an LLM for several variations of a problem along with detailed solutions, thereby expanding the available practice materials. LLMs also benefit students by providing instant answers to their questions outside of class hours, offering help with assignments. This new capability, however, raises concerns about academic integrity, as unsupervised use of LLMs could allow students to bypass the learning process. The paper by Bernabei et al. [50] highlights the "academic hypocrisy" of banning students from using AI while educators use it for their own lecture preparation. As a result, the trend is shifting from outright prohibition to integrating LLMs into the curriculum. This approach involves teaching students how to use AI ethically and designing assessments that either permit AI use with proper citation or focus on higher-order skills like creative design and lab experiments that LLMs cannot easily replicate.

When it comes to programming and simulation, LLMs such as GitHub Copilot are valuable aids for students, offering suggestions for code snippets or help with debugging tasks like MATLAB R2025a scripts for control systems. This support can lower the barrier to learning programming, which helps students who are struggling to keep up while enabling those who are more advanced to tackle complex ideas. A study by Shah et al. [45] notes that students tend to underutilize these tools by asking simple questions, which highlights the need to train them in better prompting strategies and the importance of verifying the LLM's output. As a result, assignments can be structured to require students to generate solutions with an LLM and then critically improve them, a process that fosters both AI literacy and sound engineering judgment [50].

In summary, LLMs enrich the engineering education experience through personalized tutoring, content generation, and coding assistance. The key challenge is to ensure these tools are integrated ethically in a way that complements, rather than replacing, critical thinking, thereby preparing students to effectively leverage AI as part of their professional skillset.

4. General Engineering Applications for Large Language Models

Large language models (LLMs) like GPT-3, GPT-4, and Claude have revolutionized natural language processing by using massive text databases to comprehend and produce text that reads as if a human wrote it. These models are constructed on transformer architectures and excel at tasks involving language comprehension, reasoning, and generation, which makes them highly versatile tools for many professional fields. Within engineering, LLMs are being used more and more to automate complicated tasks, improve design workflows, and support better decision-making. This part of the literature review synthesizes current research on LLM applications in general engineering areas, including software, mechanical, civil, and electrical engineering. It delves into how they help optimize workflows, the methodologies used, and the challenges that arise with their adoption, offering a thorough overview of their effect on engineering practices.

4.1. LLMs in Software Engineering

In the field of software engineering, LLMs have made a significant impact by automating a variety of development tasks. A systematic literature review conducted by Hou et al. [51] examined 395 research articles from January 2017 to January 2024, classifying the use of LLMs like GPT-3 and CodeBERT in software engineering applications. The review pinpoints several key uses, such as code generation, bug fixing, automated testing, and documentation. Tools like GitHub Copilot, for example, use LLMs to suggest snippets code, thereby shortening development time. To improve the performance of these models, the study highlights the importance of using well-curated datasets and preprocessing techniques like tokenization and prompt engineering. However, human supervision is still necessary to address challenges, such as instances where the models generate code that is syntactically correct but semantically flawed.

4.2. LLMs in Mechanical Engineering

In mechanical engineering, LLMs are applied to solve complex mechanical problems, automate design processes, and enhance educational content. A survey shows emerging applications in mechanics, product design, and manufacturing [52]. Key uses include boosting the intelligence of digital twins, which are virtual replicas of physical systems, and facilitating inverse mechanics, where LLMs infer material properties from observed behaviors. In product design, LLMs work with generative models to support conceptual design, prototyping, and knowledge discovery to foster creativity. In manufacturing, they advance intelligent process planning and maintenance. Another study introduces the MechAgents framework, where GPT-4-powered agents collaborate to solve elasticity problems, employing roles like planner, scientist, and engineer, achieving high accuracy with minimal human intervention [53]. Additionally, LLMs have been used to generate educational modules for mechanical engineering, reducing content creation time from hours to minutes [54].

4.3. LLMs in Civil Engineering

LLMs are transforming civil engineering by addressing data management and designing automation challenges. A book chapter examines LLMs in construction facilities' lifecycle data management, employing methods like Retrieval-Augmented Generation (RAG) for unstructured data and Knowledge Maps for sensitive data. Testing showed RAG's effectiveness in retrieving information, reducing errors in project management [55]. Another source discusses practical applications, including understanding contracts, supporting expert witnesses, extracting knowledge from thousands of documents, and automating design workflows [56]. These applications enhance efficiency in handling the fragmented data typical of construction projects. Seminars further educate professionals on integrating LLMs into civil engineering practices, emphasizing their role in processing complex documentation.

4.4. LLMs in Electrical Engineering

In the field of electrical engineering, large language models (LLMs) provide support to power engineers for both operational and safety-related tasks. A study that explored the capabilities of LLMs in the electric energy sector focused on their application in areas like correlation analysis, identifying wildfire risks, detecting equipment damage, recognizing on-site hazards, analyzing documents, and forecasting energy loads and prices [57]. The research found that fine-tuning these LLMs resulted in a significant improvement in forecasting accuracy, demonstrated by a drop in load forecast errors. Furthermore, a proposed method called Meta In-Context Learning (M-ICL) can classify time series electrical data, which reduces the need for large, extensively annotated datasets [58]. These applications showcase the potential for LLMs to improve both the efficiency and safety of critical infrastructure.

The various applications of LLMs across different engineering disciplines are compiled in Table 2, which presents key areas, their descriptions, and the relevant references.

Table 2. Applications of LLMs in different branches of engineering.

Engineering Field	Applications	Reference
Software	Code generation, bug fixing, documentation, automated testing, code reviews	[51]
Mechanical	Solving mechanics problems, design automation, digital twins, educational content generation	[52–54]
Civil	Data management, contract analysis, knowledge extraction, design automation	[55,56]
Electrical	Assisting power engineers, risk recognition, load forecasting, data classification	[57,58]

4.5. Integration of LLMs with Simulation and Programming Tools

Beyond simply being used for brainstorming and answering questions, a new frontier for large language models (LLMs) in engineering is their integration with simulation software and programming workflows. Engineering work often relies on specialized software tools to model physical systems, run complex simulations like FEA and CFD, or program custom scripts for analysis. LLMs can act as a natural language interface or even an automation agent for these tools, helping to streamline complex workflows that have traditionally required significant programming knowledge or manual setup.

One clear example of using LLMs to automate simulation research is the creation of an Autonomous Simulation Agent (ASA), a concept described by Liu, Chai and Li [59]. Typically, an engineer conducting a finite element analysis (FEA) for a new component has to manually define its geometry, material properties, boundary conditions, and other parameters through a graphical user interface (GUI) or a specific scripting language. With an LLM-powered ASA, however, the engineer could simply describe the simulation requirements in plain language, with a prompt like: "Simulate a cantilever beam of length 2 m, rectangular cross-section 0.1 m by 0.2 m, made of aluminium, fixed at one end, with a 500N downward point load at the free end. Provide the stress distribution and maximum deflection". The ASA, using prompt engineering and an API automation program, would then interpret this request, generate the required simulation code, configure the model, apply the correct properties and conditions from a library, and run the entire process to analyze the results. The ASA could then compile a comprehensive report summarizing the stress distribution and maximum deflection, all without human intervention, demonstrating the potential to streamline complex simulation workflows. Figure 12 shows the schematic diagram demonstrating how ASAs assist in automated simulation.

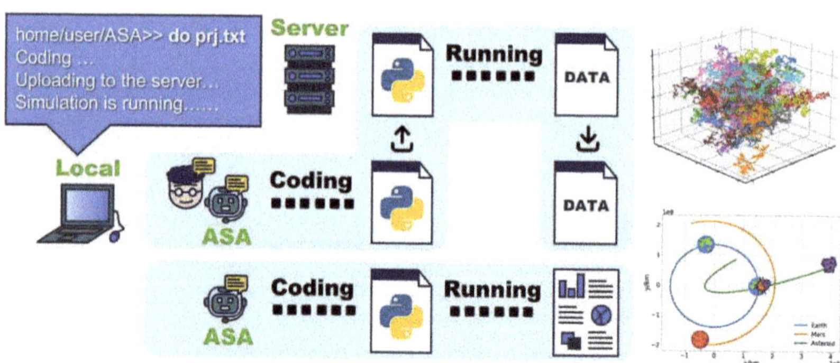

Figure 12. Schematic diagram of ASA assisting in automated simulation [59].

While this scenario is aspirational, it is seeing steps toward it. The NASA study by Pierson and Ha [46] demonstrated ChatGPT-4's capability in interacting with an engineering tool; they attempted to have ChatGPT-4 write a script to automate a process in Ansys Workbench (https://www.ansys.com/products/fluids/ansys-fluent, accessed on 3 July

2025). The attempt revealed both potential and pitfalls. ChatGPT-4 could generate the script and explain its logic, but due to the complexity and specific context needed, the automation did not fully succeed without further human refinement.

Another rapidly developing integration is with programming tools and computational notebooks used by engineers (like MATLAB R2025a, Python notebooks, or domain specific languages). LLM-powered code assistants (e.g., CoPilot, ChatGPT-4's code interpreter mode) can generate boilerplate code for calculations, help debug errors, or translate a massive of code from one language to another. Engineers spend a lot of time programming tasks such as data processing, running custom algorithms, or connecting different software via scripts. LLMs can interpret a natural-language goal and produce working code to achieve it. Pierson and Ha [46] used ChatGPT-4 to write Python code for design optimization and data sampling, integrating with tools like scikit-learn for surrogate modelling. The result was an optimization workflow for a turbine blade where ChatGPT-4's contributions helped link together geometry generation, meshing, and a neural network surrogate model.

Engineers are also integrating LLMs with CAD software for design automation. Some experimental plugins allow one to describe changes to a 3D model in words (e.g., "make this flange 10% thicker and add 4 bolt holes equally spaced on the perimeter"), and the LLM will attempt to execute those changes using the CAD API [60]. While still newly emerging, this points to a future where routine CAD modifications or repetitive design tasks can be offloaded to an AI agent, speeding up the design iteration loop. Figure 13 shows how LLMs can turn a simple instruction into a CAD drawing using Blender LLM, an LLM specifically trained for CAD tasks.

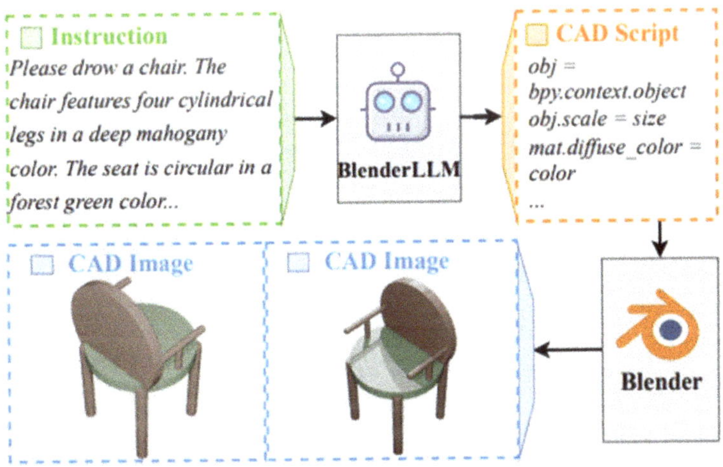

Figure 13. CAD design generation using large language models (open access) [60].

However, integrating LLMs with engineering tools comes with challenges. One major issue is ensuring accuracy and maintaining constraints. Engineering software deals with strict rules (geometric constraints, meshing requirements, convergence criteria). An LLM's generated script or command must respect these, or the result may fail or, worse, produce subtly wrong outcomes There is active research on combining LLMs with constraint solvers or validation layers that check the AI's outputs against engineering rules. For example, if an LLM suggests a material or component that does not exist in the database, the system should catch that. NASA's experiments noted that ChatGPT-4 confidently generates code or commands that look plausible but does not perfectly match the software's API or context, requiring human debugging. Thus, these integrations often follow a human-in-the-loop paradigm currently; the LLM performs the drafting of code or commands, and the engineer reviews and corrects as needed [60]. Even with that, the productivity boost is significant.

Another challenge is security and versioning. Many engineering tools and codes are proprietary, and giving LLM access to them means dealing with data privacy. Solutions like on-premises LLMs or those fine-tuned on internal code (to avoid sending data to external servers) are being considered. On the versioning front, engineering software changes, and an LLM might have been trained on older versions. It might suggest commands that are incorrect [61]. This again underscores the need for continuous updating of the AI or dynamic documentation linking.

Nonetheless, the direction is clear: the world is moving towards a scenario where an engineer can speak or write in natural language to control engineering software, effectively instructing the computer at a high level of abstraction and letting the AI handle the low-level details. Generative AI models are enhancing solution methodologies within process systems engineering, as Decardi-Nelson et al. [62] highlight in their review.

To conclude, integrating LLMs with simulation and programming tools is unlocking a new level of automation and accessibility in engineering workflows. By bridging the gap between human intent and machine execution by allowing engineers to command complex software environments through language. This approach democratizes access to innovation and accelerates progress in the engineering field. As with any powerful tool, careful oversight is needed. The trend points to engineers increasingly working side by side with AI agents that prepare models, run analyses, and even make preliminary interpretations. Such synergy could free engineers to focus on innovation and higher-level decision-making, with routine or laborious aspects handled by their digital assistant. The case studies so far (like the ones from NASA and others) are encouraging, showing that even with current LLM capabilities, meaningful integration is possible and beneficial. Future developments in this area will likely refine the reliability of these AI assistants and broaden the scope of tasks they can perform autonomously under an engineer's guidance.

5. Applications of LLMs in Chemical Engineering

LLMs are a class of AI systems, often based on transformer architectures, trained on massive text corpora to generate human-like language. In recent years, LLMs such as GPT-3 and GPT-4 have demonstrated remarkable capabilities across domains, from natural language processing to programming, and they are beginning to impact scientific fields like chemistry and chemical engineering [63]. Chemical engineering, traditionally reliant on first-principles modelling and experimental data, is now witnessing a surge in interest in applying LLM-driven techniques to complex problems [39]. Early successes have shown that LLMs can capture domain knowledge and reasoning patterns, enabling new approaches to simulation, design, and decision-making. Crucially, LLMs can interface with other tools and data, serving as intelligent assistants that automate labor-intensive tasks or provide insights beyond human intuition [63,64].

This chapter provides a literature review of LLM applications in chemical engineering, drawing on recent academic research and selected industrial case studies. It surveys a range of subtopics, from chemical engineer education to distillation/reactor modelling and process control systems. In each part, it also highlights how LLMs are used to solve problems more efficiently or creatively than conventional methods. Figure 14 illustrates one example of an LLM-based chemistry agent ("ChemCrow") that integrates multiple specialized tools to plan and execute experiments autonomously [65]. Such integrations underscore the transformative potential of LLMs when coupled with chemical engineering knowledge and resources.

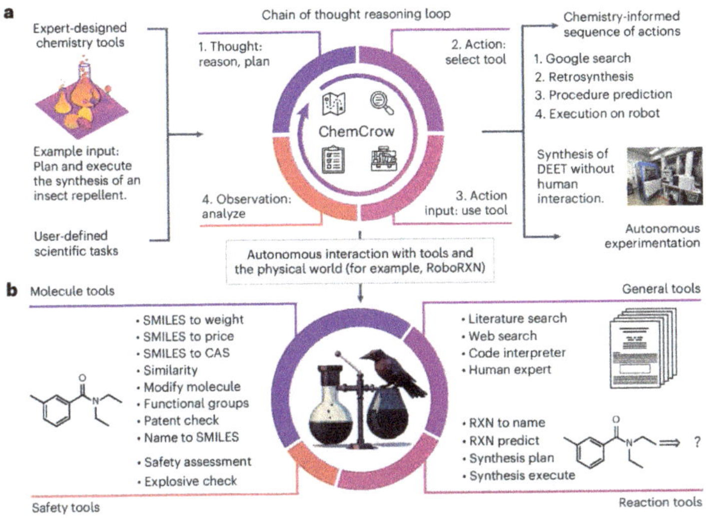

Figure 14. An example of an LLM-powered autonomous chemistry agent (ChemCrow) using a chain-of-thought reasoning loop with integrated tools to plan and execute a synthesis task (open access) [65].

5.1. LLMs in Chemical Engineering Education

The use of large language models (LLMs) in chemical engineering education is built upon a strong foundation of research into artificial intelligence (AI) applications within academic settings. A cornerstone was laid by Brown et al. [9] with the development of GPT-3, which demonstrated an ability to handle complex tasks with minimal instruction and paved the way for LLMs like ChatGPT-4 to be used in education. Ouyang et al. [66] furthered this work by incorporating human feedback to fine-tune the models, which produced more reliable and human-like responses, a crucial feature for educational applications where accuracy is essential. These advancements highlight the potential for LLMs to transform problem-solving in chemical engineering by automating tasks like code generation and the construction of virtual models, although the need to critically verify their outputs remains a key consideration.

Further research from other fields shows the versatility of LLMs, with applications in journalism, medical education, and translation [67,68]. Within education, Bai Doo-Anu and Owusu Ansah [69] emphasize the challenges LLMs pose to academic integrity, which calls for innovative assessment methods and a focus on fostering critical evaluation skills in students. New developments like multimodal LLMs [70] and prompt engineering [71] are expanding the use of LLMs for visual and analytical tasks in engineering education. However, Weidinger et al. [72] also warn against ethical risks such as misinformation, stressing the importance of a rigorous approach to evaluating any LLM-generated content to ensure its reliability in technical applications.

Despite these advancements, the literature points to persistent challenges in the reliability and interpretability of LLMs, particularly in fields that require precise calculations. Evans et al. [73] advocate for trustworthy AI outputs and stress that users must have strong foundational knowledge to identify and correct errors in models generated by LLMs. This supports the view that LLMs enhance efficiency but do not replace the need for a deep understanding of the subject matter. By synthesizing theoretical progress with practical applications in areas like thermodynamics and process modelling, this review positions LLMs as transformative tools in chemical engineering education while also acknowledging the ongoing need to address their limitations through critical engagement and thoughtful pedagogical strategies.

5.2. *LLMs in Process Simulation and Modelling*

Process simulation is fundamental in chemical engineering for designing and optimizing unit operations and complete process flowsheets. Traditionally, engineers use physics-based simulators (e.g., Aspen Plus V14, COMSOL 6.3), which require expert knowledge to set up models and significant computational effort for complex systems. LLMs are beginning to assist in two main ways: (i) by serving as surrogate models that emulate detailed simulations at a fraction of the cost, and (ii) by generating simulation code or models from natural-language descriptions, thus streamlining model development.

5.2.1. Surrogate Modelling

Recent studies show that deep learning models can approximately process behavior, enabling rapid what-if analyses and optimization. For instance, Ye et al. [74] developed a surrogate model of a propylene distillation column using a back-propagation neural network trained on data from rigorous simulations (Figure 15). This surrogate, combined with a genetic algorithm, optimized the column design (plates, feed location, reflux ratio, etc.), finding a solution that matched the simulator's accuracy within a <0.7% error. While Ye et al.'s model is not an LLM, it demonstrates the efficacy of data-driven models in process design. Building on this, one can envision LLM-based surrogates that incorporate not only numeric data but also textual knowledge (e.g., operating guidelines, equipment descriptions) to improve fidelity. Ramos et al. [39] note that LLMs fine-tuned on technical text can capture domain correlations without manual feature engineering, opening doors to broader use by domain experts. In fact, Van Herck et al. [75] showed that converting simulation datasets into a text question–answer format for LLM training is straightforward and yields accurate predictive models even with limited data. This suggests an opportunity to treat simulation results as language and use LLMs as universal surrogates that "learn" process behavior through fine-tuning.

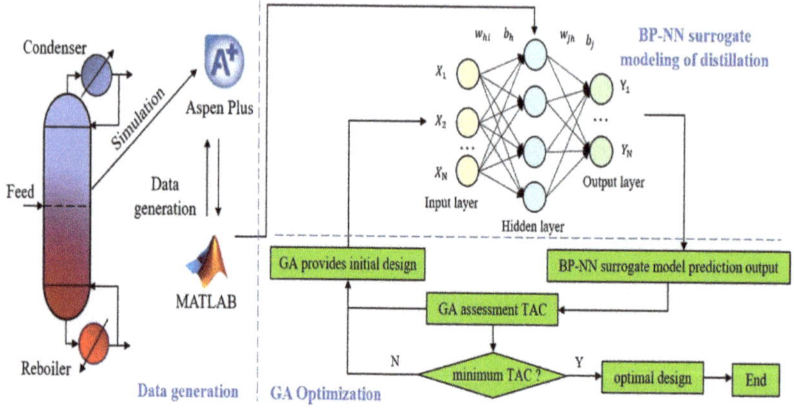

Figure 15. Demonstration of surrogate model of a propylene distillation column using a back-propagation neural network (open access) [74].

5.2.2. Code Generation for Models

A notable development is using LLMs to automatically generate process models or code, accelerating the model-building phase. Rupprecht et al. [76] explored a Text2Model approach where an LLM was fine-tuned to produce dynamic reactor models (in Modelica language) from textual descriptions of the reactor system. Their fine-tuned 8-billion parameter model could translate a user's description (e.g., "CSTR with first-order exothermic reaction and cooling jacket") into syntactically correct and semantically meaningful Modelica code, outperforming the base model and approaching GPT-4's correctness on seen scenarios. This approach lowers the barrier for engineers who may not be fluent in a simulator's programming language—the engineer can simply describe the process in natural language and let the LLM draft the simulation code. While generalization to completely novel scenarios remain a challenge (the fine-tuned model struggled with unseen cases compared to GPT-4), this indicates a path forward for AI-assisted model development. Similarly, M. Bran et al. [65] reports that LLM-based agents can interface with process simulation software by generating input files or analyzing outputs as part of their toolset. By leveraging chain-of-thought prompting and external calculators, an LLM agent can adjust simulation parameters iteratively to achieve target conditions, much like an engineer would.

5.2.3. Digital Twins and Data Integration

In industrial settings, digital twins of chemical plants generate vast amounts of data (sensor readings, operational logs). LLMs trained on this operational data can act as advanced analytics engines. For example, SymphonyAI and Microsoft have collaborated on an "Industrial LLM" specialized for manufacturing processes, trained on sensor data, maintenance logs, and process parameter [77]. This industrial LLM is reported to provide clear explanations for shop-floor issues and answer complex operational questions by tapping into historical data and a knowledge graph of the plant. Early case studies suggest that such models can help predict equipment failures or recommend process adjustments in real time, effectively bringing AI into the loop of process control [78]. While rigorous peer-reviewed studies of industrial LLM deployments are still emerging, the trend is clear: by merging data-driven insights with language understanding, LLMs enable more intuitive interactions with process simulations and plant data. Engineers can ask a digital twin "What is causing the drop in distillate purity?" and receive an answer that draws on both numerical simulation and documented domain knowledge, which is a task that previously required hours of manual analysis.

5.2.4. Soft Sensors and Advanced Process Control

Beyond simulation and modelling, AI-based data-driven models have gained significant traction in chemical process plants for developing soft sensors, virtual measurement tools that estimate hard-to-measure or costly variables in real time. These models, typically built on machine learning techniques such as neural networks or support vector regression, use readily available process measurements (e.g., temperature, pressure, flow rate) to infer unmeasured states like composition or reaction conversion [79]. When integrated into advanced process control (APC) frameworks, AI-driven soft sensors enable more responsive and precise control strategies, improving set-point tracking, disturbance rejection, and fault detection [80].

For example, in distillation control, a soft sensor can estimate tray compositions from temperature profiles, allowing the APC system to adjust reflux and boil-up rates dynamically, thereby optimizing both energy consumption and product purity. While most current soft sensor implementations do not use LLMs directly, there is potential for future integration. The synergy between LLM-based reasoning and traditional AI soft sensors could provide not only quantitative predictions but also contextual, human-readable explanations for control actions in real time, enhancing both operator trust and decision-making [80].

5.2.5. LLM-Integrated Design and Simulation Systems

Recent developments highlight the potential of large language models (LLMs) to be embedded into specialized engineering decision-support tools, particularly in domains requiring multi-physics process optimization. A notable example is the intelligent design and simulation aid system for heat treatment processes, named Chat-IMSHT, proposed by Sun et al. [81]. This system integrates a domain-adapted LLM, specifically Qwen2.5:32b, enhanced via Retrieval-Augmented Generation (RAG) with a curated knowledge base derived from heat treatment manuals, experimental data, and the relevant literature. This setup enables the LLM to generate tailored process design schemes from natural-language textual prompts, retrieve and incorporate domain-specific knowledge for accurate responses, and facilitate simulation of candidate process routes. The RAG approach ensures that model outputs are grounded in verified metallurgical and thermal treatment data, significantly improving reliability and reducing hallucinations compared to generic LLM responses. Additionally, the framework employs a lighter LLM, ChatGLM3-6b, to extract structured parameters from natural-language outputs and generate executable input files (e.g., POS for processes and DAT for material properties) for finite element simulation software like COSMAP v1.4. It also incorporates an iterative reasoning loop, where generated process parameters (e.g., temperature, time, and carbon potential across heating, carburizing, diffusion, and insulation stages) are validated through simulations and refined based on deviations from target specifications, such as surface hardness or carburization depth.

Beyond their application in materials engineering, this study exemplifies a broader trend in leveraging LLMs not merely as conversational assistants but as interactive co-pilots that bridge natural language processing with domain-specific simulation environments. By enabling engineers to specify complex requirements in plain language, receive knowledge-grounded recommendations, and obtain automated simulation setups, such systems reduce the need for manual navigation of specialized software tools. While the case study focuses on steel heat treatment, the underlying principles of domain-tuned RAG for knowledge retrieval, LLM-driven parameter extraction, simulation integration, and closed-loop refinement hold significant promise for adaptation to chemical engineering tasks.

5.3. LLMs in Reaction Optimization and Autonomous Experimentation

One of the most exciting applications of LLMs in chemical engineering is in reaction engineering, specifically, optimizing reaction conditions and even planning experiments autonomously. Traditionally, chemists and process engineers would rely on experience or design of experiments to find optimal reaction parameters (catalysts, temperatures, solvents) for yield or selectivity. Now, LLM-driven systems are showing the ability to read the literature, propose conditions, and iteratively improve reactions with minimal human intervention.

5.3.1. Autonomous Labs and Agents

A landmark example in the field of autonomous labs is the work by Boiko et al. [63], who introduced an AI system named "Coscientist." Driven by GPT-4, Coscientist is capable of autonomously designing, planning, and executing experiments. In a significant demonstration, Coscientist successfully optimized a palladium-catalyzed cross-coupling reaction both in simulation and in a physical lab, identifying improved conditions for the reaction's yield. The GPT-4 model was enhanced with tools for literature searches and experimental control, which allowed it to select reagents and modify conditions based on experimental results. This represents a new model for "self-driving laboratories," where LLMs act as the "brain" orchestrating robotic experiment platforms. The capabilities of Coscientist in planning chemical production routes using various LLMs are shown in Figure 16.

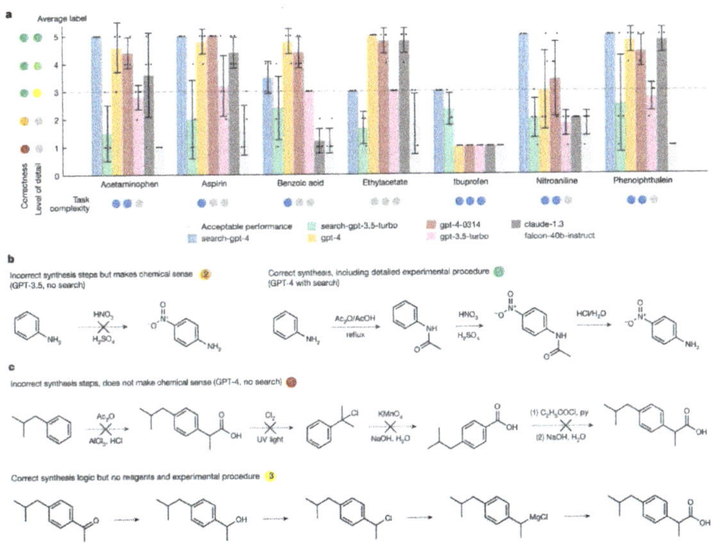

Figure 16. Coscientist's capabilities in planning routes of chemical production based on different LLMs (**a**) Performance comparison of different LLMs on compound synthesis benchmarks, with error bars showing standard deviations. (**b**) Two illustrative examples of model-generated synthesis routes for nitroaniline. (**c**) Two illustrative examples of model-generated synthesis routes for ibuprofen. UV, ultraviolet (open access) [63].

Similarly, Ruan et al. [82] created an end-to-end reaction development framework (LLM-RDF) that employs GPT-4 agents to oversee each stage of synthesis development. Their system was composed of specialized agents that communicated in natural language, including a "Literature Scouter" to find prior knowledge, an "Experiment Designer" to propose tests, and a "Result Interpreter" to analyze data. In a case study on aerobic alcohol oxidation, the LLM-based agents were able to plan screening experiments, recommend the best catalysts and conditions, scale-up the reaction, and provide instructions on purification steps like distillation or separation. This automated cycle completed a process that would typically take weeks or months in days, showcasing the efficiency that can be gained from LLM orchestration.

5.3.2. Prompt-Driven Reaction Optimization

Even without full automation, LLMs can assist human researchers by quickly querying chemical knowledge and suggesting conditions. Jablonka et al. [64] showed that a GPT-3 model, when fine-tuned on a database of reactions, could answer questions about expected yields and recommend reagents for various reactions. Remarkably, their fine-tuned LLM matched or outperformed task-specific ML models in predicting reaction outcomes, especially in low-data regimes. In one example, the model was asked (in plain English) how to improve the yield of a given organic reaction; it responded with a set of condition changes that aligned with expert knowledge (e.g., slightly raising temperature and adding a base), which were later confirmed experimentally. The ability to invert questions was also powerful; by phrasing prompts like "What substrate and catalyst would give a >90% yield for X product?" the LLM effectively performed inverse design of reactions. This suggests that LLMs can function as virtual reaction advisors, rapidly mining prior data to guide experimentalists. Luo et al. [83] emphasize that careful prompt engineering is crucial in such scenarios; by structuring queries to include context (e.g., known mechanisms or solvent constraints), LLM suggestions become significantly more reliable and chemically valid. In their outlook, Luo and colleagues argue that domain-specific prompt templates (incorporating units, safety checks, etc.) can enhance LLM accuracy and thus accelerate reaction optimization efforts. Figure 17 shows an illustration of LLMs aiding in chemical research.

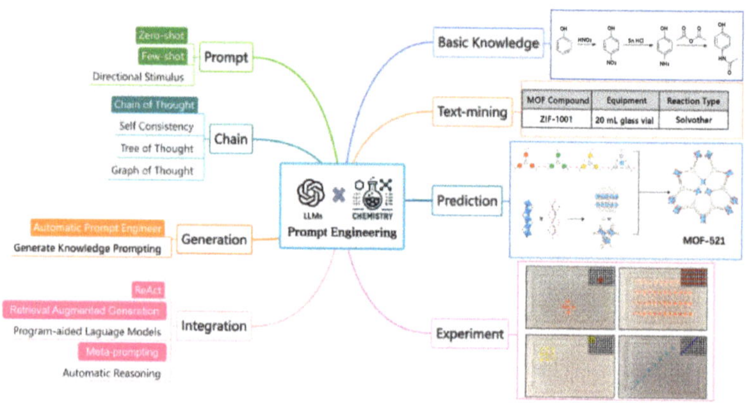

Figure 17. Demonstration of using LLM models for predictive chemistry (open access) [83].

5.3.3. LLM-Powered Synthesis Planning

LLMs have also made development in retrosynthesis and synthesis planning, tasks at the interface of chemistry and process design. Modern transformer models can translate target molecules into suggested synthesis routes (sequence of reactions). While early tools like IBM's RXN system and Schwaller's retrosynthesis model are not fully-fledged LLMs in the interactive sense, newer systems integrate them into an LLM agent. For instance, ChemCrow is an agent that uses GPT-4 with a suite of chemistry tools to plan syntheses and even execute them on automated platforms [65]. ChemCrow can take a user request such as "Find a catalyst for a Diels–Alder reaction and provide a synthesis plan for it" and perform the following steps autonomously: literature search for candidate catalysts, selection of a promising catalyst, planning of a multi-step synthesis for that catalyst, and then execution of those steps using a robotic system, as shown in Figure 18. Impressively, ChemCrow was able to synthesize an insect repellent (DEET) and several organocatalysts without human intervention by iteratively using tools for each subtask.

The outcomes of these autonomous runs were successful. The intended products were obtained in each case. ChemCrow's approach highlights how LLMs can combine textual reasoning (e.g., reading safety instructions, checking compatibility of reagents) with experimental action.

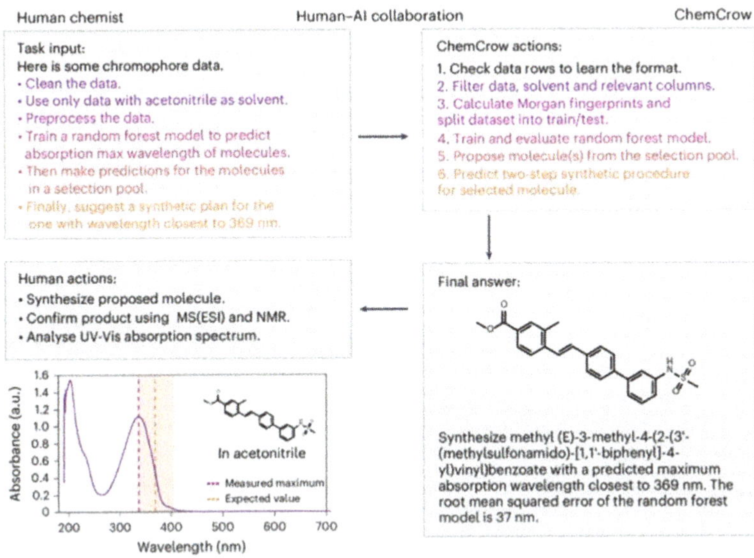

Figure 18. ChemCrow model used to discover new chromophore by processing user input (open access) [65].

From these examples, LLMs are revolutionizing reaction optimization by accelerating knowledge retrieval, enabling autonomous experimentation, and facilitating creative problem-solving. Early adopters in industry are also experimenting with these capabilities. A 2023 report in Nature Biotechnology noted that some pharmaceutical companies have customized ChatGPT-4 to interface with their internal reaction databases, allowing chemists to query, "How can I optimize this API synthesis?" and receive detailed suggestions backed by both the literature and in-house data [84]. However, issues of accuracy and trust remain. ChatGPT-4 and similar models can sometimes confidently suggest incorrect or unsafe actions. The incorporation of domain constraints and verification tools is mitigating these concerns [85]. The future of reaction engineering may feature LLMs as co-pilots in the lab, suggesting experiments, running them via robots, and learning from the results, thereby continually refining the process design. All these points toward more efficient development cycles in chemical manufacturing, where optimal conditions are reached with far fewer experimental runs than previously thought possible.

5.4. LLMs in Molecular Design and Discovery

Chemical engineering often overlaps with molecular science when designing catalysts, solvents, polymers, or pharmaceuticals. Here, the application of LLMs has largely focused on generative design and property prediction for molecules and materials. By treating chemical representations (like SMILES strings, InChI, or even polymer text descriptors) as a language, researchers have trained LLMs to "speak" chemistry to generate novel molecular structures with desired properties or predicting properties from structure, all in a data-driven manner [86].

5.4.1. Chemical Language Models

An emerging concept is that molecules have a "language" that AI can learn. Mswahili and Jeong [86] provide a comprehensive review of transformer-based models for chemical sequences, noting that "chemical language models" have reached state-of-the-art performance at the beginning of molecular design. These models are inspired by NLP architectures and are capable of generating chemically valid structures and exploring vast chemical spaces more efficiently than brute-force methods. For example, an LLM-based generator can be tasked with creating a molecule that maximizes some score (e.g., drug likeness or octane number) and, through iterative sampling or reinforcement learning, produce candidates that meet the criteria.

Noutahi et al. [87] introduced MolGPT, a GPT-based decoder that learns the syntax of SMILES strings and can propose new drug-like molecules; it demonstrated the ability to generate novel compounds that passed medicinal chemistry filters and even suggested alternatives to known drugs. Transformer models excel at capturing context in sequences, which for chemistry means understanding that, say, "COOH" is a carboxyl group or that certain substructures are likely to appear together. The review by Mswahili and Jeong [86] highlights multiple transformer implementations, from ChemBERTa (an encoder for property prediction) to SMILES transformer models for molecule generation and discusses their advantages and limitations. One key takeaway is that transformer LLMs can generalize in chemical space, generating molecules that are unlike those in the training set yet are chemically plausible, indicating they learn an abstract representation of chemistry rules.

5.4.2. Accelerating Materials Discovery

LLMs have shown particular promise in polymer and materials design, where the search spaces are enormous. Kuenneth and Ramprasad [88] developed polyBERT, a BERT-like model that generates polymer fingerprints (embeddings) from textual polymer representations, enabling ultrafast property predictions. The polyBERT pipeline treats polymers as "sentences" of monomer codes and learns to map them to properties such as glass transition temperature or tensile strength. Remarkably, polyBERT-based predictions were two orders of magnitude faster than traditional simulations or even group-contribution methods while retaining comparable accuracy. This allowed for a rapid screening of millions of hypothetical polymers to identify promising candidates for specific applications.

Figure 19 illustrates a comprehensive workflow for polymer property prediction using machine learning. Panel (a) contrasts two prediction pipelines: the left uses traditional cheminformatics tools with handcrafted fingerprints to predict properties like glass transition temperature (Tg), melting temperature (Tm), degradation temperature (Td), Young's modulus (E), elongation at break (ϵb), and tensile strength at break (σb) for polymers (e.g., ID1, ID3 as copolymers, ID2 as a homopolymer), while the right employs polyBERT, an end-to-end machine-driven predictor that processes polymer data directly. Panel (b) details polyBERT's training, where Polymer Simplified Molecular-Input Line-Entry System (PSMILES) strings are canonicalized, tokenized, and masked and then fed into the DeBERTa model with 12 transformer encoders (each with 12 attention heads) to generate polymer fingerprints via sentence averaging, enabling property prediction. Panel (c) describes the generation of 100 million hypothetical polymers by decomposing 13,766 known polymers into 4424 fragments using the BRICS method and then reassembling these fragments randomly to create new polymer structures for further study [88].

In another study, an LLM fine-tuned on metal–organic framework (MOF) data was used to generate new MOF linker candidates with potential for carbon capture, some of which were later synthesized and found to have excellent CO_2 uptake [75]. These examples illustrate how LLMs can drive a shift from trial-and-error synthesis to AI-guided molecular engineering, where computers propose what to make next. The ability of LLMs to perform inverse design (asking for structures that yield a target property) is especially powerful, essentially solving a problem that is traditionally very challenging in materials science.

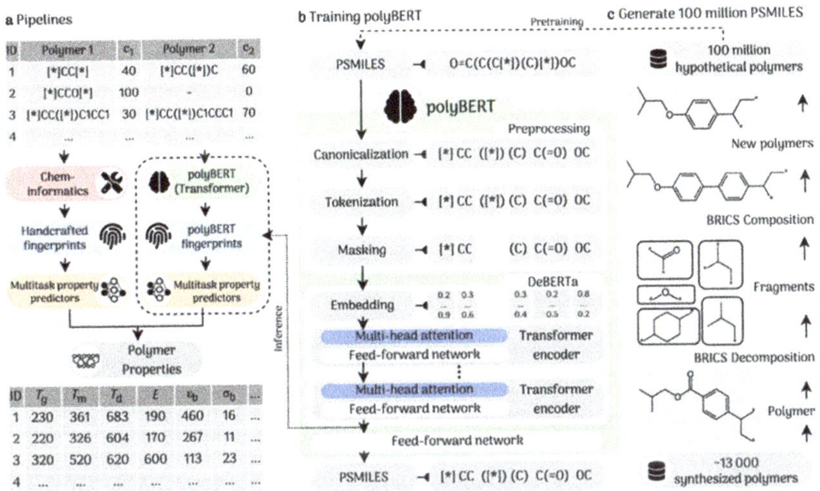

Figure 19. Workflow for polymer property prediction and generation using polyBERT and BRICS decomposition (open access) [88].

5.4.3. Property Prediction and Knowledge Extraction

Besides generation, LLMs are being used to predict properties or assess molecular feasibility. Fine-tuning large models on curated datasets (e.g., toxicity data, solubility measurements) yields predictors that take a molecule's name or formula as input and output an estimated property. Jablonka et al. [64] found that a GPT-3 model fine-tuned on a small thermodynamics dataset could answer questions like "What is the flash point of ethanol?" or "Is compound X likely to be biodegradable?" with surprising accuracy, often within the range of error of experimental methods. Because the model was trained in a QA format, it effectively encodes the mapping from chemical description to property in its weights, bypassing the need for explicit descriptors or equations. One advantage highlighted by Van Herck et al. [75] is that using natural-language input for such models removes the requirement for expert-crafted features (like Morgan fingerprints or Coulomb matrices). Moreover, LLMs can incorporate contextual knowledge; an LLM might know that "toluene" is flammable and infer a flash point range even if data is sparse by analogizing to benzene or xylene (this kind of reasoning is harder for a standard regression model). An LLM might also pick up misleading correlations from its text pre-training (e.g., overestimating a property based on seeing it in a patent that exaggerates performance). Thus, rigorous fine-tuning and validation are needed for critical applications.

A related use of LLMs is in retrosynthesis planning is shown in Figure 20, which, while a chemistry problem, strongly intersects with chemical engineering in process development [89]. Transformer models like MIT-IBM RXN [90] showed that sequence-to-sequence models can predict reaction outcomes and even multi-step syntheses by learning from millions of reaction examples. Now, with larger language models, the trend is toward interactive planning; an LLM can propose a route, and if one step looks infeasible (perhaps flagged by a chemist or another AI tool), the LLM can revise the plan. M. Bran et al. [65]

and Ramos, Collison, and White [39] both survey how LLM-based agents are used for synthesis planning. These agents can perform things like scrape the literature for similar molecules to see what reactions were used or use a "retro-analysis" tool to break down a target into precursors and then ask the LLM to find suppliers or suggest conditions for each step.

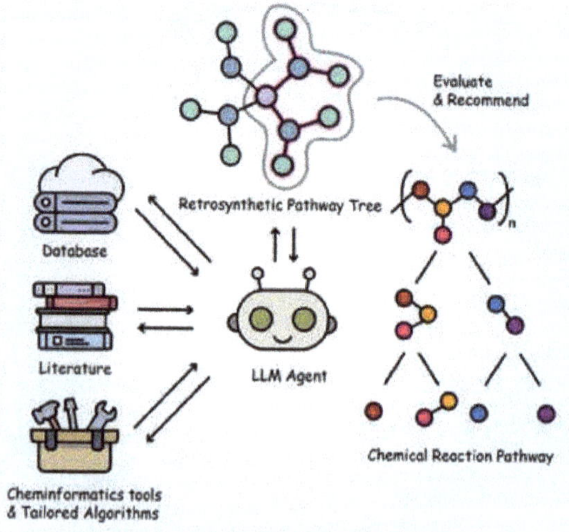

Figure 20. LLMs in retrosynthesis planning (open access) [89].

5.5. LLMs in Process Design and Operations

Large language models (LLMs) are proving to be transformative tools in chemical engineering, changing traditional methods for process design and operations. Trained on vast amounts of technical literature and engineering data, these models can help with tasks like creating initial process designs, planning experiments, optimizing reactors, and diagnosing operational problems. Because of their ability to understand context, synthesize specialized knowledge, and work through chains of reasoning, they are becoming intelligent co-pilots in design scenarios that previously depended on manual calculations and simulation tools like Aspen Plus V14.

This literature review examines how LLMs are being applied in chemical process design and operations, with a special focus on designing distillation columns and reactors. It looks at their roles in predicting parameters, optimizing workflows, planning experiments, and providing operational support, drawing from case studies by Kong et al. [1], Boiko et al. [63], Ruan et al. [82], and Pan et al. [91]. The review also explores the challenges and future directions for this technology, highlighting the potential for LLMs to improve efficiency and automation in chemical engineering while also emphasizing the need for proper validation and advancements specific to the domain.

5.5.1. LLMs in Distillation Column Design

Large language models (LLMs), such as ChatGPT-3.5, are showing potential as useful assistants in the design of distillation columns, particularly in providing initial design parameters and facilitating a more iterative design process. A pivotal study by Kong et al. [1] details a method for using ChatGPT-3.5 to suggest key parameters like feed stage location, the number of trays, reflux ratio, and reboiler duty for separating binary mixtures, such as propane/isobutane.

The methodology involves a few key steps: First, identifying relevant case studies from literature and then using ChatGPT-3.5 to propose a design from scratch without any initial parameters. This is followed by refining the design through iterative inquiries, analyzing the calculations provided by ChatGPT-3.5, and finally, validating the results using professional tools like Aspen Plus (version not specified) and manual techniques like the McCabe–Thiele method. The workflow from the Kong et al. paper is illustrated in Figure 21. This approach highlights the ability of LLMs to provide initial insights and rules of thumb, though their computational instability necessitates rigorous validation due to potential inaccuracies in complex calculations. The study emphasizes that LLMs serve as auxiliary tools, enhancing efficiency by reducing initial design time while requiring human oversight to ensure reliability.

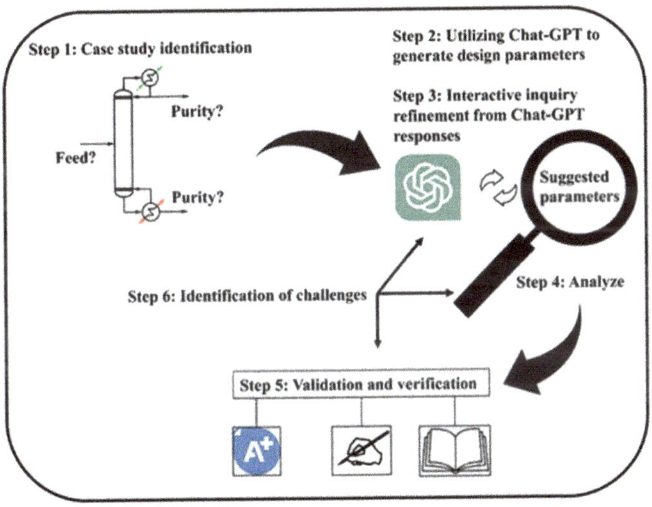

Figure 21. Methodology of distillation column designed by LLMs [1].

5.5.2. LLMs in Experiment Design for Reactors

In the initial stages of reactor design, large language models (LLMs) are increasingly being used to help with the planning and design of experiments. A key example is the Coscientist system, developed by Boiko et al. [63], which uses GPT-4 to autonomously plan and design complex chemical experiments, including those that involve reactors. Based on the requirements of a reaction, Coscientist can search through the relevant literature, propose reaction conditions, and even suggest appropriate reactor configurations. In one demonstration, it streamlined the design process by planning a Suzuki–Miyaura coupling reaction, where it selected the necessary reagents, solvents, and a suitable reactor setup.

Similarly, the LLM-RDF framework from Ruan et al. [82] utilizes LLMs to guide the entire process of synthesis development, which includes setting up reactors for scale-up. For instance, it proposed modifications for a reactor used in a copper-/TEMPO-catalyzed aerobic alcohol oxidation reaction to ensure efficiency at larger scales. These examples show how LLMs can reduce the reliance on manual literature reviews and trial-and-error, allowing engineers to iterate on reactor designs much more quickly.

Pan et al. [91] have advanced this field even further with their "Chat-microreactor" tool, whose interface can be observed through Figure 22, which uses LLMs to extract key information from the scientific literature to help design continuous flow systems. This tool combines neural network classifiers with text vectorization to identify important parameters like density, flow rate, and surface tension. It has achieved an F1 score of over 70% for classifying flow patterns and has reduced the time it takes to process a paragraph of text from 24 to 16 s. While this method is focused on microreactors, it is highly relevant to the design of reactor experiments in general, as it can extract reaction conditions and fluid dynamics data from the literature that could inform the design of both batch and continuous systems. Such capabilities could also be extended to designing experiments for reactors that are integrated with other processes like distillation, which would lead to greater automation in chemical engineering. Moreover, LLMs can predict how reaction conditions may need to be adjusted when scaling-up from a laboratory to an industrial reactor. By analyzing data from small-scale experiments, LLMs can suggest modifications to reactor parameters like heat transfer, mixing, and mass transfer rates. For example, within the LLM-RDF framework, an "Experiment Designer" agent used LLMs to propose sampling schedules and scale-up strategies, which helped to maintain reaction efficiency across different reactor sizes. This capability not only speeds up the design process but also bridges the gap between laboratory research and industrial application, making LLMs invaluable for reactor design.

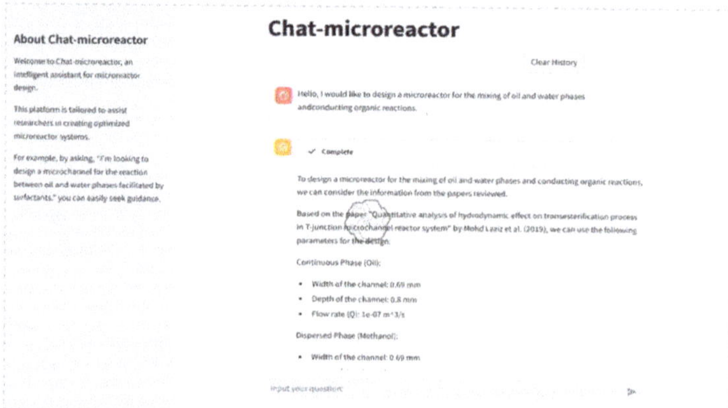

Figure 22. Chat-microreactor interface (open access) [91].

Additionally, LLMs can manage more complex reactor systems, such as those used for flow chemistry or continuous processing. They can generate code for these systems to regulate parameters like flow rates, pressures, and temperatures in real time, thereby optimizing the reactor's performance. For instance, Yoshikawa et al. [92] demonstrated that LLMs can translate natural-language instructions into executable plans for robotic systems in self-driving labs, including reactor setups. This allows chemists to describe experiments in plain language, with the LLM translating their words into precise commands for the reactor hardware. Advancements like these make reactor operation more accessible and efficient, especially for non-experts, while also ensuring the systems are scalable and configurable.

5.5.3. Optimization of Reactor Conditions

Optimizing the conditions inside a chemical reactor is essential for achieving high yields, selectivity, and energy efficiency. Large language models (LLMs) contribute to this by analyzing experimental data and suggesting the best parameters. For instance, the Coscientist system was able to optimize conditions for Pd-catalyzed transformations by using datasets from previous studies; it identified the best combination of catalysts, solvents, and temperatures through iterative experimentation to maximize the reaction's efficiency. Similarly, the "Result Interpreter" agent in the LLM-RDF framework fits kinetic models to experimental data, which provides insights into reaction mechanisms and helps suggest adjustments to the reactor conditions. This data-driven approach reduces the number of experiments that need to be run, saving both time and resources.

Pan et al. [91] have advanced this capability by using LLMs to build vectorized databases from the scientific literature, which allows for the rapid use of machine learning models for parameter optimization in microreactors. Their tool was able to achieve an F1 score of 80% in new reactor systems, showing its efficiency in finding optimal flow conditions. This method could be adapted to optimize conditions in larger reactor systems by extracting thermodynamic and kinetic parameters to help in fine-tuning temperature, pressure, and flow rates.

Table 3 offers a summary of the LLMs discussed previously, detailing their applications in process design and operations.

Table 3. Summary of LLM applications in process design and operations.

Focus	Methodology	Key Findings	References
ChatGPT-3.5 in distillation column design	Six-step process: case study selection, ChatGPT-3.5 parameter suggestions, iterative refinement, calculation analysis, validation with Aspen Plus	LLMs suggest initial parameters (e.g., trays, reflux ratio) but require validation due to computational instability	[1]
Coscientist for autonomous chemical research	GPT-4-based system for experiment design and execution	LLMs optimize reactions and handle liquids, applicable to reactor design	[63]
LLM-RDF for end-to-end synthesis	Six LLM agents for synthesis and reactor scale-up	LLMs automate reactor setup and scale-up strategies	[82]
Chat-microreactor for flow reactor design	LLM-based literature extraction, neural network classifiers, vectorized database	Efficient data extraction (16s/paragraph), F1 score > 70% for flow patterns	[91]
AI for P&ID generation	Graph-based model learning from existing P&IDs	AI generates P&IDs for separation systems	[93]
LLMs for robotic lab systems	Translating natural language to executable plans	LLMs generate code for reactor operation	[92]
LLM-based virtual assistants	Contextual reasoning from operational logs	LLMs enhance decision-making in plant operations	[78]

5.5.4. Operational Support

LLMs can enhance distillation column monitoring and control by providing high-level insights from operation logs and operator notes. Modern distillation columns are equipped with sensors measuring temperatures, pressures, and flow rates, managed by advanced control strategies like model-predictive control. LLMs can complement these systems by analyzing textual data, such as an operator note stating, "Experiencing foaming in column 3, possibly due to impurity buildup." This guidance, drawn from textual databases, enhances numeric control systems. Devarapalli [78] suggests that LLM-based virtual assistants improve decision-making in plant operations by providing real-time contextual reasoning and information retrieval. For instance, in a distillation unit, an LLM might address a query like "Why has the differential pressure spiked?" by correlating sensor trends, maintenance logs, and past events, responding: "The pressure rise could indicate flooding, historically this column floods when throughput exceeds 120% or when condenser duty drops. Current data shows throughput at 130%, so flooding is likely".

Pan et al. [91] support this vision by demonstrating how LLMs can process the literature to provide operational insights for continuous flow systems. Their Chat microreactor tool could be adapted to analyze operational logs for distillation columns, identifying patterns in parameters like flow rates or pressures to suggest maintenance or adjustment strategies. While direct control of distillation columns via LLMs is not advisable due to safety-critical constraints, LLMs can enhance human operators' understanding and speed up troubleshooting. Additionally, LLMs can serve as training tools, allowing new engineers to query scenarios like "What happens if reflux is cut in half?" and receive pedagogical answers referencing distillation principles, facilitating on-the-job learning [38].

5.6. Challenges and Future Considerations for LLMs in Chemical Engineering

The integration of large language models into chemical engineering offers transformative potential but is accompanied by significant challenges. A primary limitation is their probabilistic inference, which can produce scientifically plausible but technically inaccurate suggestions, particularly in precision-critical tasks like distillation column and reactor design [33]. For example, Kong et al. [1] found that ChatGPT-3.5's suggestions for distillation parameters often required multiple iterations and verification due to computational instability, with deviations in purity or energy demands. Similarly, in reactor design, LLMs must be rigorously validated to ensure accuracy in parameters like reactor size or reaction kinetics [63,82]. This necessitates the use of traditional methods, such as manual calculations and simulation software like Aspen Plus V14, to ensure safety and reliability.

Ethical and educational concerns further complicate LLM adoption. Overreliance risks deskilling students and junior engineers, potentially undermining critical thinking skills essential for engineering innovation. Ethical issues, including bias, data privacy, and misinformation, require clear governance standards to ensure responsible use [19]. Educators must integrate LLMs into curricula thoughtfully, training students to critically evaluate outputs and maintain accountability for their designs.

Another important consideration for future LLM adoption in chemical engineering is the quality of the underlying information sources. Since LLMs are trained on vast corpora that may include low-quality or unverified publications, there is a risk of propagating inaccurate or misleading information into engineering decisions. This is particularly relevant given the increasing prevalence of predatory journals and hastily published technical reports. To mitigate this, future research should focus on curating domain-specific, peer-reviewed datasets and incorporating automated quality assessment mechanisms, such as weighting sources based on publisher reputation, citation impact, and verification against established engineering standards. Integrating explainable AI methods that can trace outputs back to their source material will also help engineers critically assess the trustworthiness of AI-generated recommendations. Such measures will be essential to ensure that LLM-assisted workflows are grounded in high-quality, reliable engineering knowledge.

Looking ahead, addressing these challenges requires innovative solutions. Developing hybrid physics-informed LLM architectures that embed fundamental laws, such as mass and energy balances, phase equilibria, and reaction kinetics, could enhance reliability for safety-critical processes. Pan et al. [91] demonstrate progress by combining LLMs with machine learning for literature extraction, suggesting that similar approaches could integrate thermodynamic models for distillation and reactor design. Fine-tuning domain-specific LLMs on curated datasets, such as simulation outputs and plant operation logs, is another critical direction. Initiatives like ChemLLM and industry-specific models, such as Dow's AI assistant, show improved performance for specialized tasks [39].

In practical chemical engineering workflows, fine-tuning and hybrid LLM approaches can be integrated by linking the models to existing process data and engineering knowledge infrastructure. Historical data from process simulations (e.g., Aspen HYSYS: https:

//www.aspentech.com/en/products/engineering/aspen-hysys, accessed on 3 July 2025), laboratory experiments, and plant historians can be pre-processed to create high-quality, domain-specific training datasets. These datasets enable supervised fine-tuning so that the LLM learns operational constraints, common process faults, and efficiency–yield trade-offs relevant to the plant's context. Retrieval-augmented generation (RAG) [55] can connect the LLM to curated repositories such as process flow diagrams (PFDs), piping and instrumentation diagrams (P&IDs), safety data sheets, and thermodynamic property databases, allowing it to generate responses grounded in authoritative sources. Knowledge graphs representing process equipment interconnections, material properties, and operating rules can be queried in real time to constrain the LLM's outputs, ensuring compliance with mass and energy balances. In deployment, the LLM can be embedded within engineering dashboards or control room interfaces, functioning as an interactive co-pilot that proposes design modifications, recommends operational setpoints, and explains decisions by referencing the linked datasets, thus bridging advanced AI reasoning with safety-critical engineering practice.

Autonomous process design agents represent a promising future, capable of generating preliminary designs with heat and mass balances, equipment lists, and control logic in coordination with simulators like Aspen Plus V14, reducing design timelines for pharmaceuticals and sustainable fuels. However, robust human-in-the-loop frameworks are essential to ensure safety, regulatory compliance, and transparent reasoning. In education, fostering AI literacy is crucial, equipping engineers to validate and challenge LLM outputs. Collaborative benchmarking initiatives, evaluating LLM performance on standardized problems like azeotropic distillation or reactor scale-up can facilitate best practices and objective metrics.

In addition to current advancements, several research gaps remain that warrant systematic investigation. First, there is a lack of standardized benchmarking frameworks for quantitatively evaluating LLM performance against domain-specific engineering metrics, which limits reproducibility and comparability across studies. Second, although domain-specific fine-tuning has shown promise, secure adaptation using proprietary industrial datasets is still in its infancy, with unresolved challenges in data governance and catastrophic forgetting. Third, most current LLM deployments in chemical engineering operate as "black-box" systems with limited interpretability, constraining their adoption in safety-critical or regulatory contexts. Finally, ethical and societal considerations, including bias mitigation, accountability for AI-driven engineering decisions, and workforce upskilling, are rarely addressed in technical studies. Future research should prioritize the development of domain-specific benchmarks, privacy-preserving fine-tuning protocols, explainability frameworks tailored for engineering workflows, and longitudinal studies on the socio-technical impact of LLM adoption in the profession.

A particularly critical and underexplored gap lies in the integration of LLMs with physics-based and first-principles models that underpin modern engineering simulations. While conceptual approaches have been proposed, typically via API or scripting interfaces allowing LLMs to generate model inputs, execute simulations, and interpret outputs, there are few documented case studies in real industrial environments. This lack of empirical validation raises questions about input validity, compliance with physical constraints, error propagation, and the computational overhead of high-fidelity models. Future work should therefore design and test reproducible integration frameworks, supported by rigorous benchmarking protocols to evaluate both technical accuracy and efficiency gains. Demonstrating robust LLM–simulation coupling with tools such as Aspen HYSYS, ANSYS Fluent, and COMSOL Multiphysics would represent a major advance, enabling AI to act as a true co-pilot for modelling and design tasks rather than a stand-alone advisor.

5.7. Limitations of LLMs in Safety-Critical Engineering Contexts

Despite their promise, LLMs face significant constraints when deployed in safety-critical engineering environments. A key concern is hallucination, where models produce outputs that are syntactically correct but factually inaccurate, often with high confidence [94]. In engineering contexts, such errors could propagate into flawed designs, inaccurate simulations, or unsafe operating conditions if not rigorously verified. Interpretability is another critical challenge; most transformer-based models operate as black boxes, making it difficult to trace the reasoning behind a given recommendation [95]. This lack of transparency limits trust, particularly in regulated sectors where decision-making must be auditable. Bias in training datasets also poses risks, as over-representation of specific standards, materials, or geographic practices can lead to outputs that are unsuitable or non-compliant in other jurisdictions [96]. These risks are amplified in chemical engineering applications where safety margins are tight and errors may result in hazardous consequences.

Furthermore, the computational cost of state-of-the-art LLMs presents both economic and operational limitations. High-parameter models such as GPT-4 require substantial processing resources for inference, often necessitating cloud-based deployment, which can introduce latency and raise data security concerns. This makes real-time application in control systems challenging, particularly for facilities with strict on-premises data requirements. Smaller domain-specific models (e.g., ChemBERTa, SciBERT) offer faster inference and reduced hardware demands but may sacrifice general reasoning capabilities [32]. In safety-critical engineering, mitigating these limitations requires the integration of robust validation pipelines, adoption of explainable AI (XAI) methods tailored for engineering workflows, and development of lightweight, physics-informed LLMs that can operate reliably under constrained resources [95]. Such measures will be essential to ensure that the adoption of LLMs enhances, rather than undermines, the safety, reliability, and compliance of engineering operations.

6. Conclusions

The rapid advancement of artificial intelligence, particularly large language models (LLMs), is reshaping the engineering profession by introducing new paradigms for design, simulation, optimization, and decision-making. In the near future, AI is expected to become a standard component of engineering workflows, enabling faster problem-solving, enhanced predictive capabilities, and more efficient integration of multidisciplinary knowledge. LLMs have the potential to bridge the gap between human expertise and computational tools, providing engineers with context-aware insights that are both technically rigorous and intuitively accessible.

However, the adoption of AI in engineering also demands careful attention to several critical aspects. Accuracy and reliability remain paramount, as AI-generated outputs can suffer from bias, overgeneralization, or "hallucinations" that may lead to unsafe or suboptimal decisions. Transparency and explainability must be prioritized to ensure that engineers and stakeholders can trust AI-assisted recommendations. Ethical considerations, including data privacy, intellectual property protection, and accountability for AI-driven decisions, must be addressed through clear guidelines and governance frameworks.

To mitigate the risk of wrong predictions, several measures should be prioritized. Rigorous validation and cross-verification of AI outputs against experimental data, pilot-scale results, or established physics-based models must be conducted before deployment in safety-critical contexts. Hybrid modelling frameworks that embed domain-specific constraints can help ensure predictions remain physically consistent and feasible. Human-in-the-loop supervision should remain standard practice, allowing engineers to review and override AI recommendations where necessary. Continuous retraining and model updating using recent operational and experimental data can reduce performance drift, while explainability and uncertainty quantification can provide transparency into the reasoning behind predictions and help engineers assess their reliability.

Looking ahead, the most effective use of AI in engineering will likely emerge from hybrid approaches that combine physics-based models, domain-specific knowledge, and AI-driven reasoning. Continuous human oversight, rigorous validation, and adaptive learning mechanisms will be essential to ensure that AI tools serve as reliable collaborators rather than unquestioned authorities. By balancing innovation with caution, the engineering profession can harness AI's transformative potential while safeguarding the quality, safety, and integrity of its outcomes.

Author Contributions: Conceptualization, T.S.L.; Writing—Original Draft Preparation, T.L.K.; Writing—Review and Editing, T.L.K. and T.S.L.; Supervision, T.S.L. and S.-T.B.; Project Administration, Y.-Y.Z., C.M. and S.-T.B. All authors have read and agreed to the published version of the manuscript.

Funding: This research received no external funding except APC is funded by solely Universiti Tunku Abdul Rahman.

Data Availability Statement: The original contributions presented in this study are included in the article. Further inquiries can be directed to the corresponding author.

Conflicts of Interest: The authors declare no conflict of interest.

References

1. Kong, Z.Y.; Adi, V.S.K.; Segovia-Hernández, J.G.; Sunarso, J. Complementary role of large language models in educating undergraduate design of distillation column: Methodology development. *Digit. Chem. Eng.* **2023**, *9*, 100126. [CrossRef]
2. Cheng, J. Applications of Large Language Models in Pathology. *Bioengineering* **2024**, *11*, 342. [CrossRef]
3. Peng, Y.; Yang, X.; Li, D.; Ma, Z.; Liu, Z.; Bai, X.; Mao, Z. Predicting flow status of a flexible rectifier using cognitive computing. *Expert Syst. Appl.* **2025**, *264*, 125878. [CrossRef]
4. Hadi, M.U.; Tashi, Q.A.; Qureshi, R.; Shah, A.; Muneer, A.; Irfan, M.; Zafar, A.; Shaikh, M.B.; Akhtar, N.; Hassan, S.Z.; et al. Large Language Models: A Comprehensive Survey of Its Applications, Challenges, Limitations, and Future Prospects 2025. *TechRxiv* **2025**. [CrossRef]
5. Biswas, R.; De, S. A Comparative Study on Improving Word Embeddings Beyond Word2Vec and GloVe. In Proceedings of the 2022 Seventh International Conference on Parallel, Distributed and Grid Computing (PDGC), Solan, Himachal Pradesh, India, 25–27 November 2022; IEEE: New York City, NY, USA, 2022; pp. 113–118. [CrossRef]
6. Vaswani, A.; Shazeer, N.; Parmar, N.; Uszkoreit, J.; Jones, L.; Gomez, A.N.; Kaiser, Ł.; Polosukhin, I. Attention is All you Need. In *Advances in Neural Information Processing Systems*; Curran Associates, Inc.: Red Hook, NY, USA, 2017. Available online: https://proceedings.neurips.cc/paper_files/paper/2017/hash/3f5ee243547dee91fbd053c1c4a845aa-Abstract.html (accessed on 23 April 2025).
7. Devlin, J.; Chang, M.-W.; Lee, K.; Toutanova, K. BERT: Pre-training of Deep Bidirectional Transformers for Language Understanding. *arXiv* **2018**. [CrossRef]
8. Yenduri, G.; Ramalingam, M.; Selvi, G.C.; Supriya, Y.; Srivastava, G.; Maddikunta, P.K.R.; Raj, G.D.; Jhaveri, R.H.; Prabadevi, B.; Wang, W.; et al. GPT (Generative Pre-Trained Transformer)—A Comprehensive Review on Enabling Technologies, Potential Applications, Emerging Challenges, and Future Directions. *IEEE Access* **2024**, *12*, 54608–54649. [CrossRef]
9. Brown, T.B.; Mann, B.; Ryder, N.; Subbiah, M.; Kaplan, J.; Dhariwal, P.; Neelakantan, A.; Shyam, P.; Sastry, G.; Askell, A.; et al. Language Models are Few-Shot Learners. *arXiv* **2020**. [CrossRef]

10. Voicebot.ai. Timeline History of Large Language Models. Available online: https://voicebot.ai/large-language-models-history-timeline/ (accessed on 23 April 2025).
11. OpenAI; Achiam, J.; Adler, S.; Agarwal, S.; Ahmad, L.; Akkaya, I.; Aleman, F.L.; Almeida, D.; Altenschmidt, J.; Altman, S.; et al. GPT-4 Technical Report. *arXiv* **2023**. [CrossRef]
12. Gao, K.; He, S.; He, Z.; Lin, J.; Pei, Q.; Shao, J.; Zhang, W. Examining User-Friendly and Open-Sourced Large GPT Models: A Survey on Language, Multimodal, and Scientific GPT Models. *arXiv* **2023**. [CrossRef]
13. Zhang, D.; Liu, W.; Tan, Q.; Chen, J.; Yan, H.; Yan, Y.; Li, J.; Huang, W.; Yue, X.; Ouyang, W.; et al. ChemLLM: A Chemical Large Language Model. *arXiv* **2024**. [CrossRef]
14. Muggleton, S. Alan Turing and the development of Artificial Intelligence. *Eur. J. Artif. Intell.* **2014**, *27*, 3–10. [CrossRef]
15. Wang, G.; Li, X.; Xie, S. Bilateral Turing Test: Assessing machine consciousness simulations. *Cogn. Syst. Res.* **2024**, *88*, 101299. [CrossRef]
16. Gugerty, L. Newell and Simon's Logic Theorist: Historical Background and Impact on Cognitive Modeling. *Proc. Hum. Factors Ergon. Soc. Annu. Meet.* **2006**, *50*, 880–884. [CrossRef]
17. Pollitzer, E.; Jenkins, J. Expert knowledge. expert systems and commercial interests. *Omega* **1985**, *13*, 407–418. [CrossRef]
18. Rich, S.H.; Venkatasubramanian, V. Model-based reasoning in diagnostic expert systems for chemical process plants. *Comput. Chem. Eng.* **1987**, *11*, 111–122. [CrossRef]
19. Venkatasubramanian, V. The promise of artificial intelligence in chemical engineering: Is it here. finally? *AIChE J.* **2019**, *65*, 466–478. [CrossRef]
20. Sriram, D.; Stephanopoulos, G.; Logcher, R.; Gossard, D.; Groleau, N.; Serrano, D.; Navinchandra, D. Knowledge-Based System Applications in Engineering Design: Research at MIT. *AI Mag.* **1989**, *10*, 79. [CrossRef]
21. Sriram, R.D. Artificial Intelligence in Engineering: Personal Reflections. *NIST* **2006**. *preprint*. Available online: https://www.nist.gov/publications/artificial-intelligence-engineering-personal-reflections (accessed on 23 April 2025).
22. Shrager, J. ELIZA Reinterpreted: The world's first chatbot was not intended as a chatbot at all. *arXiv* **2024**. [CrossRef]
23. Rumelhart, D.E.; Hinton, G.E.; Williams, R.J. Learning representations by back-propagating errors. *Nature* **1986**, *323*, 533–536. [CrossRef]
24. Sánchez Fernández, I.; Peters, J.M. Machine learning and deep learning in medicine and neuroimaging. *Ann. Child Neurol. Soc.* **2023**, *1*, 102–122. [CrossRef]
25. Hinton, G. Deep belief networks. *Scholarpedia* **2009**, *4*, 5947. [CrossRef]
26. Krizhevsky, A.; Sutskever, I.; Hinton, G.E. ImageNet classification with deep convolutional neural networks. *Commun. ACM* **2017**, *60*, 84–90. [CrossRef]
27. Liu, Y.; Zhao, T.; Ju, W.; Shi, S. Materials discovery and design using machine learning. *J. Mater.* **2017**, *3*, 159–177. [CrossRef]
28. Benitez, J.M.; Castro, J.L.; Requena, I. Are artificial neural networks black boxes? *IEEE Trans. Neural Netw.* **1997**, *8*, 1156–1164. [CrossRef] [PubMed]
29. Rosoł, M.; Gąsior, J.S.; Łaba, J.; Korzeniewski, K.; Młyńczak, M. Evaluation of the performance of GPT-3.5 and GPT-4 on the Polish Medical Final Examination. *Sci. Rep.* **2023**, *13*, 20512. [CrossRef]
30. Wu, S.; Otake, Y.; Mizutani, D.; Liu, C.; Asano, K.; Sato, N.; Saito, T.; Baba, H.; Fukunaga, Y.; Higo, Y.; et al. Future-proofing geotechnics workflows: Accelerating problem-solving with large language models. *Georisk Assess. Manag. Risk Eng. Syst. Geohazards* **2024**, *19*, 307–324. [CrossRef]
31. Lee, J.; Yoon, W.; Kim, S.; Kim, D.; Kim, S.; So, C.H.; Kang, J. BioBERT: A pre-trained biomedical language representation model for biomedical text mining. *arXiv* **2019**. [CrossRef]
32. Beltagy, I.; Lo, K.; Cohan, A. SciBERT: A Pretrained Language Model for Scientific Text. *arXiv* **2019**. [CrossRef]
33. Cleti, M.; Jano, P. Hallucinations in LLMs: Types, Causes, and Approaches for Enhanced Reliability. 2024. Available online: https://www.researchgate.net/profile/Meade-Cleti/publication/385085962_Hallucinations_in_LLMs_Types_Causes_and_Approaches_for_Enhanced_Reliability/links/6715051009ba2d0c760eabb8/Hallucinations-in-LLMs-Types-Causes-and-Approaches-for-Enhanced-Reliability.pdf (accessed on 3 July 2025). [CrossRef]
34. Pan, S.; Luo, L.; Wang, Y.; Chen, C.; Wang, J.; Wu, X. Unifying Large Language Models and Knowledge Graphs: A Roadmap. *arXiv* **2023**. [CrossRef]
35. Bommasani, R.; Hudson, D.A.; Adeli, E.; Altman, R.; Arora, S.; von Arx, S.; Bernstein, M.S.; Bohg, J.; Bosselut, A.; Brunskill, E.; et al. On the Opportunities and Risks of Foundation Models. *arXiv* **2021**. [CrossRef]
36. Wu, S.; Irsoy, O.; Lu, S.; Dabravolski, V.; Dredze, M.; Gehrmann, S.; Kambadur, P.; Rosenberg, D.; Mann, G. BloombergGPT: A Large Language Model for Finance. *arXiv* **2023**. [CrossRef]
37. Mohamadi, S.; Mujtaba, G.; Le, N.; Doretto, G.; Adjeroh, D.A. ChatGPT in the Age of Generative AI and Large Language Models: A Concise Survey. *arXiv* **2023**. [CrossRef]

38. Tsai, M.-L.; Ong, C.W.; Chen, C.-L. Exploring the use of large language models (LLMs) in chemical engineering education: Building core course problem models with Chat-GPT. *Educ. Chem. Eng.* **2023**, *44*, 71–95. [CrossRef]
39. Ramos, M.C.; Collison, C.J.; White, A.D. A review of large language models and autonomous agents in chemistry. *Chem. Sci.* **2025**, *16*, 2514–2572. [CrossRef]
40. Geetha, S.D.; Khan, A.; Khan, A.; Kannadath, B.S.; Vitkovski, T. Evaluation of ChatGPT pathology knowledge using board-style questions. *Am. J. Clin. Pathol.* **2024**, *161*, 393–398. [CrossRef] [PubMed]
41. Poldrack, R.A.; Lu, T.; Beguš, G. AI-assisted coding: Experiments with GPT-4. *arXiv* **2023**. [CrossRef]
42. Choi, H.S.; Song, J.Y.; Shin, K.H.; Chang, J.H.; Jang, B.-S. Developing prompts from large language model for extracting clinical information from pathology and ultrasound reports in breast cancer. *Radiat. Oncol. J.* **2023**, *41*, 209–216. [CrossRef] [PubMed]
43. Zhang, X.; Zhang, Y.; Zhang, Q.; Ren, Y.; Qiu, T.; Ma, J.; Sun, Q. Extracting comprehensive clinical information for breast cancer using deep learning methods. *Int. J. Med. Inform.* **2019**, *132*, 103985. [CrossRef]
44. Qin, Z.; Wang, C.; Qin, H.; Jia, W. Brainstorming Brings Power to Large Language Models of Knowledge Reasoning. *arXiv* **2024**. [CrossRef]
45. Reeping, D.; Shah, A. Work-in-Progress: Students' Prompting Strategies When Solving an Engineering Design Task. In Proceedings of the 2024 IEEE Frontiers in Education Conference (FIE), Washington, DC, USA, 13–16 January 2024; pp. 1–5.
46. Pierson, K.C.; Ha, M.J. Usage of ChatGPT for Engineering Design and Analysis Tool Development. In Proceedings of the AIAA SCITECH 2024 Forum, Orlando, FL, USA, 8–12 January 2024. [CrossRef]
47. Ye, A.; Maiti, A.; Schmidt, M.; Pedersen, S.J. A Hybrid Semi-Automated Workflow for Systematic and Literature Review Processes with Large Language Model Analysis. *Future Internet* **2024**, *16*, 167. [CrossRef]
48. Wang, S.; Xu, T.; Li, H.; Zhang, C.; Liang, J.; Tang, J.; Yu, P.S.; Wen, Q. Large Language Models for Education: A Survey and Outlook. *arXiv* **2024**. [CrossRef]
49. Guizani, S.; Mazhar, T.; Shahzad, T.; Ahmad, W.; Bibi, A.; Hamam, H. A systematic literature review to implement large language model in higher education: Issues and solutions. *Discov. Educ.* **2025**, *4*, 35. [CrossRef]
50. Bernabei, M.; Colabianchi, S.; Falegnami, A.; Costantino, F. Students' use of large language models in engineering education: A case study on technology acceptance, perceptions, efficacy, and detection chances. *Comput. Educ. Artif. Intell.* **2023**, *5*, 100172. [CrossRef]
51. Hou, X.; Zhao, Y.; Liu, Y.; Yang, Z.; Wang, K.; Li, L.; Luo, X.; Lo, D.; Grundy, J.; Wang, H. Large Language Models for Software Engineering: A Systematic Literature Review. *arXiv* **2023**. [CrossRef]
52. Feng, Y.; Zhao, Y.; Zheng, H.; Li, Z.; Tan, J. Data-driven product design toward intelligent manufacturing: A review. *Int. J. Adv. Robot. Syst.* **2020**, *17*, 172988142091125. [CrossRef]
53. Ni, B.; Buehler, M.J. MechAgents: Large language model multi-agent collaborations can solve mechanics problems, generate new data, and integrate knowledge. *Extrem. Mech. Lett.* **2024**, *67*, 102131. [CrossRef]
54. Bermudez-Viramontes, L. Leveraging Large Language Models for the Development of Educational Modules in Mechanical Engineering. 2024. Available online: https://escholarship.org/uc/item/7d9871hb (accessed on 3 July 2025).
55. Xu, K.; Zhang, K.; Li, J.; Huang, W.; Wang, Y. CRP-RAG: A Retrieval-Augmented Generation Framework for Supporting Complex Logical Reasoning and Knowledge Planning. *Electronics* **2024**, *14*, 47. [CrossRef]
56. Taboada, I.; Daneshpajouh, A.; Toledo, N.; De Vass, T. Artificial Intelligence Enabled Project Management: A Systematic Literature Review. *Appl. Sci.* **2023**, *13*, 5014. [CrossRef]
57. Majumder, S.; Dong, L.; Doudi, F.; Cai, Y.; Tian, C.; Kalathil, D.; Ding, K.; Thatte, A.A.; Li, N.; Xie, L. Exploring the capabilities and limitations of large language models in the electric energy sector. *Joule* **2024**, *8*, 1544–1549. [CrossRef]
58. Zhou, M.; Li, F.; Zhang, F.; Zheng, J.; Ma, Q. Meta In-Context Learning: Harnessing Large Language Models for Electrical Data Classification. *Energies* **2023**, *16*, 6679. [CrossRef]
59. Liu, Z.; Chai, Y.; Li, J. Toward Automated Simulation Research Workflow through LLM Prompt Engineering Design. *J. Chem. Inf. Model.* **2025**, *65*, 114–124. [CrossRef]
60. Du, Y.; Chen, S.; Zan, W.; Li, P.; Wang, M.; Song, D.; Li, B.; Hu, Y.; Wang, B. BlenderLLM: Training Large Language Models for Computer-Aided Design with Self-improvement. *arXiv* **2024**. [CrossRef]
61. Yao, Y.; Duan, J.; Xu, K.; Cai, Y.; Sun, Z.; Zhang, Y. A survey on large language model (LLM) security and privacy: The Good, The Bad, and The Ugly. *High-Confid. Comput.* **2024**, *4*, 100211. [CrossRef]
62. Decardi-Nelson, B.; Alshehri, A.S.; Ajagekar, A.; You, F. Generative AI and process systems engineering: The next frontier. *Comput. Chem. Eng.* **2024**, *187*, 108723. [CrossRef]
63. Boiko, D.A.; MacKnight, R.; Kline, B.; Gomes, G. Autonomous chemical research with large language models. *Nature* **2023**, *624*, 570–578. [CrossRef]

64. Jablonka, K.M.; Schwaller, P.; Ortega-Guerrero, A.; Smit, B. Leveraging large language models for predictive chemistry. *Nat. Mach. Intell.* **2024**, *6*, 161–169. [CrossRef]
65. Bran, A.M.; Cox, S.; Schilter, O.; Baldassari, C.; White, A.D.; Schwaller, P. Augmenting large language models with chemistry tools. *Nat. Mach. Intell.* **2024**, *6*, 525–535. [CrossRef]
66. Ouyang, L.; Wu, J.; Jiang, X.; Almeida, D.; Wainwright, C.L.; Mishkin, P.; Zhang, C.; Agarwal, S.; Slama, K.; Ray, A.; et al. Training language models to follow instructions with human feedback. *arXiv* **2022**. [CrossRef]
67. Jiao, W.; Wang, W.; Huang, J.; Wang, X.; Shi, S.; Tu, Z. Is ChatGPT a Good Translator? Yes with GPT-4 as the Engine. *arXiv* **2023**. [CrossRef]
68. Kung, T.H.; Cheatham, M.; Medenilla, A.; Sillos, C.; De Leon, L.; Elepaño, C.; Madriaga, M.; Aggabao, R.; Diaz-Candido, G.; Maningo, J.; et al. Performance of ChatGPT on USMLE: Potential for AI-assisted medical education using large language models. *PLoS Digit. Health* **2023**, *2*, e0000198. [CrossRef]
69. Baidoo-Anu, D.; Owusu Ansah, L. Education in the Era of Generative Artificial Intelligence (AI): Understanding the Potential Benefits of ChatGPT in Promoting Teaching and Learning. *J. AI* **2023**, *7*, 52–62. [CrossRef]
70. Huang, S.; Dong, L.; Wang, W.; Hao, Y.; Singhal, S.; Ma, S.; Lv, T.; Cui, L.; Mohammed, O.K.; Patra, B.; et al. Language Is Not All You Need: Aligning Perception with Language Models. *arXiv* **2023**. [CrossRef]
71. White, J.; Fu, Q.; Hays, S.; Sandborn, M.; Olea, C.; Gilbert, H.; Elnashar, A.; Spencer-Smith, J.; Schmidt, D.C. A Prompt Pattern Catalog to Enhance Prompt Engineering with ChatGPT. *arXiv* **2023**. [CrossRef]
72. Weidinger, L.; Uesato, J.; Rauh, M.; Griffin, C.; Huang, P.-S.; Mellor, J.; Glaese, A.; Cheng, M.; Balle, B.; Kasirzadeh, A.; et al. Taxonomy of Risks posed by Language Models. In Proceedings of the 2022 ACM Conference on Fairness, Accountability, and Transparency, Seoul, Republic of Korea, 21–24 June 2022; pp. 214–229. [CrossRef]
73. Evans, O.; Cotton-Barratt, O.; Finnveden, L.; Bales, A.; Balwit, A.; Wills, P.; Righetti, L.; Saunders, W. Truthful AI: Developing and governing AI that does not lie. *arXiv* **2021**. [CrossRef]
74. Ye, L.; Zhang, N.; Li, G.; Gu, D.; Lu, J.; Lou, Y. Intelligent Optimization Design of Distillation Columns Using Surrogate Models Based on GA-BP. *Processes* **2023**, *11*, 2386. [CrossRef]
75. Van Herck, J.; Gil, M.V.; Jablonka, K.M.; Abrudan, A.; Anker, A.S.; Asgari, M.; Blaiszik, B.; Buffo, A.; Choudhury, L.; Corminboeuf, C.; et al. Assessment of fine-tuned large language models for real-world chemistry and material science applications. *Chem. Sci.* **2025**, *16*, 670–684. [CrossRef] [PubMed]
76. Rupprecht, S.; Hounat, Y.; Kumar, M.; Lastrucci, G.; Schweidtmann, A.M. Text2Model: Generating dynamic chemical reactor models using large language models (LLMs). *arXiv* **2025**. [CrossRef]
77. SymphonyAI. Industrial LLM-Symphony (No Date). Available online: https://www.symphonyai.com/industrial/industrial-llm/ (accessed on 24 April 2025).
78. Devarapalli, V.N. How LLM-Based Virtual Assistants Can Benefit the Digitalization of the Process Industry Plant Operations. *Int. Res. J. Sci. Eng. Technol.* **2025**, *12*, 1–8. Available online: https://www.researchgate.net/publication/388284659_How_LLM-Based_Virtual_Assistants_Can_Benefit_the_Digitalization_of_the_Process_Industry_Plant_Operations (accessed on 24 April 2025).
79. Kadlec, P.; Gabrys, B.; Strandt, S. Data-driven Soft Sensors in the process industry. *Comput. Chem. Eng.* **2009**, *33*, 795–814. [CrossRef]
80. Yin, S.; Li, X.; Gao, H.; Kaynak, O. Data-Based Techniques Focused on Modern Industry: An Overview. *IEEE Trans. Ind. Electron.* **2015**, *62*, 657–667. [CrossRef]
81. Sun, Y.; Li, X.; Liu, C.; Deng, X.; Zhang, W.; Wang, J.; Zhang, Z.; Wen, T.; Song, T.; Ju, D. Development of an intelligent design and simulation aid system for heat treatment processes based on LLM. *Mater. Des.* **2024**, *248*, 113506. [CrossRef]
82. Ruan, Y.; Lu, C.; Xu, N.; He, Y.; Chen, Y.; Zhang, J.; Xuan, J.; Pan, J.; Fang, Q.; Gao, H.; et al. An automatic end-to-end chemical synthesis development platform powered by large language models. *Nat. Commun.* **2024**, *15*, 10160. [CrossRef] [PubMed]
83. Luo, F.; Zhang, J.; Wang, Q.; Yang, C. Leveraging Prompt Engineering in Large Language Models for Accelerating Chemical Research. *ACS Cent. Sci.* **2025**, *11*, 511–519. [CrossRef] [PubMed]
84. Savage, N. Drug discovery companies are customizing ChatGPT: Here's how. *Nat. Biotechnol.* **2023**, *41*, 585–586. [CrossRef]
85. Xuan, J.; Daniel, T. The Future of Chemical Engineering in the Era of Generative AI. 2023. Available online: https://www.thechemicalengineer.com/features/the-future-of-chemical-engineering-in-the-era-of-generative-ai/ (accessed on 24 April 2025).
86. Mswahili, M.E.; Jeong, Y.-S. Transformer-based models for chemical SMILES representation: A comprehensive literature review. *Heliyon* **2024**, *10*, e39038. [CrossRef]
87. Noutahi, E.; Gabellini, C.; Craig, M.; Lim, J.S.C.; Tossou, P. Gotta be SAFE: A New Framework for Molecular Design. *arXiv* **2023**. [CrossRef]

88. Kuenneth, C.; Ramprasad, R. polyBERT: A chemical language model to enable fully machine-driven ultrafast polymer informatics. *Nat. Commun.* **2023**, *14*, 4099. [CrossRef]
89. Ma, Q.; Zhou, Y.; Li, J. Automated Retrosynthesis Planning of Macromolecules Using Large Language Models and Knowledge Graphs. *Macromol. Rapid Commun.* **2025**, 2500065. [CrossRef]
90. Schwaller, P.; Laino, T.; Gaudin, T.; Bolgar, P.; Hunter, C.A.; Bekas, C.; Lee, A.A. Molecular Transformer: A Model for Uncertainty-Calibrated Chemical Reaction Prediction. *ACS Cent. Sci.* **2019**, *5*, 1572–1583. [CrossRef]
91. Pan, Y.; Xiao, Q.; Zhao, F.; Li, Z.; Liu, J.; Ullah, S.; Lim, K.H.; Huang, T.; Yu, Z.; Li, C.; et al. Chat-microreactor: A large-language-model-based assistant for designing continuous flow systems. *Chem. Eng. Sci.* **2025**, *311*, 121567. [CrossRef]
92. Yoshikawa, N.; Skreta, M.; Darvish, K.; Arellano-Rubach, S.; Ji, Z.; Bjørn Kristensen, L.; Li, A.Z.; Zhao, Y.; Xu, H.; Kuramshin, A.; et al. Large language models for chemistry robotics. *Auton. Robot.* **2023**, *47*, 1057–1086. [CrossRef]
93. Hirtreiter, E.; Schulze Balhorn, L.; Schweidtmann, A.M. Toward automatic generation of control structures for process flow diagrams with large language models. *AIChE J.* **2024**, *70*, e18259. [CrossRef]
94. Ji, Z.; Lee, N.; Frieske, R.; Yu, T.; Su, D.; Xu, Y.; Ishii, E.; Bang, Y.J.; Madotto, A.; Fung, P. Survey of Hallucination in Natural Language Generation. *ACM Comput. Surv.* **2023**, *55*, 1–38. [CrossRef]
95. Rai, A. Explainable AI: From black box to glass box. *J. Acad. Mark. Sci.* **2020**, *48*, 137–141. [CrossRef]
96. Mehrabi, N.; Morstatter, F.; Saxena, N.; Lerman, K.; Galstyan, A. A Survey on Bias and Fairness in Machine Learning. *ACM Comput. Surv.* **2022**, *54*, 1–35. [CrossRef]

processes

MDPI

Article

Investigation of a Cogeneration System Combining a Solid Oxide Fuel Cell and the Organic Rankine Cycle: Parametric Analysis and Multi-Objective Optimization

Sheng Yang, Anman Liang, Zhengpeng Jin and Nan Xie *

School of Energy Science and Engineering, Central South University, No. 932 South Lushan Road, Changsha 410083, China; ceshyang@csu.edu.cn (S.Y.); lianganman@csu.edu.cn (A.L.); zhengpengjinyh@163.com (Z.J.)

* Correspondence: xienan@csu.edu.cn; Tel.: +86-731-8887-9863

Abstract: A novel solid oxide fuel cell (SOFC)-based cogeneration system is proposed here, integrating an organic Rankine cycle for waste heat recovery. Technical–economic and parametric analyses are conducted, and a multi-objective optimization is carried out. The results reveal that the net electrical efficiency, investment cost, and payback time are 56.6%, USD 2,408,256, and 3.27 years, respectively. The parametric analysis indicates that the current density should be limited between 0.3 A/cm^2 and 0.9 A/cm^2, and the stack temperature should be controlled between 675 °C and 875 °C. After the operational optimization of η_{ele}-$Cost_{TCI}$, the investment cost and the net electrical efficiency are obtained as USD 2,164,742 and 62.1%. After the η_{ele}-PBT optimization, the payback period and the net electrical efficiency are 3.22 years and 58.9%. The heat transfer network optimization achieves the highest efficiency and reduces the cold utilities by 43 kW, but three additional heat exchangers should be added to the system. This research provides practical reference and pragmatic guidance for the integration, analysis, operation, and heat transfer network optimization of SOFC-based cogeneration systems.

Keywords: SOFC; waste heat recovery; cogeneration system; multi-objective optimization; heat exchange network

1. Introduction

The depletion of fossil fuels and environmental pollution are becoming increasingly serious [1]. Against this background, it is particularly important to explore green energy generation contributing to sustainability and SDGs [2]. As a novel type of clean and efficient power generation equipment, fuel cells have received wide attention. With the maturity of technology and cost reduction, fuel cells can play a role in more areas [3,4].

Fuel cells can be categorized according to fuel type, electrolyte type, and operating temperature [5]. Among the many types of fuel cells, solid oxide fuel cells (SOFCs) are more adaptable to various fuels and are more efficient [6]. The high waste heat temperature of SOFCs can be effectively recovered through the establishment of a cogeneration system, further improving the efficiency of energy utilization [7]. The SOFC-based cogeneration system has become a very promising green energy-saving technology [8]. Many scholars have proposed novel SOFC-based cogeneration systems. Comprehensive performance evaluations and key parameter analyses of the systems have been carried out [9,10]. Zhu et al. [11] proposed a coupled system based on SOFCs, engine generation, and ARC. Parametric analyses showed that increasing the steam-to-biomass ratio or the fuel utilization contributed to an improvement in efficiency. Gholamian et al. [12] investigated SOFC-based hybrid systems coupled with either the organic Rankine cycle (ORC) or the Kalina cycle. It was found that the system coupled with the ORC had a higher energy efficiency of 62% than the energy efficiency of 60% of the system coupled with the KC cycle. Zeng et al. [13]

designed waste-heat-driven hybrid systems with different configurations, using the ORC and absorption refrigeration cycle.

System optimization is a necessary step to further improve the operation performance of the proposed novel systems [14,15]. The combined processes need to be comprehensively evaluated from multiple aspects such as output, cost, and efficiency [16]. In most energy systems, the performance of the system is affected by multiple parameters, and it is difficult to achieve a trade-off between multiple objectives using a parametric analysis [17]. Therefore, many scholars often use algorithms to conduct multi-objective optimization. Wang et al. [18] used a genetic algorithm (NSGA-II) for the multi-objective optimization of a novel hybrid system. They indicated that the efficiency and the total cost rate of the atmospheric and pressurized systems reached 56.1% and 16.82 USD/h and 66.83% and 12.02 USD/h, respectively. Mahdi et al. [19] investigated a novel cogeneration system using the Gray Wolf optimization algorithm. The energy efficiency and exergy efficiency were obtained as 49.48% and 47.21%, respectively, in their objective optimization research. Abedinia et al. [20] conducted optimization research using a particle swarm approach coupled with an artificial neural network. Under optimal operating conditions, the highest exergy efficiency and the lowest total product cost were obtained as 42.71% and 6.29 USD/GJ. It can be seen that a lot of work has been carried out for the optimization of fuel-cell-based cogeneration systems, but the heat exchanger network is rarely considered in the system optimization.

Heat exchanger network optimization is one way to improve industrial energy efficiency [21]. Heat exchangers are the most common equipment in energy systems, especially cogeneration systems, which involve complex heat exchanger networks [22]. The optimization of heat exchanger network design is necessary to maximize energy efficiency and reduce costs [23]. Lei et al. [24] proposed an optimization model in which the HEN scheme was optimized with the operating parameters and working fluids at the same time. The HEN scheme was found to be an important reason for the different Pareto optimal solutions. Wang et al. [25] combined tri-reforming into an SOFC system and constructed four different heat transfer network configurations for the system. The results showed that configuration A, Reformer I-AOGR, gave the best performance. Huber et al. [26] optimized the coupling of a heat exchange network and operational parameters for a 1 MW hydroelectric unit. Zhang et al. [27] developed a multi-objective optimization method for the heat exchanger networks of a coal-to-methanol system, which reduced the power consumption by 29.6%.

The published literature mentioned above provides a solid foundation for fuel-cell-based systems, while fewer articles have investigated the interconnections among parameters and objectives after a thorough analysis of the system. In addition, the heat exchange network imposes an obvious impact on the system's performance. However, the optimization of the heat exchange network is seldom considered with a parametric analysis at the same time. In this research, thermodynamic and economic analyses and a parametric analysis of the key parameters are conducted for a novel SOFC-based cogeneration system. Based on the analysis results of the proposed system, the heat exchange network is also considered for the operational optimization of this system. Then, the interconnections among the optimal parameters and objectives are investigated. The novel contributions can be concluded as the simultaneous optimization of the operational parameters and the heat transfer network of the system and the investigation of intrinsic connections among objectives and variables based on the parametric analysis. This work provides pragmatic information for the system integration, technical–economic and parametric analyses, operation, and heat transfer network optimization of SOFC-based energy utilization systems.

2. Methodology

2.1. System Description

A flowchart of the SOFC-based cogeneration system is depicted in Figure 1, as a combination of an SOFC and the ORC. In the SOFC subsystem, the natural gas is pressurized and mixed with the anode recycle steam, and then it flows into the reformer for the reforming process. Pre-reforming can prevent carbon accumulation caused by improper operating conditions, which directly affects the performance and life of the stack. The SOFC stack has a certain tolerance to CO, so the obtained syngas (rich in hydrogen and carbon monoxide) directly enters the anode, and the air enters the cathode. In the stack, efficient electrochemical reactions generate electricity and heat. A part of the exhaust flows back to preheat the anode inlet syngas and then mixes with the fuel. Because the SOFC type is ionic conduction, the anode exhaust contains not only unreacted fuel but also water vapor, and the returned anode exhaust also provides the required water vapor needed for the reaction in the reformer. The other part of the anode exhaust and the cathode exhaust are burned in the burner. The heat released during combustion is fed to the reformer to maintain the heat required for the reforming reaction.

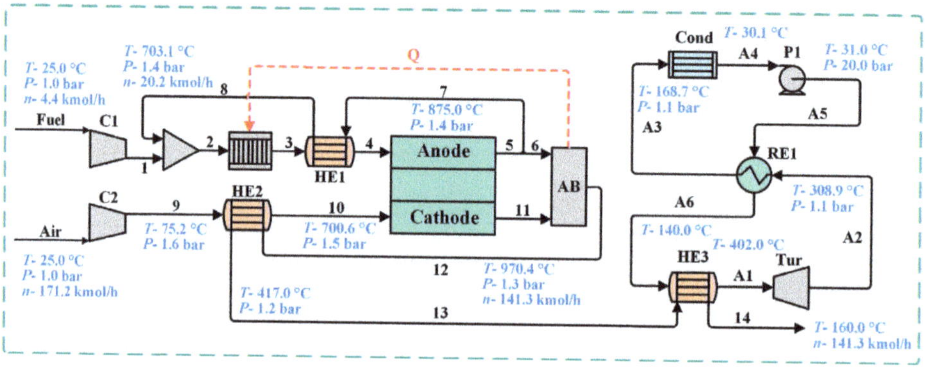

Figure 1. Flowchart of the proposed system.

The high-temperature exhaust gases preheat the air and then drive the ORC subsystem. The temperature is still high after preheating the air. The ORC cycle is more suitable for higher heat source temperatures than the KC cycle. The heat source between 300 °C and 450 °C can drive the trans-critical cycle or the supercritical cycle. However, the heat source is a gas, which has a low heat capacity and a large temperature drop. And the operating pressure of trans-critical or supercritical cycles is high. Therefore, the ORC subsystem is used to recover waste heat from the system. In the ORC subsystem, the organic steam is first heated in a heat exchanger (HE5) by the waste heat from the SOFC subsystem. Then, these high-temperature and high-pressure vapors expand in the expander, converting thermal energy into electrical energy in the turbine. Finally, the organic steam is condensed into a liquid, and then it flows back to the heater exchanger. Thus, the SOFC and ORC subsystems provide electrical power to the load.

2.2. Simulation Models

(1) Basic assumptions

The process simulation is conducted using the Aspen Plus software. The block specifications and descriptions are shown in Table A1 [28]. Peng-Rob is employed as the global physical property method [29]. The calculation of electrochemical models and multi-objective optimization are conducted in MATLAB 2019a. The basic assumptions are given as follows [30,31]: i. The chemical reaction is in equilibrium; ii. All gases are considered ideal; iii. The exhaust is completely consumed in the afterburner; iv. The pre-reformer and afterburner are adiabatic.

(2) Pre-reformer reactor model

RGibbs is adopted for the CH_4 reforming, and the relevant reactions are given as Equations (1)–(3) [32]:

$$CH_4 + H_2O \Leftrightarrow CO + 3H_2 \quad (1)$$

$$CO + H_2O \Leftrightarrow CO_2 + H_2 \quad (2)$$

$$CH_4 + 2H_2O \Leftrightarrow CO_2 + 4H_2 \quad (3)$$

(3) SOFC stack model

The residual CH_4 in the reformer is still useful in the stack because of the high temperature and catalyst in this stack, as in Equations (1)–(3). RGibbs and Splitter blocks are employed for the anode and cathode simulation, respectively [28]. Electrochemical reactions in the stack are given as follows. The process specifications of the stack are shown in Table 1 [33–35].

$$O_2 + 4e^- \rightarrow 2O^{2-} \quad (4)$$

$$H_2 + O^{2-} \rightarrow H_2O + 2e^- \quad (5)$$

$$CO + O^{2-} \rightarrow CO_2 + 2e^- \quad (6)$$

Table 1. Process parameters of SOFC [3,4,33].

Parameters	Value
N_{cell}	2054
A_{cell} (cm^2)	625
γ_a (A/m^2)	7×10^9
γ_c (A/m^2)	7×10^9
E_0 (V)	$E_0 = 1.2723–2.7645 \times 10^{-4} \times T$
ASR_{ohm} (Ωcm^2)	0.04
E^{an} (J/mol)	110,000
E^{ca} (J/mol)	160,000
T_{air} (°C)	25
T_{fuel} (°C)	25
T_{atm} (°C)	25
P_{atm} (bar)	1
U_f/U_a	0.85/0.2
P_{SOFC} (bar)	1.1

The stack power, current density, and voltage are calculated as follows [33,36,37]:

$$U_f = \frac{n_{H_{2,consumed}}}{n_{H_{2,equ}}} \quad (7)$$

$$i = \frac{I}{A} = \frac{2n_{H_{2,equ}} F U_f}{A} \quad (8)$$

$$n_{H_{2,equ}} = nH_2 + nCO + 4nCH_4 \quad (9)$$

$$W_{SOFC} = I \cdot V \quad (10)$$

where i refers to the current density; A_{cell} refers to the stack active area; F refers to Faraday's constant; and U_f refers to the fuel utilization factor.

$$V = V_N - V_{act} - V_{ohm} - V_{con} \quad (11)$$

$$V_N = E_0 + \frac{RT_{avg}}{4F} \ln\left(\frac{P_{H_2} P_{O_2}}{P_{H_2O} P_{atm}^{0.5}}\right) \quad (12)$$

where V represents the single-cell voltage; V_N refers to the Nernst voltage; $E_0 = 1.2723–2.7645 \times 10^{-4} \times T$; V_{act} refers to the activation loss; V_{ohm} is the ohmic loss; and

V_{con} refers to the concentration loss. The activation loss (V_{act}) of the stack can be determined by the Butler–Volmer equation.

$$i = i_0[\exp(\alpha \frac{nF}{RT} V_{act}) - \exp(-(1-\alpha)\frac{nF}{RT} V_{act})] \quad (13)$$

$$i_{0,c} = \gamma_c (\frac{P_{O_2}}{P_{atm}})^{0.25} \exp(-\frac{E^{ca}}{RT}) \quad (14)$$

$$i_{0,a} = \gamma_a (\frac{P_{H_2}}{P_{atm}})(\frac{P_{H_2O}}{P_{atm}}) \exp(-\frac{E^{an}}{RT}) \quad (15)$$

where i_0 is the pre-exponential factor that is specific to each electrode, γ is an activation overpotential factor, and E_{act} is the activation energy.

The ohmic loss (V_{ohm}) depends on the resistivity and thickness of the stacked elements. The stack ohmic loss can be approximated using the area-specific resistance by the following equation.

$$V_{ohm} = i \cdot ASR_{ohm} \quad (16)$$

where ASR_{ohm} is the ohmic area-specific resistance estimated, 0.04 Ω cm^2.

Concentration loss (V_{con}) occurs mainly at high current densities, where the electrochemical reaction rate is fast and the reactants at the electrode are not replenished in a timely manner. It can be expressed by the following equation:

$$V_{con} = \frac{RT}{nF} \ln(1 - \frac{i}{i_L}) \quad (17)$$

(4) ORC subsystem

The corresponding models of ORC are provided in Table A1. The models follow the principle of energy and mass conservation [13]:

$$Q + \sum m_{in} h_{in} = W_T + \sum m_{out} h_{out} \quad (18)$$

$$\sum m_{in} = \sum m_{out} \quad (19)$$

2.3. Performance Indicators

(1) Energy performance indicators

The stack net electrical efficiency η_{SOFC}:

$$\eta_{SOFC} = \frac{W_{SOFC} - W_{Comp1} - W_{Comp2}}{m_{fuel} \cdot LHV_{fuel}} \times 100\% \quad (20)$$

where W_{comp1} and W_{comp2} are the consumed electricity in Comp1 and Comp2 and LHV_{fuel} represents the lower heating value of the fuel.

The system net electrical efficiency η_{ele} is as follows:

$$\eta_{ele} = \frac{W_{SOFC} + W_{Tur} - \Sigma(W_{Comp} + W_P)}{m_{fuel} \cdot LHV_{fuel}} \times 100\% \quad (21)$$

(2) Exergy performance indicators

Exergy refers to the qualitative value of energy, which includes the physical and chemical parts [38,39]:

$$Ex = Ex^{ph} + Ex^{ch} \quad (22)$$

$$Ex^{ph} = m[h_{in} - h_{out} - T_0(s_{in} - s_{out})] \quad (23)$$

$$Ex^{ch} = \sum_i^n n_i \overline{ex}_i^{ch,0} + RT_0 \sum_i^n n_i \ln x_i \quad (24)$$

(3) Economic performance indicators

Economic indexes reveal the economic performance and clarify the investment and benefits of a project [40]. The total cost of this system ($Cost_{TCI}$) consists of two parts: the total depreciation cost ($Cost_{TDC}$) and startup cost ($Cost_{start}$). Among them, the total depreciation cost includes the fixed investment cost ($Cost_{fix}$), maintenance cost ($Cost_{maint}$), contingency cost ($Cost_{cont}$), and fuel cost ($Cost_{fuel}$) [41]. The cost equations are presented in Table 2.

Table 2. Cogeneration system economic equations.

Cost	Equation
Fixed investment cost $Cost_{fix}$	$Cost_{fix}$
Maintenance cost $Cost_{maint}$	$Cost_{maint} = 0.1\ Cost_{fix}$
Contingency cost $Cost_{cont}$	$Cost_{cont} = 0.198\ Cost_{fix}$
Fuel cost $Cost_{fuel}$	$Cost_{fuel}$
Total depreciation cost $Cost_{TDC}$	$Cost_{TDC} = Cost_{fix} + Cost_{maint} + Cost_{cont} + Cost_{fuel}$
Startup cost $Cost_{start}$	$Cost_{start} = 0.1\ Cost_{TDC}$
Total cost of the system $Cost_{TCI}$	$Cost_{TCI} = Cost_{TDC} + Cost_{start}$

The fixed investment cost is the cost of the system components (C_K). To estimate the system cost more accurately, this paper introduces the Chemical Engineering Plant Cost Index ($CEPCI$), which converts the calculated component costs for the year 2022 ($CEPCI$: 541.7) [42]. Detailed formulas for C_K and $CEPCI_{ref}$ are shown in Table A2 [43].

$$C_{current} = C_{ref} \times \frac{CEPCI_{current}}{CEPCI_{ref}} \tag{25}$$

The fuel cost is obtained by using the following equation [44]:

$$C_{current} = C_{ref} \times \frac{CEPCI_{current}}{CEPCI_{ref}} \tag{26}$$

where c_f represents the natural gas unit cost of 3.5 USD/MMBTU.

Payback time (PBT) is a very important indicator in the economic evaluation of a project. PBT is the length of time it takes for a project to recover the full cost of investment from the start of the investment. PBT can be calculated by the following formula [45]:

$$PBT = \frac{Cost_{TDC}}{W_{net} t_w C_{elec}} \tag{27}$$

where t_w represents the operating time of 7884 h, the system life is 15 years, W_{net} is the system net electrical output, and C_{elec} refers to the grid electricity price of 0.15 USD/kWh.

2.4. Model Verification

The modeling of the ORC subsystem is well-developed and does not involve complex physical processes. As the core component of a cogeneration system, the performance and reliability of the SOFC stack are critical to the effective operation of the entire system. Therefore, accurate stack modeling is key to the simulation of fuel cell cogeneration systems. During the SOFC stack model validation, it is common to compare the stack polarization curves obtained from the simulation with those from experimental data or other reliable references. Under the same parameter conditions (atmospheric pressure, operating temperature of 725 °C, and consistent feed), the polarization curves of the stack obtained in this research are compared with the polarization curves in reference [33], and the results are shown in Figure 2. As can be seen from the comparison, the error between the two is within 5%. This indicates that the established SOFC stack model has high reliability and accuracy.

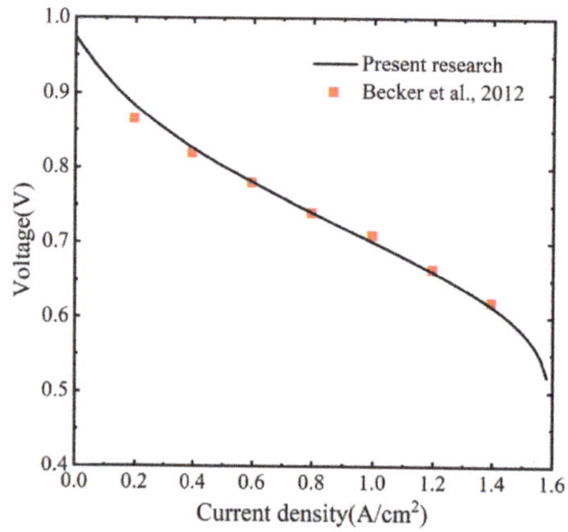

Figure 2. Verification of stack polarization curves [33].

3. Parametric and Performance Analyses

3.1. Current Density

The polarization curve in Figure 3a shows that the voltage of the SOFC decreases gradually as the current density increases. At a low current density, the voltage decreases faster, which is caused by the activation loss. And at high current density, the voltage drop tends to slow down, which is mainly caused by ohmic losses. Then, the stack power output improves with the rising current density. The change in current density is achieved by varying the fuel flow rates at a constant total polarization area. The improved system output is less than the fuel consumption, resulting in a decrease in efficiency.

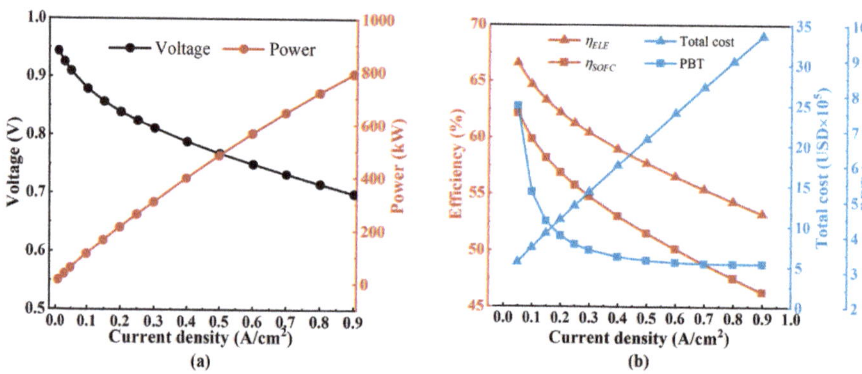

Figure 3. Effect of current density on the system.

In Figure 3b, the net efficiency of the SOFC stack goes down from 55% to 46%, and the system electrical efficiency goes down from 61% to 53.3% with the current density varying from 0.3 A/cm^2 to 0.9 A/cm^2. The total cost of the system increases linearly with increasing current density. As the current density rises to 0.4 A/cm^2, the *PBT* drops rapidly. At a current density from near 0.4 A/cm^2 to 0.9 A/cm^2, the *PBT* also decreases, but slower. With increasing current density at a low current density, the improvement in output power is more significant than the increase in total cost. As the current density increases, the increase in output power slows down. Therefore, the *PBT* decreases with increasing current density, but the trend is gradually smaller. The *PBT* increases even at current densities close to 0.9 A/cm^2. In summary, although the stack efficiency is as high as 60% at the current density of 0.1 A/cm^2, the electrical output is only 110 kW, and the *PBT* is high. This power is as high as 792 kW at the current density of 0.9 A/cm^2, but the efficiency of the stack drops to 46%, and the cost is high. Therefore, to ensure a proper stack operation, the current density should be limited in the range of 0.3 A/cm^2–0.9 A/cm^2.

3.2. Average Stack Temperature

Figure 4a shows the polarization curves at different average stack temperatures. At a high current density, the effect of temperature on voltage is greater. The increase in temperature directly accelerates the rate of chemical reaction. At temperatures close to 900 °C, the water–gas reaction is limited, and the voltage increase will slow down. Figure 4b shows that the η_{SOFC} has increased but gradually slows down with the improving average temperature. The stack electrical output accounts for a major part of the system, so the trend of the η_{ele} is similar to the η_{SOFC}. As the average temperature increases from 675 °C to 950 °C, the η_{SOFC} increases from 45% to 52%, and the η_{ele} increases from 52% to 58%. The total cost increases as the average temperature increases but is not significant. The cost of each component and the output power both increase with the increase in temperature. Under the combined effect of the two, the *PBT* of the system is shortened from 3.40 years to 3.26 years, with the average temperature varying from 675 °C to 850 °C. However, the *PBT* is slightly increased when the average temperature improves from 850 °C to 875 °C. In conclusion, increasing the temperature of the stack will have a positive effect on the system. However, the increase is limited at high temperatures, and it will put higher demands on the materials of the components. Therefore, the average temperature of the stack should be moderate. The average temperature is taken to be between 675 °C and 875 °C in this paper.

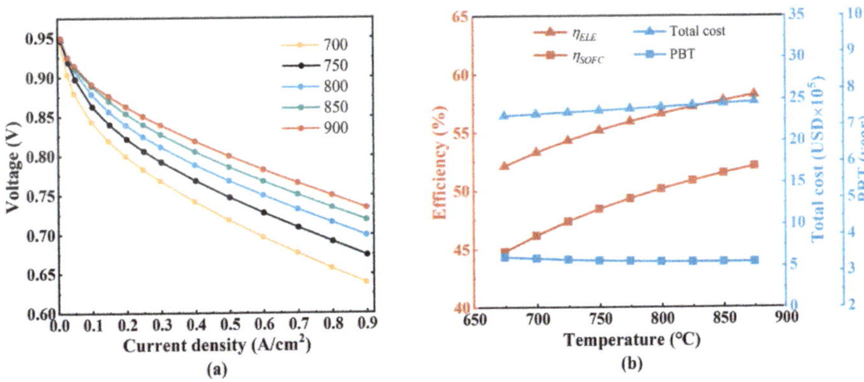

Figure 4. Effect of the average temperature of the stack on the system.

3.3. Anode Recycle Ratio

This parameter directly affects the reformer inlet temperature and water–carbon ratio. Figure 5a shows the variations in the reformer outlet composition for different recycle ratios. When the recycle ratio varies from 0.5 to 0.75, the mole fraction of methane drops rapidly, while the amount of H_2 and CO increases. If this ratio is further increased, the methane conversion is approaching the limit, and the residual water vapor gradually accumulates, resulting in a decrease in the mole fraction of H_2. The effect of the recycle ratio on the reforming conditions in the cogeneration system would influence the entire system. As shown in Figure 5b, the stack electrical efficiency and system electrical efficiency rise at first and then decrease as the recycle ratio increases from 0.5 to 0.85. The anode recycle ratio of the stack also affects the system economics, but the magnitude of the effect is small compared to the current density and stack temperature. The total cost consistently increases when increasing the recycle ratio, but the increase is less than 1×10^5 USD; the system *PBT* decreases and then increases as the recycle ratio improves, but the change is less than 0.5 years. In summary, too small a recycle ratio increases the risk of carbon accumulation, and too large a recycle ratio leads to a decrease in the proportion of H_2, which results in lower efficiency, higher costs, and longer payback periods. Therefore, the recycle ratio of the cogeneration system in this paper should be in the range of 0.5 to 0.8.

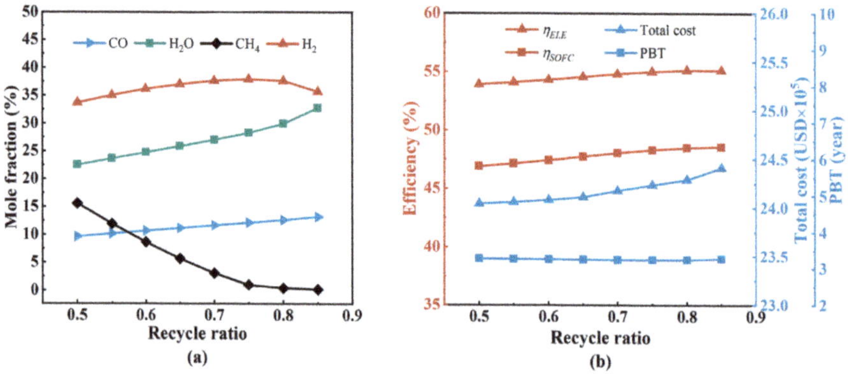

Figure 5. Effect of anode recycle ratio on the system.

3.4. System Performance

Based on the input parameters, the physical properties of each stream of this cogeneration system are obtained, as shown in Tables A3 and A4. The performance of the cogeneration system under the initial operating conditions is shown in Tables 3 and 4. In the cogeneration system, the SOFC stack output is 566 kW with a net stack efficiency of 50.1%, the turbine output is 65 kW, and the system total electrical power is 631 kW with a net electrical efficiency of 56.6%. The total investment in the cogeneration system amounted to USD 2,408,256, of which the total depreciation costs amounted to USD 2,189,324. The fuel cost is the main part of the total depreciation cost, which is 58.9%. The payback period is less than 4 years, indicating that the proposed system is profitable. Figure 6 illustrates the exergy destruction of each component. Significant exergy losses are found in the air preheater, afterburner, and the SOFC stack. The logarithmic mean temperature difference of the air preheater is calculated as higher than 300 °C. Hence, the huge exergy destruction occurs in this component because of its large load and heat transfer temperature difference. The afterburner burns the residual substances in the exhaust gas and converts the chemical energy of fuel into thermal energy, resulting in considerable exergy destruction because of the chemical reactions and substances mixing. Since the SOFC stack is the main fuel-consuming component in the system, it causes large exergy damage along with high energy consumption and output.

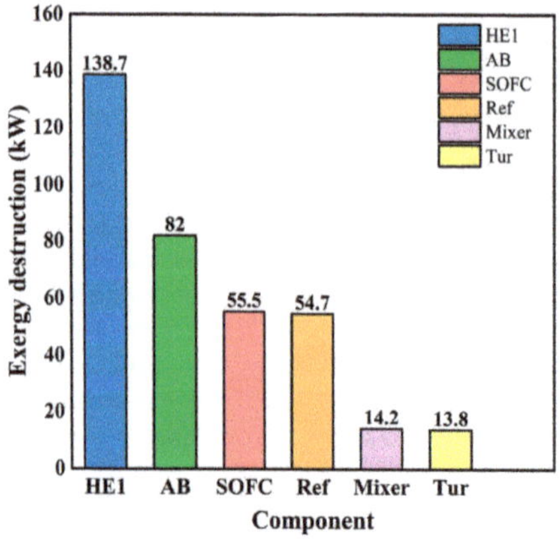

Figure 6. Exergy destruction of each component.

Table 3. Cogeneration system performance.

Parameters	Value	Parameters	Value
Current density i (A/m^2)	6000	Single-cell voltage V_{cell} (V)	0.75
SOFC stack power W_{SOFC} (kW)	566	Turbine power W_{Tur} (kW)	65
SOFC electrical efficiency η_{SOFC} (%)	50.1	System electrical efficiency η_{ele} (%)	56.6

Table 4. Cogeneration system economic equations.

Item	Value	Percentage
Fixed investment cost $Cost_{fix}$ (USD)	593,377.5	24.6%
Maintenance cost $Cost_{maint}$ (USD)	59,337.75	2.5%
Contingency cost $Cost_{cont}$ (USD)	117,488.7	4.9%
Fuel cost $Cost_{fuel}$ (USD)	1,419,120	58.9%
Total depreciation cost $Cost_{TDC}$ (USD)	2,189,324	90.9%
Startup cost $Cost_{start}$ (USD)	218,932.4	9.1%
Total cost of the system $Cost_{TCI}$ (USD)	2,408,256	100%
Payback Time PBT (Year)	3.27	-

4. System Optimization

4.1. Optimization Model

For cogeneration systems, efficiency and cost are often important criteria for evaluating a system [46]. Both the total cost and PBT can describe the system's economics. And in the system analysis, the total cost and the PBT will vary with the parameters. Therefore, the multi-objective optimization will make two cases: case 1, η_{ele}-$Cost_{TCI}$, and case 2, η_{ele} -PBT. To optimize the cogeneration system, the decision variables should be selected and specified [47]. Based on the system parameters and performance analysis, the current density, average stack temperature, and anode recycle ratio are selected as decision variables, as shown in Table 5.

Table 5. Decision variables and ranges of values.

Decision Variables	Value Range	Unit
Current density	0.3–0.9	A/cm^2
Average stack temperature	675–750	°C
Anode recycle ratio	0.5–0.8	-

Multiple heat exchangers are inevitably used in cogeneration systems, including air preheating, fuel preheating, and multiple waste heat recovery heat exchangers. The flow rates and temperatures vary greatly among the streams. Inefficient network configurations of heat exchange will lead to energy waste and cost increases. The heat exchanger network of a cogeneration system consists of cold streams (CS) and hot streams (HS), and then the heat exchanger network can be expressed as [24]:

$$Q_{HEN} = \sum_{i=1}^{n} Q_{HS,i} - Q_{CU} = \sum_{i=1}^{n} Q_{CS,i} - Q_{HU} \quad (28)$$

$$Q_{HS,i} = cp_{HS,i} m_{HS,i} (T_{HS,i,in} - T_{HS,i,out}) \quad (29)$$

$$Q_{CS,i} = cp_{CS,i} m_{CS,i} (T_{CS,i,in} - T_{CS,i,out}) \quad (30)$$

Multi-objective optimization can deal with multiple objectives simultaneously and find the best trade-off solutions between these objectives, which can be achieved by genetic algorithms. A genetic algorithm is often utilized to achieve the optimal combination in terms of energy and exergy efficiencies, cost, and environmental performance. NSGA-II, with less computational complexity, is a common fast tool for solving multi-objective problems. The specifications for the genetic algorithm are listed in Table 6.

Table 6. Genetic algorithm parameter values.

Algorithm Parameters	Values
Population size	100
Number of variables	3
Crossover ratio	80%
Proportion of variation	20%
Maximum number of iterations	100

Based on the model establishment described above, the genetic algorithm is employed for multi-objective optimization, and the flowchart of this algorithm is described in Figure 7. The Technique for Order Preference by Similarity to an Ideal Solution (TOPSIS) is an available decision-making method for final optimal solution selection [48]. It finds the optimal solution according to the relative proximities of each solution in the Pareto solution set computed by the distance from the positive ideal solution and the distance from the negative ideal solution. The final optimal solution is selected with the shortest distance from the ideal point and the farthest distance from the non-ideal point.

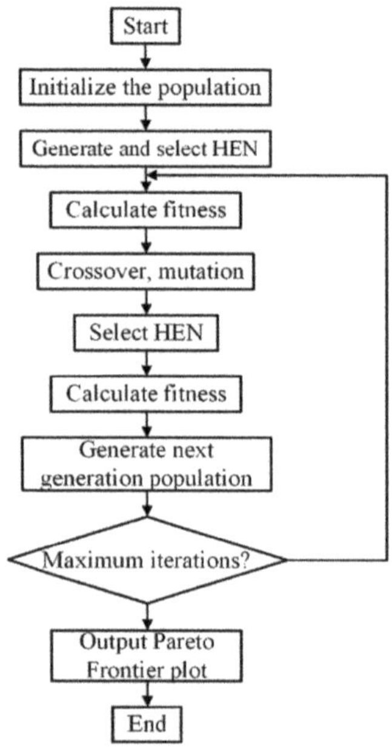

Figure 7. Algorithm operation flowchart.

4.2. Optimization Results of Case 1 (ηele-CostTCI)

The Pareto frontier plot of case 1 (η_{ele}-$Cost_{TCI}$) is shown in Figure 8. A1 is the lowest point of total system cost, USD 1,998,885. But in this case, A1 has the lowest net electric efficiency of 55.6%. C1 is the highest point of net electric efficiency of the system at 63.3%, but simultaneously, it has the highest total cost of USD 2,215,737. B1 is the optimal point with a total cost of USD 2,164,742 and a net electrical efficiency of 62.1%.

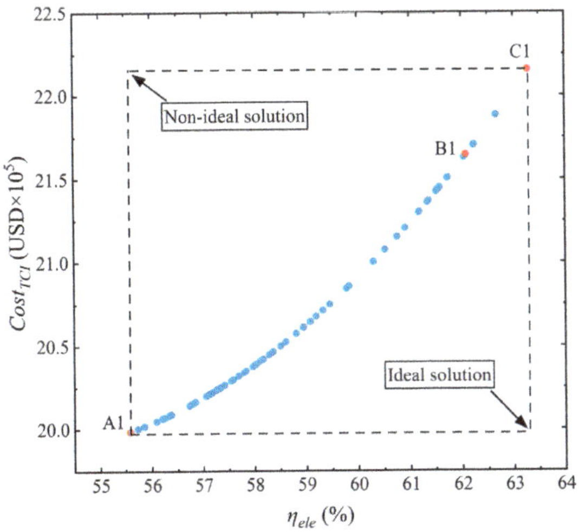

Figure 8. Pareto frontier plot for η_{ele}-$Cost_{TCI}$.

The current density at A1, B1, and C1 is set to the lowest value. The reason for this is that as the current density increases, not only does the efficiency decrease, but also the total cost increases due to increased fuel consumption. Therefore, to achieve the goal of low cost and high efficiency, the current density chosen should be as low as possible. Secondly, the operating temperature of the stack at the three points increases sequentially. The operating temperature at point A1 is close to the minimum limit, while the operating temperature at point C1 is close to the maximum limit. The reason for this is that the system efficiency increases as the temperature rises, but accordingly, the cost of the system components increases. Therefore, a balance needs to be found between efficiency improvement and cost control. Finally, the anode recycle ratio at the three points is also gradually increased. However, unlike the operating temperature, the recycle ratio at point C1 does not reach the maximum limit. This is because when the recycle ratio is too high, it will reduce the system efficiency instead. The corresponding specifications are listed in Table 7.

Table 7. Trade-off points for η_{ele}-*PBT*.

	Unit	A1	B1	C1
Current density	A/cm^2	0.3	0.3	0.3
Stack temperature (average)	°C	677	843	875
Anode recycle ratio	-	0.51	0.60	0.78
η_{ele}	%	55.6	62.1	63.3
$Cost_{TCI}$	USD	1,998,885	2,164,742	2,215,737
PBT	Year	3.58	3.57	3.67

4.3. Optimization Results of Case2 (η_{ele}-*PBT*)

In the optimization results in case 1, although the three points of case 1 have high net electrical efficiency, the current density is only at the minimum limit, which leads to insufficient output power of the stack and a long *PBT*.

In Figure 9, the Pareto frontier plot of case 2 (η_{ele}-*PBT*) is shown, where A2 is the lowest point of the *PBT* at 3.17 years, but at this time, the lowest net electricity efficiency is 56.7%; C2 is the highest point of the system net electricity efficiency at 63.3%, but at this time, the *PBT* is the largest at 3.67 years; and B2 is the optimal point achieved by the entropy power method of trade-offs, with a recovery cycle of 3.22 and a net electricity efficiency of 58.9%. Points C1 and C2 both aim to achieve the highest net electrical efficiency, so they take the same values for the decision variables. At higher current densities, the output power of the system increases, which results in shorter payback periods. Therefore, in case 2, the current density settings at points A and B are higher than those in case 1. The increase in the operating temperature of the power stack will continue to increase the net power efficiency, but the recovery cycle will be shortened and then lengthened. Therefore, the SOFC stack temperatures of points A2 and B2 are set lower than C2. The corresponding specifications are given in Table 8.

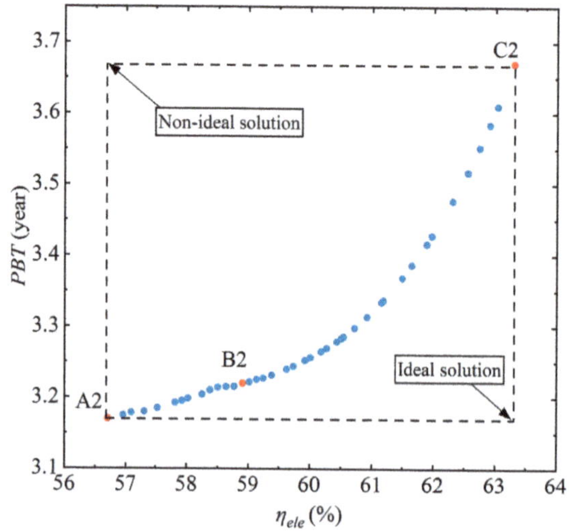

Figure 9. Pareto frontier plot for η_{ele}- *PBT*.

Table 8. Trade-off points for η_{ele}-*PBT*.

	Unit	A2	B2	C2
Current density	A/cm^2	0.8	0.57	0.3
Stack temperature (average)	°C	850	850	875
Anode recycle ratio	-	0.65	0.62	0.78
η_{ele}	%	56.7	58.9	63.3
PBT	Year	3.17	3.22	3.67
$Cost_{TCI}$	USD	3,136,668	2,363,673	2,215,737

4.4. Optimization Results of Heat Exchange Network

The points in the Pareto frontier plot have different heat transfer network arrangements. Take point A1 as an example. As shown in Figure 10a, the heat exchanger network configuration at point A1 is extremely simplified, containing only two heat exchangers: the cathode preheater and waste heat recovery heat exchanger. This configuration ensures the basic operational requirements of the fuel cell system while minimizing the cost. Point C1 is then the most efficient configuration. As shown in Figure 10b, the heat transfer network at point C1 is more complex and contains five heat exchangers to improve system efficiency. The additional HE1 is used to avoid mixing the high-temperature anode recycle gas and the fresh fuel with large temperature differences and to reduce energy loss during mixing. By adding HE2 to achieve segmented preheated air, the temperature in the air preheater can be better matched and the turbine inlet temperature can be increased. Finally, RE1 was also added to the ORC cycle to recover the higher-temperature waste heat at the turbine outlet and reduce system energy loss. However, it is very practical that although the additional heat exchangers can improve the efficiency of this system, their addition inevitably increases the initial investment in equipment. Subsequently, auxiliary devices such as pipeline components will also be multiplied in the actual process. This puts forward higher requirements for the steady operation, sealing, and durability of the system.

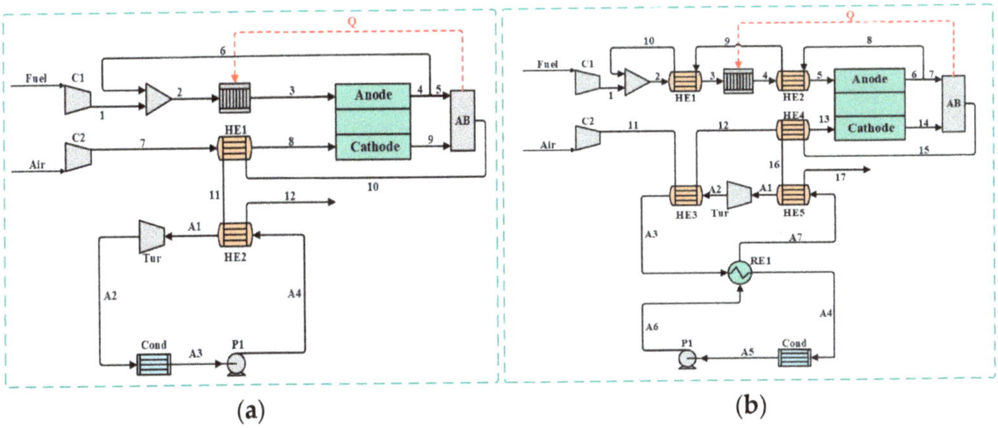

Figure 10. Flowchart corresponding to (**a**) A1 and (**b**) C1.

The variation curves of the cold and hot flows of the heat transfer network corresponding to points A1 and C1 are shown in Figure 11. The minimum heat transfer temperature difference in the heat transfer network of A1 is large. The heat recovery of the system is 440 kW, and the cold utility consumption is 129 kW. The minimum heat transfer temperature difference in the heat transfer network of C1 is only 10 °C. The heat recovery of the system is 802 kW, and the cold utility consumption is 86 kW. In summary, the matching of the heat exchange temperature of the cogeneration system is achieved through the optimization of the heat exchanger network, thus improving the energy efficiency of the whole system. In the process of heat exchanger network optimization, additional heat exchangers are added to the system, and the layout of the existing heat exchangers is adjusted, which will lead to an increase in the initial investment cost.

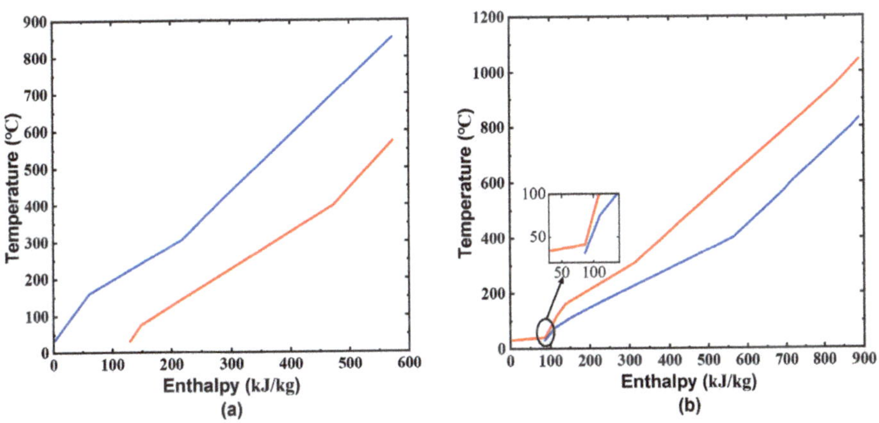

Figure 11. The combination curves of the cold and hot flow of (**a**) A1 and (**b**) C1.

5. Conclusions

This research achieves the simultaneous optimization of operational parameters and the heat transfer network of a novel SOFC-based cogeneration system and carried out an investigation of the intrinsic connections among the objectives and variables based on the parametric analysis results. A technical–economic analysis, a parametric analysis, and an operational and heat transfer network optimization were conducted. The main conclusions are summarized as follows:

(1) The variations in current density and stack temperature lead to an imbalance between efficiency and economy. The current density is recommended to be between 0.3 A/cm^2 and 0.9 A/cm^2. The operating temperature of the SOFC stack should be limited between 675 °C and 875 °C. A lower recycle ratio would improve the risk of carbon accumulation, and a higher recycle ratio would reduce the efficiency and increase the total cost, so the recycle ratio of this system is recommended to be within 0.5–0.8.
(2) Under initial conditions, the system net power efficiency, investment cost, and payback period are 56.6%, USD 2,408,256, and 3.27 years, respectively. In the case 1 (η_{ele}-$Cost_{TCI}$) optimization, the cost and the electrical efficiency of the optimal point are USD 2,164,742 and 62.1%. In the case 2 (η_{ele}-PBT) optimization, the PBT and the electrical efficiency of B2 are 3.22 years and 58.9%.
(3) Comparing the two configurations of the heat exchange network for the lowest-cost purpose and highest-efficiency purpose, the former requires only two heat exchangers, while the latter requires five heat exchangers. The heat exchange network optimization reduces the consumption of cold utilities by 43 kW.

However, some limitations need to be clarified in this research: (1) only three operational parameters were selected for the decision variables; (2) the waste heat temperature would be different under different operating parameters, while the different cycles for waste heat recovery and the corresponding change in work quality were not investigated; (3) an uncertainty analysis is necessary as well, considering the variations and fluctuations of all the factors; (4) the impacts of operating environments (i.e., ambient temperature, atmospheric pressure, air composition, fuel composition, etc.) were not described in detail; (5) this research also neglects the requirements of the user. For future efforts, it is suggested that energy management strategies should be considered in the proposed system to achieve a fast and accurate response to the changing demands of the users.

Author Contributions: Conceptualization, S.Y.; methodology, S.Y.; software, A.L. and Z.J.; validation, Z.J.; formal analysis, Z.J.; investigation, Z.J.; resources, Z.J.; data curation, A.L.; writing—original draft preparation, A.L.; writing—review and editing, N.X.; visualization, Z.J.; supervision, S.Y.; project administration, N.X.; funding acquisition, N.X. All authors have read and agreed to the published version of the manuscript.

Funding: This work was supported by the National Natural Science Foundation of China [grant number 22478443, 52206037, 22008265] and the Natural Science Foundation of Hunan Province [grant number 2022JJ40574, 2023JJ10077].

Data Availability Statement: Date will be made available on request.

Conflicts of Interest: The authors declare no conflict of interest.

Nomenclature

Abbreviations	
AB	Afterburner
COND	Condenser
COP	Coefficient of performance
HE	Humidity efficiency/heat exchanger
KC	Karina cycle
LHV	Lower heating value
ORC	Organic Rankine cycle
RE	Recycle heat exchange
REF	Reformer
STCR	Steam-to-carbon ratio
SOFC	Solid oxide fuel cell
TUR	Turbine
TOPSIS	Technique for Order Preference by Similarity to an Ideal Solution

Symbols

A	Area
F	Faraday's constant
G	Gibbs free energy
h	Specific enthalpy
i	Current density
I	Current
L	Length
n	Molar flow rate
P	Power/pressure
T	Temperature
U_a	Air utilization factor
U_f	Fuel utilization factor
V	Voltage
V_{act}	Activation loss
V_{con}	Concentration loss
V_N	Nernst voltage
V_{ohm}	Ohmic loss
Greek letters	
γ	Pre-exponential coefficient
η	Efficiency
Subscripts	
an	Anode
avg	Average
ca	Cathode
ele	Electricity
p	Pump
T	Turbine

Appendix A

See Tables A1–A4 as below.

Table A1. Blocks used in establishing the systems.

Systems	Block type	Block ID	Description
SOFC subsystem	RGibbs	Ref	Simulate methane reforming process
	RGibbs	anode	Simulate the electrochemical reaction process
	Sep	cathode	Simulate oxygen ion transport
	RStoic	Burner	Simulate combustion reaction
	Compr	Comp1	Increase the pressure of the fuel
	Compr	Comp2	Increase the pressure of the air
	Heater	HE1	Preheated air
	Heater	HE2	Preheated air
	Mixer	Mix	Mix the stream
ORC subsystem	Compr	Tur	Convert the energy into mechanical work
	Heater	HE3	Heat recovery
	Heater	RE1	Heat the working fluid
	Heater	Cond1	Condense the working fluid
	Pump	P1	Increase the pressure of the working fluid

Table A2. Component cost equations.

Components	Cost (USD)	Year	CEPCI
SOFC stack	$Z_{SOFC} = A_a N_{cell}(2.96T_{SOFC} - 1907)$	2002	395.6
SOFC auxiliaries	$Z_{aux} = 0.1 \times Z_{SOFC}$	2002	395.6
SOFC inverter	$Z_{inv} = 10^5 \times (W_{SOFC}/500)^{0.7}$	2002	395.6
Afterburner	$Z_{AB} = \frac{46.08m_6}{0.995 - P_{in}/P_{out}}(1 + \exp(0.018T_{13} - 26.4))$	1994	368.1
Compressor	$Z_{Comp} = 91562(W_{Comp}/445)^{0.67}$	2003	402.3
Vale	$Z_{Vale} = 114.5 \times \dot{m}$	2001	394.3
Pump	$Z_P = 705.48 \times (1 + 0.2/(1 - \eta_P))W_P^{0.71}$	2001	394.3
Turbine	$Z_{Tur} = W_{Tur}(1318.5 - 98.328\ln(W_{Tur}))$	2001	394.3
Preheater	$Z_{RE} = 130 \times (A_{RE}/0.093)^{0.78}$	2005	468.2
Heat exchanger	$Z_{HE} = 130 \times (A_{HE}/0.093)^{0.78}$	2000	394.1
Condenser	$Z_{Cond} = 516.62 \times A_{Cond}^{0.6}$	2005	468.2

Table A3. Detailed stream data for SOFC subsystem.

Stream	Temperature (°C)	Pressure (bar)	Mole Flow (kmol/h)	Mole Fraction (%)						
				N_2	O_2	CO_2	CO	H_2O	CH_4	H_2
Fuel	25.0	1.0	4.4	0	0	0	0	0	1	0
Air	25.0	1.0.	171.2	79.0	21.0	0.0	0	0	0	0
1	64.7	1.6	4.5	0	0	0	0	0	1	0
2	562.8	1.6	24.7	0	0	19.4	7.8	40.5	18.2	14.0
3	600.0	1.5	28.7	0	0	19.8	10.7	24.8	8.6	36.1
4	725.0	1.5	28.7	0	0	19.8	10.7	24.8	8.6	36.1
5	875.0	1.4	33.7	0	0	23.8	9.6	49.6	0	17.1
6	875.0	1.4	13.5	0	0	23.8	9.6	49.6	0	17.1
7	875.0	1.4	20.2	0	0	23.8	9.6	49.6	0	17.1
8	703.1	1.4	20.2	0	0	23.8	9.6	49.6	0	17.1
9	75.2	1.6	136.9	0.79	0.21	0	0	0	0	0
10	700.6	1.5	136.9	0.79	0.21	0	0	0	0	0
11	875.0	1.4	129.8	83.4	16.6	0	0	0	0	0
12	970.4	1.3	141.3	76.5	14.0	3.2	0	6.4	0	0
13	417.0	1.2	141.3	76.5	14.0	3.2	0	6.4	0	0
14	160.0	1.4	141.3	76.5	14.0	3.2	0	6.4	0	0

Table A4. Detailed stream data for ORC subsystem.

Stream	Temperature (°C)	Pressure (bar)	Mole Flow (kmol/h)	R123 Mole Fraction (%)
A1	402.0	20.0	20.86	100
A2	308.9	1.1	20.86	100
A3	168.7	1.1	20.86	100
A4	30.1	1.1	20.86	100
A5	31.0	20.0	20.86	100
A6	140.0	20.0	20.86	100

References

1. Wang, L.; Guo, W.; Zhang, Z.; Wang, F.; Yuan, J. Performance analysis and optimization of syngas composition for reversible solid oxide fuel cells in dual-mode operation based on extreme learning machine. *J. Power Sources* **2024**, *614*, 234982. [CrossRef]
2. Vengatesan, S.; Jayakumar, A.; Sadasivuni, K.K. FCEV vs. BEV—A short overview on identifying the key contributors to af-fordable & clean energy (SDG-7). *Energy Strategy Rev.* **2024**, *53*, 101380. [CrossRef]
3. Song, M.; Zhuang, Y.; Zhang, L.; Li, W.; Du, J.; Shen, S. Thermodynamic performance assessment of SOFC-RC-KC system for multiple waste heat recovery. *Energy Convers. Manag.* **2021**, *245*, 114579. [CrossRef]
4. Song, M.; Zhuang, Y.; Zhang, L.; Wang, C.; Du, J.; Shen, S. Advanced exergy analysis for the solid oxide fuel cell system combined with a kinetic-based modeling pre-reformer. *Energy Convers. Manag.* **2021**, *245*, 114560. [CrossRef]

5. Zhao, J.; Cai, S.; Huang, X.; Luo, X.; Tu, Z. 4E analysis and multiobjective optimization of a PEMFC-based CCHP system with dehumidification. *Energy Convers. Manag.* **2021**, *248*, 114789. [CrossRef]
6. Tu, B.; Wen, H.; Yin, Y.; Zhang, F.; Su, X.; Cui, D.; Cheng, M. Thermodynamic analysis and experimental study of electrode reactions and open circuit voltages for methane-fuelled SOFC. *Int. J. Hydrog. Energy* **2020**, *45*, 34069–34079. [CrossRef]
7. Abbasi, H.R.; Pourrahmani, H.; Chitgar, N.; Van herle, J. Thermodynamic analysis of a tri-generation system using SOFC and HDH desalination unit. *Int. J. Hydrog. Energy* **2021**, *72*, 1204–1215. [CrossRef]
8. Singh, M.; Zappa, D.; Comini, E. Solid oxide fuel cell: Decade of progress, future perspectives and challenges. *Int. J. Hydrog. Energy* **2021**, *46*, 27643–27674. [CrossRef]
9. Ding, A.; Sun, H.; Zhang, S.; Dai, X.; Pan, Y.; Zhang, X.; Rahman, E.; Guo, J. Thermodynamic analysis and parameter optimization of a hybrid system based on SOFC and graphene-collector thermionic energy converter. *Energy Convers. Manag.* **2023**, *291*, 117327. [CrossRef]
10. Li, C.; Wang, Z.; Liu, H.; Guo, F.; Li, C.; Xiu, X.; Wang, C.; Qin, J.; Wei, L. Exergetic and exergoeconomic evaluation of an SOFC-Engine-ORC hybrid power generation system with methanol for ship application. *Fuel* **2024**, *357*, 129944. [CrossRef]
11. Zhu, P.; Wu, Z.; Guo, L.; Yao, J.; Dai, M.; Ren, J.; Kurko, S.; Yan, H.; Yang, F.; Zhang, Z. Achieving high-efficiency conversion and poly-generation of cooling, heating, and power based on biomass-fueled SOFC hybrid system: Performance assessment and multi-objective optimization. *Energy Convers. Manag.* **2021**, *240*, 114245. [CrossRef]
12. Gholamian, E.; Zare, V. A comparative thermodynamic investigation with environmental analysis of SOFC waste heat to power conversion employing Kalina and Organic Rankine Cycles. *Energy Convers. Manag.* **2016**, *117*, 150–161. [CrossRef]
13. Zeng, R.; Guo, B.; Zhang, X.; Li, H.; Zhang, G. Study on thermodynamic performance of SOFC-CCHP system integrating ORC and double-effect ARC. *Energy Convers. Manag.* **2021**, *242*, 114326. [CrossRef]
14. You, H.; Han, J.; Liu, Y.; Chen, C.; Ge, Y. 4E analysis and multi-objective optimization of a micro poly-generation system based on SOFC/MGT/MED and organic steam ejector refrigerator. *Energy* **2020**, *206*, 118122. [CrossRef]
15. Yuan, X.; Liu, Y.; Bucknall, R. Optimised MOPSO with the grey relationship analysis for the multi-criteria objective energy dispatch of a novel SOFC-solar hybrid CCHP residential system in the UK. *Energy Convers. Manag.* **2021**, *243*, 114406. [CrossRef]
16. Liang, W.; Yu, Z.; Liu, W.; Ji, S. Investigation of a novel near-zero emission poly-generation system based on biomass gasification and SOFC: A thermodynamic and exergoeconomic evaluation. *Energy* **2023**, *282*, 128822. [CrossRef]
17. Guo, Y.; Guo, X.; Wang, J.; Guan, Z.; Wang, Z.; Zhang, Y.; Wu, W.; Wang, X. Performance analysis and multi-objective optimization for a hybrid system based on solid oxide fuel cell and supercritical CO_2 Brayton cycle with energetic and ecological objective approaches. *Appl. Therm. Eng.* **2023**, *221*, 119871. [CrossRef]
18. Wang, H.; Yu, Z.; Wang, D.; Li, G.; Xu, G. Energy, exergetic and economic analysis and multi-objective optimization of atmospheric and pressurized SOFC based trigeneration systems. *Energy Convers. Manag.* **2021**, *239*, 114183. [CrossRef]
19. Forootan, M.M.; Ahmadi, A. Machine learning-based optimization and 4E analysis of renewable-based polygeneration system by integration of GT-SRC-ORC-SOFC-PEME-MED-RO using multi-objective grey wolf optimization algorithm and neural networks. *Renew. Sustain. Energy Rev.* **2024**, *200*, 114616. [CrossRef]
20. Abedinia, O.; Shakibi, H.; Shokri, A.; Sobhani, B.; Sobhani, B.; Yari, M.; Bagheri, M. Optimization of a syngas-fueled SOFC-based multigeneration system: Enhanced performance with biomass and gasification agent selection. *Renew. Sustain. Energy Rev.* **2024**, *199*, 114460. [CrossRef]
21. Oliva, D.G.; Francesconi, J.A.; Mussati, M.C.; Aguirre, P.A. Ethanol and glycerin processor systems coupled to solid oxide fuel cells (SOFCs). Optimal operation and heat exchangers network synthesis. *Int. J. Hydrog. Energy* **2013**, *38*, 7140–7158. [CrossRef]
22. Fei, Z.; Yang, H.; Du, L.; Guerrero, J.M.; Meng, K.; Li, Z. Two-Stage Coordinated Operation of A Green Multi-Energy Ship Microgrid With Underwater Radiated Noise by Distributed Stochastic Approach. *IEEE Trans. Smart Grid* **2024**, 1. [CrossRef]
23. Xu, Y.; Lei, Y.; Xu, C.; Chen, Y.; Tan, Q.; Ye, S.; Li, J.; Huang, W. Multi-objective optimization of a solid oxide fuel cell-based integrated system to select the optimal closed thermodynamic cycle and heat coupling scheme simultaneously. *Int. J. Hydrog. Energy* **2021**, *46*, 31828–31853. [CrossRef]
24. Lei, Y.; Ye, S.; Xu, Y.; Kong, C.; Xu, C.; Chen, Y.; Huang, W.; Xiao, H. Multi-objective optimization and algorithm improvement on thermal coupling of SOFC-GT-ORC integrated system. *Comput. Chem. Eng.* **2022**, *164*, 107903. [CrossRef]
25. Xuan Wang, R.H. Comparison and analysis of heat exchange and offgas recycle strategies in tri-reforming-SOFC system. *Int. J. Hydrog. Energy* **2023**, *48*, 34069–34079. [CrossRef]
26. Huber, D.; Birkelbach, F.; Hofmann, R. Unlocking the potential of synthetic fuel production: Coupled optimization of heat exchanger network and operating parameters of a 1 MW power-to-liquid plant. *Chem. Eng. Sci.* **2024**, *284*, 119506. [CrossRef]
27. Zhang, Z.; Zhao, L.; Tera, I.; Liu, G. Multi-objective optimization of heat exchanger network with disturbances based on graph theory and decoupling. *Chem. Eng. Sci.* **2024**, *287*, 119763. [CrossRef]
28. Doherty, W.; Reynolds, A.; Kennedy, D. Computer simulation of a biomass gasification-solid oxide fuel cell power system using Aspen Plus. *Energy* **2010**, *35*, 4545–4555. [CrossRef]
29. Anderson, T.; Vijay, P.; Tade, M.O. An adaptable steady state Aspen Hysys model for the methane fuelled solid oxide fuel cell. *Chem. Eng. Res. Des.* **2014**, *92*, 295–307. [CrossRef]
30. Wu, Z.; Tan, P.; Chen, B.; Cai, W.; Chen, M.; Xu, X.; Zhang, Z.; Ni, M. Dynamic modeling and operation strategy of an NG-fueled SOFC-WGS-TSA-PEMFC hybrid energy conversion system for fuel cell vehicle by using MATLAB/SIMULINK. *Energy* **2019**, *175*, 567–579. [CrossRef]

31. Alsarraf, J.; Alnaqi, A.A.; Al-Rashed, A.A.A.A. Thermodynamic modeling and exergy investigation of a hydrogen-based integrated system consisting of SOFC and CO_2 capture option. *Int. J. Hydrog. Energy* **2022**, *47*, 26654–26664. [CrossRef]
32. Diglio, G.; Hanak, D.P.; Bareschino, P.; Pepe, F.; Montagnaro, F.; Manovic, V. Modelling of sorption-enhanced steam methane reforming in a fixed bed reactor network integrated with fuel cell. *Appl. Energy* **2018**, *210*, 1–15. [CrossRef]
33. Becker, W.L.; Braun, R.J.; Penev, M.; Melaina, M. Design and technoeconomic performance analysis of a 1MW solid oxide fuel cell polygeneration system for combined production of heat, hydrogen, and power. *J. Power Sources* **2012**, *200*, 34–44. [CrossRef]
34. Huang, Y.; Turan, A. Mechanical equilibrium operation integrated modelling of hybrid SOFC–GT systems: Design analyses and off-design optimization. *Energy* **2020**, *208*, 118334. [CrossRef]
35. Yang, S.; Jin, Z.; Ji, F.; Deng, C.; Liu, Z. Proposal and analysis of a combined cooling, heating, and power system with humidity control based on solid oxide fuel cell. *Energy* **2023**, *284*, 129233. [CrossRef]
36. Pirkandi, J.; Ghassemi, M.; Hamedi, M.H.; Mohammadi, R. Electrochemical and thermodynamic modeling of a CHP system using tubular solid oxide fuel cell (SOFC-CHP). *J. Clean. Prod.* **2012**, *29–30*, 151–162. [CrossRef]
37. Ma, L.; Ru, X.; Wang, J.; Lin, Z. System model and performance analysis of a solid oxide fuel cell system self-humidified with anode off-gas recycling. *Int. J. Hydrog. Energy* **2024**, *57*, 1164–1173. [CrossRef]
38. Sinha, A.A.; Saini, G.; Sanjay; Shukla, A.K.; Ansari, M.Z.; Dwivedi, G.; Choudhary, T. A novel comparison of energy-exergy, and sustainability analysis for biomass-fueled solid oxide fuel cell integrated gas turbine hybrid configuration. *Energy Convers. Manag.* **2023**, *283*, 116923. [CrossRef]
39. Wei, S.; Ni, S.; Ma, W.; Du, Z.; Lu, P. Exergy analysis and parameters optimization for zero-carbon/low-carbon fuels on SOFC-engine hybrid system. *Fuel* **2024**, *364*, 131076. [CrossRef]
40. Meng, N.; Li, T.; Jia, Y.; Qin, H.; Liu, Q.; Zhao, W.; Lei, G. Techno-economic performance comparison of enhanced geothermal system with typical cycle configurations for combined heating and power. *Energy Convers. Manag.* **2020**, *205*, 112409. [CrossRef]
41. Zhao, Z.; Shi, X.; Zhang, M.; Ouyang, T. Multi-scale assessment and multi-objective optimization of a novel solid oxide fuel cell hybrid power system fed by bio-syngas. *J. Power Sources* **2022**, *524*, 231047. [CrossRef]
42. Comidy, L.J.F.; Staples, M.D.; Barrett, S.R.H. Technical, economic, and environmental assessment of liquid fuel production on aircraft carriers. *Appl. Energy* **2019**, *256*, 113810. [CrossRef]
43. Su, Z.; Ouyang, T.; Chen, J.; Xu, P.; Tan, J.; Chen, N.; Huang, H. Green and efficient configuration of integrated waste heat and cold energy recovery for marine natural gas/diesel dual-fuel engine. *Energy Convers. Manag.* **2020**, *209*, 112650. [CrossRef]
44. Chitgar, N.; Emadi, M.A.; Chitsaz, A.; Rosen, M.A. Investigation of a novel multigeneration system driven by a SOFC for electricity and fresh water production. *Energy Convers. Manag.* **2019**, *196*, 296–310. [CrossRef]
45. Ouyang, T.; Zhao, Z.; Lu, J.; Su, Z.; Li, J.; Huang, H. Waste heat cascade utilisation of solid oxide fuel cell for marine applications. *J. Clean. Prod.* **2020**, *275*, 124133. [CrossRef]
46. Xia, W.; Ren, Z.; Li, H.; Pan, Z. A data-driven probabilistic evaluation method of hydrogen fuel cell vehicles hosting capacity for integrated hydrogen-electricity network. *Appl. Energy* **2024**, *376*, 123895. [CrossRef]
47. Yan, Z.; He, A.; Hara, S.; Shikazono, N. Modeling of solid oxide fuel cell (SOFC) electrodes from fabrication to operation: Microstructure optimization via artificial neural networks and multi-objective genetic algorithms. *Energy Convers. Manag.* **2019**, *198*, 111916. [CrossRef]
48. Xie, N.; Liu, Z.; Luo, Z.; Ren, J.; Deng, C.; Yang, S. Multi-objective optimization and life cycle assessment of an integrated system combining LiBr/H_2O absorption chiller and Kalina cycle. *Energy Convers. Manag.* **2020**, *225*, 113448. [CrossRef]

processes

MDPI

Article

Corrosion State Monitoring Based on Multi-Granularity Synergistic Learning of Acoustic Emission and Electrochemical Noise Signals

Rui Wang [1,†], Guangbin Shan [2,†], Feng Qiu [2], Linqi Zhu [2], Kang Wang [3], Xianglong Meng [3], Ruiqin Li [3], Kai Song [1,*] and Xu Chen [1]

1 School of Chemical Engineering and Technology, Tianjin University, Tianjin 300072, China; ruiw@tju.edu.cn (R.W.); xchen@tju.edu.cn (X.C.)

2 State Key Laboratory of Chemical Safety, SINOPEC Research Institute of Safety Engineering Co., Ltd., Qingdao 266071, China; shangb.qday@sinopec.com (G.S.); qiuf.qday@sinopec.com (F.Q.); zhulq.qday@sinopec.com (L.Z.)

3 Shanxi Beifang XingAn Chemical Industry Co., Ltd., Taiyuan 030021, China; jokensj@163.com (K.W.); 15735109769@163.com (X.M.); lrq072019@163.com (R.L.)

* Correspondence: ksong@tju.edu.cn

† These authors contributed equally to this work.

Abstract: Corrosion monitoring is crucial for ensuring the structural integrity of equipment. Acoustic emission (AE) and electrochemical noise (EN) have been proven to be highly effective for the detection of corrosion. Due to the complementary nature of these two techniques, previous studies have demonstrated that combining both signals can facilitate research on corrosion monitoring. However, current machine learning models have not yet been able to effectively integrate these two different modal types of signals. Therefore, a new deep learning framework, CorroNet, is designed to synergistically integrate AE and EN signals at the algorithmic level for the first time. The CorroNet leverages multimodal learning, enhances accuracy, and automates the monitoring process. During training, paired AE-EN data and unpaired EN data are used, with AE signals serving as anchors to help the model better align EN signals with the same corrosion stage. A new feature alignment loss function and a probability distribution consistency loss function are designed to facilitate more effective feature learning to improve classification performance. Experimental results demonstrate that CorroNet achieves superior accuracy in corrosion stage classification compared to other state-of-the-art models, with an overall accuracy of 97.01%. Importantly, CorroNet requires only EN signals during the testing phase, making it suitable for stable and continuous monitoring applications. This framework offers a promising solution for real-time corrosion detection and structural health monitoring.

Keywords: corrosion monitoring; process automation and monitoring; artificial neural networks; deep learning; acoustic emission; electrochemical noise

1. Introduction

The corrosion of equipment can result in facility damage, production interruptions, environmental pollution, and even explosion. It is one of the primary factors causing the structural integrity degradation in equipment. Therefore, developing effective automated corrosion monitoring methods is crucial to ensure process safety, equipment reliability, and environment protection [1,2].

Acoustic emission (AE) and electrochemical noise (EN) technologies are two types of technologies proven to be effective for corrosion monitoring. AE involves detecting high-frequency elastic waves generated by the rapid release of energy within a material [3–5]. Such signals primarily reflect physical phenomena occurring during crack propagation, pitting initiation, hydrogen embrittlement, and other microscopic damage behaviors. On the other hand, EN signals primarily reflect electrochemical reactions during corrosion, with random or semi-random fluctuations in current and potential due to uneven reactions, revealing changes in the corrosion process mechanisms. [6–9]. Because EN signals are sensitive to the early stages of corrosion, they have been applied in the research of the corrosion behavior of metallic materials [10], the corrosion inhibition of steel reinforcement in concrete [11], stress corrosion cracking [12], and several other corrosion areas.

AE and EN signals for pitting corrosion are chaotic, non-linear, and complex, requiring advanced non-linear methods for processing. Deep learning algorithms effectively address this need. For example, Homborg et al. interpreted EN time–frequency spectra as an image classification problem. They explored the application of convolutional neural networks (CNNs) for the deep learning-based image classification of electrochemical noise time–frequency transient information [13]. Rubio et al. utilized a deep autoencoder model to analyze local electrochemical noise for PEM fuel cell fault diagnosis [14]. Guo et al. applied the InceptionTime model for damage classification in carbon fiber-reinforced composites using AE time-series data [15]. Liu et al. proposed a transfer learning method based on convolutional neural networks (CNN-TL) for pipeline leak detection under various working conditions [16]. Qiu et al. established a CNN-based hydrogen defect recognition (HDR) model from AE signals [17].

As a passive method, however, AE relies on damage events, lacks early warning capabilities, and struggles with noise interference and signal interpretation, while EN signals can directly represent the chemical reactions occurring during corrosion. But EN signals' characteristics are complex, especially with the indistinct features across different corrosion stages. This complexity makes stage-specific feature analysis challenging. Due to the complementary nature of AE and EN signals in corrosion monitoring, their combined use offers significant potential for advancing corrosion research. In fact, the integrated analysis of these signals has traditionally been conducted manually [18–22]. But such extensive manual analyses are time consuming and labor intensive.

Recently developed multimodal techniques provide a way to integrate different types of data at the feature and/or algorithmic levels to enhance the model's performance [23–27]. Inspired by these state-of-the-art (SOTA) methods, in this paper, a new network, called CorroNet, was proposed to synergistically integrate AE and EN signals at the algorithmic level for the first time. In CorroNet, AE and EN are treated as two distinct modalities with different granularities. Each modality reflects the different aspects of corrosion process while sharing semantically related features of the corrosion stages. By employing multimodal deep learning algorithms to jointly process these two modalities, more accurate corrosion state monitoring may be achieved. The contributions of this paper are summarized as follows:

(1) A new deep learning framework, named CorroNet, for corrosion monitoring is proposed. It achieves multi-granularity synergistic learning between AE and EN signals at the model learning level;
(2) A feature alignment method for corrosion events and a probability distribution alignment loss function are designed for CorroNet to significantly improve the performance in the corrosion monitoring task.

This paper is organized as follows: Section 2 introduces the proposed CorroNet model, the experimental setup, and the corrosion dataset. Section 3 presents an analysis of the model's corrosion monitoring results. Section 4 provides the conclusions.

2. Methods

The goal of the acoustic-electrochemical synergistic corrosion monitoring is to achieve synergistic learning between the two modalities of acoustic emission (AE) and electrochemical noise (EN) data and to learn an accurate mapping function $F(x)$ from the training set X_{train} in order to minimize the prediction error on the test set X_{test}.

As shown in Figure 1, CorroNet leverages both AE and EN signals to enhance corrosion monitoring. AE, as a passive method, captures signals only during active damage events, resulting in irregular and discontinuous data. In contrast, EN provides continuous-time data with uniform sampling, making it suitable for stable monitoring. CorroNet integrates AE and EN signals during training to leverage both modalities. During testing, it uses only EN signals for consistent and reliable online monitoring.

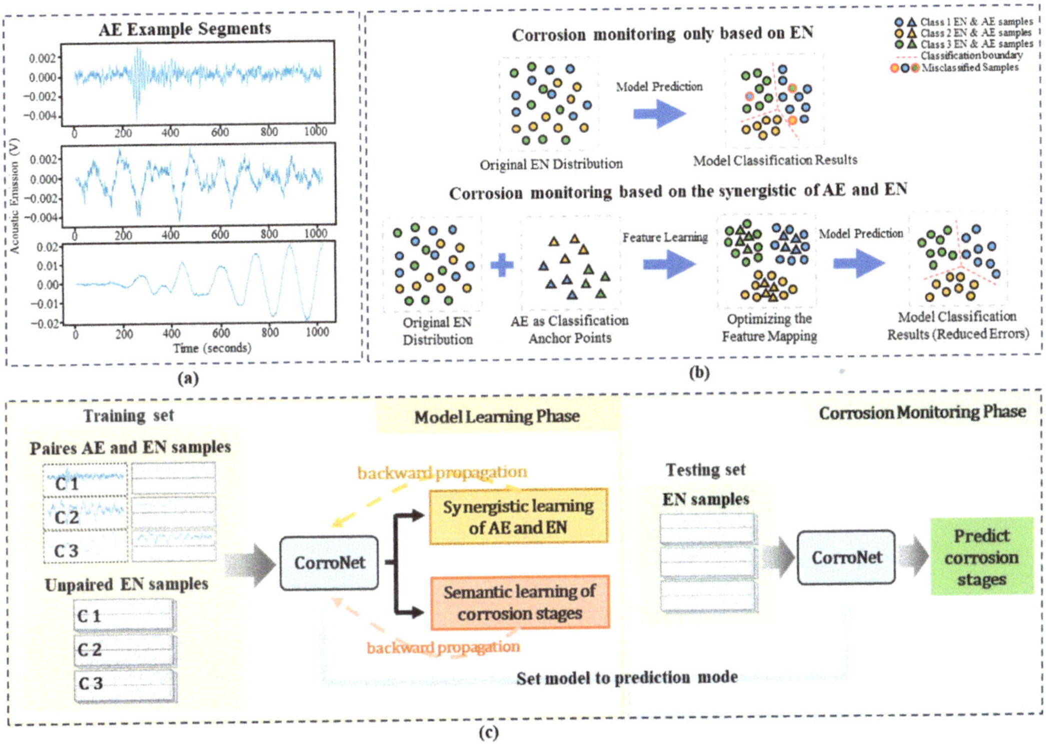

Figure 1. Explanation of AE and EN synergistic learning; (**a**) is an example of AE signals with significant corrosion features; (**b**) is the comparison between the synergistic learning model using AE and EN signals as well as existing models that use only EN signals; (**c**) is the overview of the proposed synergistic learning-based corrosion monitoring method.

The training set consists of paired AE-EN data and additional unpaired EN samples. AE signals are selected based on significant corrosion stage features (as shown in Figure 1a), serving as anchors during the learning process. These anchors assist in aligning EN signals of the same category, enhancing the model's ability to recognize corrosion stages and improving the classification accuracy of the EN data (as shown in Figure 1b).

As illustrated in Figure 1c, AE signals support synergistic learning with EN during training, enriching CorroNet's understanding of corrosion mechanisms. Once fully trained, CorroNet can effectively combine AE's intermittent insights with EN's stable monitoring capability. Therefore, it can accurately predict corrosion stages using only EN signals. The subsequent experimental results (Section 3) provide validation and support for this.

2.1. The Overview of the CorroNet

The Transformer encoder, known for strong sequential modeling, is used as CorroNet's backbone to extract features from EN and AE time-series data. CorroNet has two branches to process AE and EN signals separately and eventually performs synergistic learning. Different modules are connected after the Transformer encoders in each branch to accomplish various functions. The detailed architecture of CorroNet is shown in Figure 2:

(1) Most importantly, to achieve multi-granularity synergistic learning, a feature alignment loss L_{InfoNCE} and a probability distribution consistency loss L_{KL} are designed. After the Transformer encoders pass f_{A} and f_{E} in the two branches, projection heads g_{A} and g_{E}^{a} (represented in orange and blue, respectively) are introduced to align the representations p_{A} and p_{E} of the two modalities in feature space. Additionally, supervised contrastive learning is applied to the EN data using a second projection head g_{E}^{b} (green) after the EN Transformer encoder with the loss function L_{SC};
(2) Secondly, since AE signals serve as anchor points to optimize the feature mapping of EN signals, a classification head q_{A} is connected after the Transformer encoder f_{A}. It is used to initially train the model for accurate AE classification, for more precise anchor points in the feature space, and for the subsequent optimization of EN signals' mapping. Similarly, a classification head q_{E} is connected after the EN branch's Transformer encoder f_{E};
(3) Additionally, due to the multi-granularity feature learning, the complex multi-level feature alignment and loss constraints make precise training and pre-training even more essential. Therefore, a specific training strategy is proposed for CorroNet, dividing the training process into three stages. The details can be found in Section 2.4.

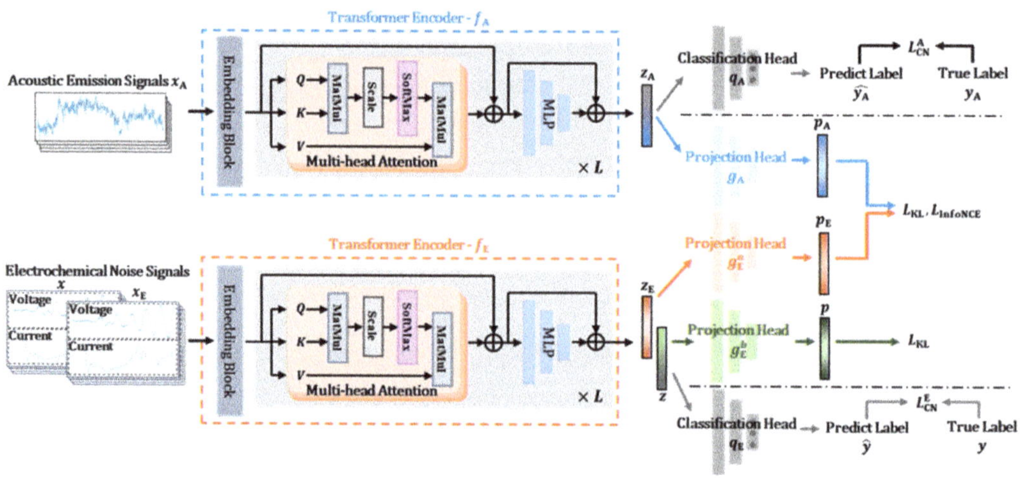

Figure 2. The structure of CorroNet.

Each component of the CorroNet is discussed in detail in the following section.

2.2. The Transformer Encoder

AE and EN signals are both time-series data characterized by temporal dependencies. Previous studies have demonstrated that Transformer models have achieved remarkable success in temporal modeling tasks [28–32]. As illustrated in Figure 2, the Transformer encoder is introduced as the backbone of CorroNet to leverage the attention mechanism for efficient feature extraction from AE and EN signals, represented as f_{A} and f_{E}, respectively. Specifically, the Transformer encoder consists of three main components: the embedding block, multi-head attention, and multi-layer perceptron (MLP). The embedding block transforms the raw time-series data into multiple tokens; multi-head attention captures temporal dependencies, and the MLP performs non-linear transformations and feature fusion to enhance expressiveness. The details of the Transformer encoder are in Appendix A.1.

2.3. The AE-EN Alignment Framework

The core component of CorroNet is the multi-granularity synergistic learning between AE and EN signals. It includes three main parts, feature alignment learning, probability distribution alignment learning, and supervised contrastive learning. Each of these components addresses different aspects of the model's learning process, and their integration enhances the overall corrosion monitoring capabilities of the model. Feature alignment captures mutual information between AE and EN signals; distribution alignment minimizes modality mismatches, and supervised contrastive learning uses labels to enhance corrosion stage feature extraction in EN. The following sections provide a detailed explanation of each module within CorroNet.

The specific structure of the multi-granularity synergistic learning consists of Transformer encoders f_A and f_E and projection heads g_A, g_E^a, and g_E^b. Transformer encoder f_A is responsible for feature extraction from AE signals, followed by the projection head g_A, which comprises an MLP to further map the extracted features. The features z_A extracted by f_A serve as the class anchors for guiding the feature learning of EN signals. Another branch extracts features from EN signals using the Transformer encoder f_E. Two parallel projection heads, g_E^a and g_E^b, map these features into different subspaces. Specifically, g_E^a works together with g_A to achieve a joint mapping for the AE and EN modalities. The mapping produced by g_E^b is used for supervised contrastive learning to further enhance the model's learning of EN signals.

In the synergistic learning modules, all samples from X_{train} are utilized, including the paired AE-EN dataset $X_{AE-EN} = \left\{(x_E, x_A, y_A)_{i=1}^{N_1}\right\}$ and additional EN training data $X_{EN} = \left\{(x, y)_{i=1}^{N_2}\right\}$. As illustrated in Figure 2, AE samples x_A are fed into the Transformer encoder f_A to generate the features z_A. The features z_A are then passed through the projection head g_A, resulting in the representations p_A. The paired EN samples x_E are fed into the Transformer encoder f_E, which outputs the features z_E. Subsequently, the features z_E are further mapped by the projection head g_E^a to generate the representations p_E. As shown in Figure 2, the model uses feature alignment learning $L_{InfoNCE}$ and distribution alignment learning L_{KL} to treat AE representations p_A as class anchors. This aligns EN representations p_E with p_A, training f_E to learn corrosion stage semantics.

The feature alignment learning $L_{InfoNCE}$ is based on the InfoNCE loss [33], which enhances the similarity between paired representations (p_A, p_E) while pushing apart unpaired samples, as formulated in Equation (1):

$$L_{InfoNCE} = -\sum_{i=1}^{N_1} \log\left(\frac{\exp\left(p_E^i \cdot p_A^i/\tau\right)}{\exp\left(p_E^i \cdot p_A^i/\tau\right) + \sum_{j\neq i}^{N_1} \exp\left(p_E^i \cdot p_A^j/\tau\right)}\right), \tag{1}$$

where τ is the temperature parameter [34], used to control the smoothness during training.

Distribution alignment learning L_{KL} is designed based on the Kullback–Leibler divergence. L_{KL} minimizes the distribution mismatch between AE and EN modalities. Specifically, it first computes the individual distribution probability scores s_{AE} and s_{EN} for AE and EN samples. Then, an exponential kernel is applied to fit the distribution probability scores of the two modalities to compute their similarity. Taking AE as an example, the distribution probability score $s_{AE}^{i,j}$ for the (i, j)-th pair of AE samples is calculated using Equation (2):

$$s_{AE}^{i,j} = \frac{\exp\left(p_A^i \cdot p_A^j/\tau\right)}{\sum_{k=1}^{N_1} \exp\left(p_A^i \cdot p_A^k/\tau\right)} \tag{2}$$

The probability score of EN is obtained in the same way and is denoted as $s_{\mathrm{EN}}^{i,j}$. The probability distribution similarity alignment loss L_{KL} is given by Equation (3):

$$L_{\mathrm{KL}} = \sum_{i=1}^{N_1}\sum_{j=1}^{N_1} s_{\mathrm{EN}}^{i,j} \cdot \log\left(\frac{s_{\mathrm{EN}}^{i,j}}{s_{\mathrm{AE}}^{i,j}}\right). \tag{3}$$

Unpaired EN samples x are processed by f_{E}, producing features z, which are mapped by g_{E}^{b} into representations p. The model is further trained using supervised contrastive learning to enhance the model's ability to learn discriminative features for different stages of EN signals. Supervised contrastive learning L_{SC} encourages the model to pull closer the feature representations of samples belonging to the same class, while pushing apart the feature representations of samples from different classes, as described in Equation (4):

$$L_{\mathrm{SC}} = \sum_{i \in I} \frac{-1}{|P(i)|} \sum_{j \in P(i)} \log\left(\frac{\exp(z_i \cdot z_j/\tau)}{\sum_{a \epsilon A(i)} \exp(z_i \cdot z_a/\tau)}\right), \tag{4}$$

where I denotes the set of indices for all samples in a batch, and $P(i)$ represents the set of positive samples that share the same label as sample i. $A(i)$ is the set of all other samples in the training batch except for the i-th sample itself. τ is the temperature parameter that controls the smoothness of contrastive learning. In summary, the training loss function for multi-granularity synergistic learning consists of three components: L_{InfoNCE}, L_{KL}, and L_{SC}. The detailed optimization process for the parameters of each module in the model is provided in Section 2.4.

Additionally, since AE signals serve as anchors to assist in EN signal learning, it is essential to first obtain a feature extractor which is capable of extracting effective corrosion stage discriminative features (denoted as z_{A}) from AE signals. Therefore, before the AE-EN synergistic learning, the Transformer encoder f_{A} is pre-trained with a supervised classification task. After encoding AE signals with f_{A}, a classification head q_{A} composed of an MLP is added to classify the corrosion stages of AE signals. This outputs the predicted stage label $\hat{y_{\mathrm{A}}} = q_{\mathrm{A}}(f_{\mathrm{A}}(x_{\mathrm{A}}))$. The AE portion $\left\{(x_{\mathrm{A}}, y_{\mathrm{A}})_{i=1}^{N_1}\right\}$ of the $X_{\mathrm{AE-EN}}$ is used as the training data. The AE signals x_{A} are input into the classifier, and the predicted outputs $\hat{y_{\mathrm{A}}}$ are compared with the true labels y_{A} to compute the cross-entropy loss $L_{\mathrm{CN}}^{\mathrm{A}}$, as shown in Equation (5):

$$L_{\mathrm{CN}}^{\mathrm{A}} = -\sum_{i=1}^{N_1} {y_{\mathrm{A}}}^i \log\left(\hat{y_{\mathrm{A}}}^i\right), \tag{5}$$

where i represents the sample index, and N_1 is the number of AE samples.

After multi-granularity AE-EN synergistic training, the Transformer encoder f_{E} gains the ability to effectively extract corrosion-related semantic information from EN signals. Finally, f_{E} is fine-tuned to perform the final corrosion stage classification task. As shown in Figure 2, a classification head q_{E}, composed of an MLP, is attached to f_{E}. During the fine-tuning stage, only EN data are used to train the model. The samples x from X_{EN} are fed into f_{E} and q_{E}, which output the predicted labels $\hat{y}$. Similar to Equation (5), the cross-entropy loss $L_{\mathrm{CN}}^{\mathrm{E}}$ is computed between $\hat{y}$ and the true labels y and used to fine-tune the model, further improving its classification performance.

$$L_{\mathrm{CN}}^{\mathrm{E}} = -\sum_{i=1}^{N_2} y^i \log\left(\hat{y}^i\right). \tag{6}$$

2.4. Model Training and Optimization

This section describes the training strategy proposed for CorroNet. The proposed CorroNet model undergoes a three-stage training process to maximize its ability to extract semantic corrosion stage features from EN signals, thereby improving its performance in corrosion stage monitoring. Stage 1 aims to pre-train the transformer encoder of the AE branch to extract highly discriminative representations of corrosion stages from AE signals to provide anchor points for feature mapping of EN signals. Stage 2 focuses on achieving synergistic learning between AE and EN signals to enhance the model's ability to recognize features of different corrosion stages through multi-granularity synergistic learning. Stage 3 further utilizes corrosion label information to enable the model to fully learn the corrosion stage features from EN signals. The parameters of the CorroNet model are optimized using gradient backpropagation, with the specific process detailed in Equations (7)–(16). The parameters for each component f_{A}, f_{E}, q_{A}, q_{E}, g_{A}, g_{E}^{a}, and g_{E}^{b} are denoted as $W_{f_{\mathrm{A}}}$, $W_{f_{\mathrm{E}}}$, $W_{q_{\mathrm{A}}}$, $W_{q_{\mathrm{E}}}$, $W_{g_{\mathrm{A}}}$, $W_{g_{\mathrm{E}}^{a}}$, and $W_{g_{\mathrm{E}}^{b}}$, respectively.

(1) Stage 1 Parameter Optimization:

AE signals serve as anchors to optimize the feature mapping of EN signals, so the AE branch's Transformer encoder f_{A} is first trained to extract corrosion stage features. Therefore, only the parameters of f_{A} and q_{A} are updated in Stage 1, while other modules are frozen.

$$W_{f_{\mathrm{A}}} \leftarrow W_{f_{\mathrm{A}}} - \eta_1 \left(\frac{\partial L_{\mathrm{CN}}^{\mathrm{A}}}{\partial W_{f_{\mathrm{A}}}} \right) \tag{7}$$

$$W_{q_A} \leftarrow W_{q_A} - \eta_1 \left(\frac{\partial L_{CN}^{A}}{\partial W_{q_A}} \right) \tag{8}$$

where η_1 represents the learning rate for Stage 1.

(2) Stage 2 Parameter Optimization:

In the Stage 2, multi-granularity synergistic learning between AE and EN, as well as supervised contrastive learning for EN signals, is performed. Thus, the parameters of f_{A}, g_{A}, f_{E}, g_{E}^{a}, and g_{E}^{b} are updated, while other modules remain frozen.

$$W_{f_{\mathrm{A}}} \leftarrow W_{f_{\mathrm{A}}} - \eta_2 \left(\lambda \frac{\partial L_{\mathrm{KL}}}{\partial W_{f_{\mathrm{A}}}} + (1-\lambda) \frac{\partial L_{\mathrm{InfoNCE}}}{\partial W_{f_{\mathrm{A}}}} \right) \tag{9}$$

$$W_{g_{\mathrm{A}}} \leftarrow W_{g_{\mathrm{A}}} - \eta_2 \left(\lambda \frac{\partial L_{\mathrm{KL}}}{\partial W_{f_{\mathrm{A}}}} + (1-\lambda) \frac{\partial L_{\mathrm{InfoNCE}}}{\partial W_{f_{\mathrm{A}}}} \right) \tag{10}$$

$$W_{f_{\mathrm{E}}} \leftarrow W_{f_{\mathrm{E}}} - \eta_2 \left(\lambda \frac{\partial L_{\mathrm{KL}}}{\partial W_{f_{\mathrm{E}}}} + (1-\lambda) \frac{\partial L_{\mathrm{InfoNCE}}}{\partial W_{f_{\mathrm{E}}}} \right) \tag{11}$$

$$W_{g_{\mathrm{E}}^{a}} \leftarrow W_{g_{\mathrm{E}}^{a}} - \eta_2 \left(\lambda \frac{\partial L_{\mathrm{KL}}}{\partial W_{g_{\mathrm{E}}^{a}}} + (1-\lambda) \frac{\partial L_{\mathrm{InfoNCE}}}{\partial W_{g_{\mathrm{E}}^{a}}} \right) \tag{12}$$

$$W_{f_{\mathrm{E}}} \leftarrow W_{f_{\mathrm{E}}} - \eta_2 \left(\frac{\partial L_{\mathrm{SC}}}{\partial W_{f_{\mathrm{E}}}} \right) \tag{13}$$

$$W_{g_{\mathrm{E}}^{b}} \leftarrow W_{g_{\mathrm{E}}^{b}} - \eta_2 \left(\frac{\partial L_{\mathrm{SC}}}{\partial W_{f_{\mathrm{E}}}} \right) \tag{14}$$

where η_2 represents the learning rate for Stage 2, and λ is the hyperparameter that balances the loss function components.

(3) Stage 3 Parameter Optimization:

In the final stage, fine-tuning is conducted to achieve the final corrosion stage classification. In this stage, only the parameters of f_{E} and q_{E} are updated, and all other modules are frozen.

$$W_{f_{\mathrm{E}}} \leftarrow W_{f_{\mathrm{E}}} - \eta_3 \left(\frac{\partial L_{\mathrm{CN}}^{\mathrm{E}}}{\partial W_{f_{\mathrm{E}}}} \right), \tag{15}$$

$$W_{q_{\mathrm{E}}} \leftarrow W_{q_{\mathrm{E}}} - \eta_3 \left(\frac{\partial L_{\mathrm{CN}}^{\mathrm{E}}}{\partial W_{q_{\mathrm{E}}}} \right), \tag{16}$$

where η_3 represents the learning rate for Stage 1.

Based on the above formulas and descriptions, the pseudocode for the training and inference process can be found in Algorithm 1.

Algorithm 1: Training and Inference Procedure of CorroNet

Input: Training sample sets X_{train} consist of the paired AE-EN dataset $X_{\mathrm{AE-EN}} = \left\{ (x_{\mathrm{E}}, x_{\mathrm{A}}, y_{\mathrm{A}})_{i=1}^{N_1} \right\}$ and the unpaired EN dataset $X_{\mathrm{EN}} = \left\{ (x, y)_{i=1}^{N_2} \right\}$. Testing sample sets $X_{test} = \left\{ (x_t, y_t)_{i=1}^{N_3} \right\}$ consist of the EN data.
Output: The predicted corrosion stage label $\hat{y}_t$ for the sample x_t from the testing set.

Initialize randomly the network weight $W_{f_{\mathrm{A}}}$, $W_{f_{\mathrm{E}}}$, $W_{q_{\mathrm{A}}}$, $W_{q_{\mathrm{E}}}$, $W_{g_{\mathrm{A}}}$, $W_{g_{\mathrm{E}}^a}$, and $W_{g_{\mathrm{E}}^b}$. Set the number of iterations for each stage as n_1, n_2, and n_3.
Training:
  for $(i = 1\ to\ n_1)$ **do:** # Stage 1 training process
    Frozen $W_{f_{\mathrm{E}}}$, $W_{q_{\mathrm{E}}}$, $W_{g_{\mathrm{A}}}$, $W_{g_{\mathrm{E}}^a}$, and $W_{g_{\mathrm{E}}^b}$.
    obtain AE data and labels $\{(x_{\mathrm{A}}, y_{\mathrm{A}})\}$ from $X_{\mathrm{AE-EN}}$.
    Calculate $L_{\mathrm{CN}}^{\mathrm{A}}$ by solving Equation (5).
    Update $W_{f_{\mathrm{A}}}$ and $W_{q_{\mathrm{A}}}$ by solving Equations (7) and (8).
  end for.
  for $(i = 1\ to\ n_2)$ **do:** # Stage 2 training process
    for obtain a batch of paired AE-EN data $\{(x_{\mathrm{E}}, x_{\mathrm{A}})\}$ from $Dataloader(X_{\mathrm{AE-EN}})$ **do:**
      Frozen $W_{q_{\mathrm{A}}}$, $W_{q_{\mathrm{E}}}$, and $W_{g_{\mathrm{E}}^b}$.
      Calculate L_{InfoNCE} and L_{KL} by solving Equations (1) and (3).
      Update $W_{f_{\mathrm{A}}}$, $W_{g_{\mathrm{A}}}$, $W_{f_{\mathrm{E}}}$, and $W_{g_{\mathrm{E}}^a}$ by solving Equations (9)–(12).
    end for.
    for obtain a batch of EN data $\{x\}$ from $Dataloader(X_{EN})$ **do:**
      Frozen $W_{f_{\mathrm{A}}}$, $W_{q_{\mathrm{A}}}$, $W_{q_{\mathrm{E}}}$, $W_{g_{\mathrm{A}}}$, and $W_{g_{\mathrm{E}}^a}$.
      Calculate L_{SC} by solving Equation (4).
      Update $W_{f_{\mathrm{E}}}$ and $W_{g_{\mathrm{E}}^b}$ by solving Equations (13) and (14).
    end for.
  for $(i = 1\ to\ n_3)$ **do:** # Stage 3 training process
    for obtain a batch of EN data $\{x\}$ from $Dataloader(X_{EN})$ **do:**
      Frozen $W_{f_{\mathrm{A}}}$, $W_{q_{\mathrm{A}}}$, $W_{g_{\mathrm{A}}}$, $W_{g_{\mathrm{E}}^a}$, and $W_{g_{\mathrm{E}}^b}$.
      Calculate $L_{\mathrm{CN}}^{\mathrm{E}}$ by solving Equation (6).
      Update $W_{f_{\mathrm{E}}}$ and $W_{q_{\mathrm{E}}}$ by solving Equations (15) and (16).
    end for.
Inference:
  Obtain a test sample x_t from X_{test}.
  Calculate the predicted label $\hat{y}_t = q_{\mathrm{E}}(f_{\mathrm{E}}(x_t))$.
  Provide the model's predicted result for the corresponding corrosion stage of the sample.

2.5. Evaluation Metrics

To evaluate the effectiveness of the model in corrosion monitoring, Accuracy and F1 score are chosen as evaluation metrics. Accuracy measures the proportion of correctly classified samples out of the total number of samples. It provides a straightforward measure of the model's overall performance by indicating the percentage of correct predictions. The

F1 score is the harmonic mean of precision and recall, balancing the trade-off between these two metrics. Precision measures the proportion of true positives out of all predicted positives, and recall measures the proportion of true positives out of all actual positives. Specifically, in addition to calculating the Accuracy (Acc.) and F1 score for each class individually, the Average Accuracy (Ave. Acc.) and the average Macro-F1 score on the test set were also calculated.

For each class *c*, Accuracy and F1 score are calculated using Equations (17) and (18):

$$Acc_c = \frac{TP_c + TN_c}{TP_c + TN_c + FP_c + FN_c} \tag{17}$$

$$F1_c = 2 \times \frac{Precision_c \times Recall_c}{Precision_c + Recall_c} \tag{18}$$

$$Precision_c = \frac{TP_c}{TP_c + FP_c} \tag{19}$$

$$Recall_c = \frac{TP_c}{TP_c + FN_c} \tag{20}$$

TP_c is the true positives for class c; TN_c is true negatives for class c; FP_c is false positives for class c; FN_c is false negatives for class c. The Average Accuracy (Ave. Acc) across all classes is the mean of the per-class Accuracies, and the Macro-F1 score is the unweighted average of the per-class F1 scores.

2.6. Experimental Setup and Data Description

Pitting corrosion damage is a major issue affecting material strength as it can lead to catastrophic failure of metallic systems and structures, which is often difficult to predict. Therefore, this paper uses pitting corrosion as an example to evaluate the performance of CorroNet and other comparative methods. To obtain AE and EN data for model training and testing, pitting corrosion experiments were conducted on 304 stainless steels. The experimental setup is shown in Figure 3, with Figure 3a depicting a schematic diagram and Figure 3b showing the physical experimental setup. The experiments were conducted at room temperature, using 3.5% NaCl (Sinopharm Chemical Reagent Co. Ltd., Shanghai, China) as the corrosive medium. The number of 304 stainless steel specimens (Shen Xin Ke Ji Co. Ltd., Tengzhou, Shandong, China) used in this study was 20. The working surface of the 304 stainless steel specimens (25 × 50 × 3 mm) was ground with 2000-grit silicon carbide paper, cleaned with ethanol, and then air-dried. A circular area of 1 cm^2 was exposed for corrosion, with the rest sealed using D-31 epoxy/polyurethane primer. Except for the NaCl solution, all other materials were supplied by Shen Xin Ke Ji Co. Ltd. (Tengzhou, Shandong, China). Sampling details are in Appendix A.2.

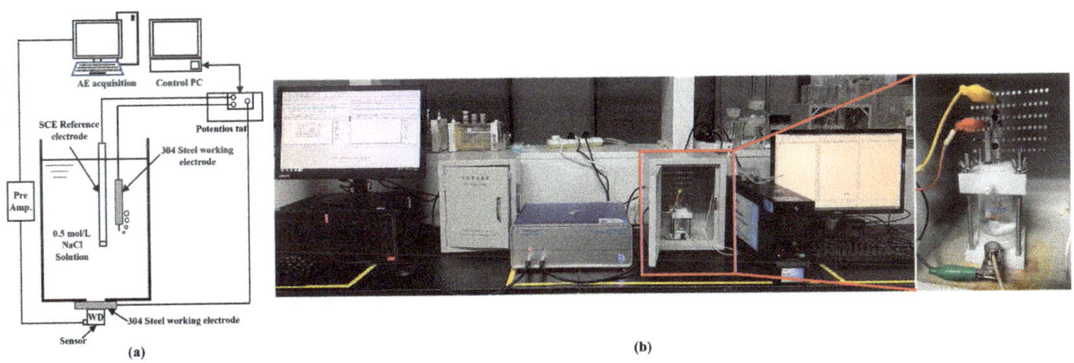

Figure 3. Corrosion experimental setup; (**a**) is a schematic and (**b**) is the physical setup.

The EN and AE data were labeled into three categories according to the corrosion states: no corrosion, metastable corrosion, and stable corrosion, denoted as C1, C2, and C3, respectively. In the no corrosion stage, the metal remains in a healthy state with no obvious signs of corrosion. In the metastable corrosion stage, corrosion has initiated but remains unstable, where corrosion reactions may be localized or intermittent. In the stable corrosion stage, the corrosion process enters a stable phase, with a steady rate and intensified damage [22,35].

Similar to previous studies [36], the synergistic AE-EN data pairs were generated by capturing both signals occurring simultaneously at the same time step. Each AE event comprises 1024 sampling points, and EN signals within 5 min before and after the event were paired as synergistic data. Given the small sample size of AE data, the goal was to achieve effective learning with a limited dataset. As a result, approximately 90 AE-EN aligned data segments were sampled.

In addition, we conducted multiple corrosion experiments with the same experimental setup as described above and collected additional unpaired EN data to expand the dataset. Each experiment lasted approximately 100 h in total, metastable corrosion began after about 10 h and stable corrosion after about 70 h. At the end of the experiment, the long-sequence data collected throughout the entire experiment were segmented. Each segment consists of 3600 sampling points, corresponding to a 1 h duration. In total, the EN dataset comprised 324 samples, with 108 samples for each corrosion stage. To ensure a fair comparison between models and consistency with previous studies, the EN data were randomly split into a train set and a test set in an 8:2 ratio. Using a sliding window approach (window length: 600, stride: 30), the original 3600-time-step sample segments were further divided into smaller segments. The details of the EN dataset are shown in Table 1.

Table 1. Composition of electrochemical noise dataset.

	Training Set		Testing Set	
	Original Count	Count After Sliding Window	Original Count	Count After Sliding Window
C1	86	8686	22	2222
C2	86	8686	22	2222
C3	86	8686	22	2222

The proposed CorroNet method was implemented in PyTorch 12.1, which was built using Anaconda (Anaconda Inc., Austin, TX, USA), and the code editor used was Visual Studio Code (Microsoft Corporation, Redmond, WA, USA). The Transformer encoders, f_A and f_E, were both configured with two layers ($L = 2$) and the multi-head attention mechanism included two heads. The optimizer was configured as AdamW with a learning rate of 0.003, the temperature parameter τ of 0.07, and the weight λ set at 0.5. All tasks used the same preprocessing method, and all models were trained and tested using the same workstation (Windows, Intel Xeon Gold 6248R, NVIDIA GeForce RTX 4090, and 256 GB RAM).

3. Results and Discussion

3.1. Comparison of Corrosion Monitoring Performance

To evaluate the corrosion monitoring performance of CorroNet, it was compared against commonly used deep learning models, i.e., CNN and LSTM. The dataset described in Section 2.6 was used to train and test all models. CNN is a representative model widely utilized in corrosion analysis [16,17,37]. LSTM is a deep learning model extensively employed for analyzing and modeling time-series data [38]. Previous studies have demonstrated their effectiveness in corrosion classification tasks.

In prior research, CNN and LSTM have predominantly been applied to single-modal signal corrosion classification tasks. For a fair comparison, experiments using CNN and

LSTM within the proposed CorroNet multimodal framework were also conducted. Specifically, the Transformer backbone in CorroNet was replaced with CNN and LSTM, respectively. This created two variants: Corro-CNN and Corro-LSTM. This setup was intended to demonstrate the robustness and effectiveness of the proposed acoustic-electrochemical alignment framework in improving corrosion classification performance across different backbones. The comparison results are presented in Table 2.

Table 2. Results for corrosion monitoring. (The bolded values in the table indicate the optimal values).

	Ave. Acc.	Macro-F1	C1		C2		C3	
			Acc.	F1	Acc.	F1	Acc.	F1
CNN	0.9310	0.9307	0.8573	0.9221	0.9356	0.9037	**1.0000**	0.9663
Corro-CNN	0.9548	0.9547	0.8816	0.9351	0.9847	0.9362	0.9982	0.9928
LSTM	0.9389	0.9385	0.8807	0.9275	0.9361	0.9271	**1.0000**	0.9609
Corro-LSTM	0.9602	0.9601	**0.9761**	0.9445	0.9046	0.9466	**1.0000**	0.9893
CorroNet	**0.9701**	**0.9701**	0.9203	**0.9549**	**0.9928**	**0.9600**	0.9973	**0.9953**

As shown in Table 2, CorroNet achieved the highest Average Accuracy (Ave. Acc.) of 0.9701 and Macro-F1 score of 0.9701 among all models. For specific classes, C1 samples were the most challenging to classify for all models. This may be attributed to the less distinctive features and higher noise interference in the "no corrosion" stage. Notably, CorroNet achieved an Accuracy of 0.9203 and an F1 score of 0.9549 for C1, which are 11.28% higher in Accuracy and 4.8% higher in F1 score than the CNN and 8.94% higher in Accuracy and 4.26% higher in F1 score than the LSTM. Additionally, CorroNet maintained remarkable Accuracies and F1 scores for C2 and C3 with minimal variance while CNN and LSTM exhibited comparatively larger performance drops. This superior performance is attributed to CorroNet's Transformer backbone for effective feature extraction. Its synergistic learning framework that aligns AE and EN signals enables more precise corrosion stage classification compared to the single-modal CNN and LSTM models.

The comparison between the regular models (CNN and LSTM) and their variants (Corro-CNN and Corro-LSTM) also demonstrated the effectiveness of synergistic learning. The regular CNN achieved the lowest performance with an Ave. Acc. of 0.9310 and a Macro-F1 of 0.9307. This suggested that CNN may have encountered difficulties when handling time-series data with complex temporal dependencies. LSTM performed slightly better than CNN (Ave. Acc. 0.9389, Macro-F1 0.9385). This result was expected, as LSTMs were generally more capable of handling time-series information compared to CNNs. However, LSTM still fell short compared to CorroNet, particularly in C1 (Accuracy 0.8807), where it struggled to capture the necessary features for accurate classification.

Integrating CNN into the CorroNet framework (Corro-CNN) significantly improved its performance, with a 2.38% increase in Ave. Acc. And a 2.4% increase in Macro-F1 score. The enhanced performance, especially in C1 (Accuracy +2.43%) and C2 (Accuracy +4.91%), indicated that the CorroNet framework helped CNN handle complex temporal tasks. Similarly, Corro-LSTM outperformed the LSTM, with a 2.13% increase in Ave. Acc. And a 2.16% increase in Macro-F1 score. The most significant improvement was in C1 (Accuracy +9.54%), where Corro-LSTM achieved the highest Accuracy (0.9761) among all the models. These results suggested that the CorroNet framework enhanced the ability to capture and align features from dual-modal data to further improve the overall performance.

To further analyze each model's corrosion monitoring performance and ability to learn corrosion-related semantic features, Figure 4 presents the confusion matrixes of the predicted results on the test set. The horizontal axis represents the predicted labels, and the vertical axis represents the true labels.

The analysis of the confusion matrices for all five models revealed a general pattern, samples from adjacent corrosion stages exhibited more similar features, resulting in a relatively higher misclassification rate. The non-corroded stage (C1) and the metastable corrosion stage (C2) were misclassified by most of the models. For instance, in CNN model (Figure 4a), 13.5% of C1 samples were misclassified as C2, while in LSTM (Figure 4c), 8.33% of C1 samples were also misclassified as C2. Similarly, 6.21% of C2 samples in the CNN model were also misclassified as C3, and 4.55% in the LSTM model. This indicated that most models exhibited weaker recognition capabilities for early-stage corrosion data (C1 and C2). Despite these challenges, CorroNet (Figure 4e) maintained strong corrosion monitoring performance, with only 0.72% of C2 samples misclassified as C1.

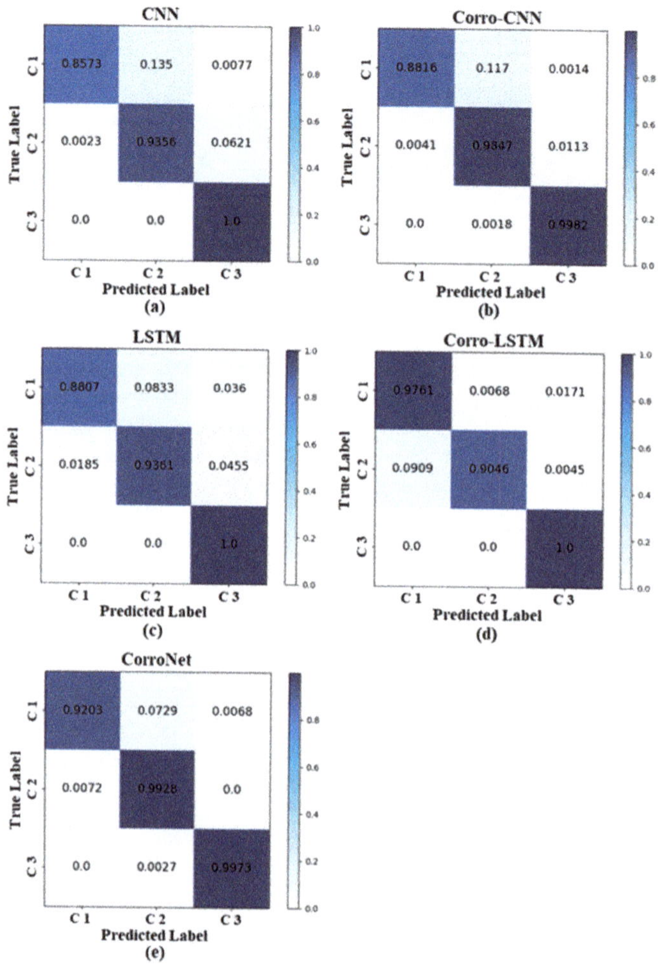

Figure 4. The confusion matrixes of different models; (**a**–**e**) are the confusion matrices of the prediction results obtained by CNN, Corro-CNN, LSTM, Corro-LSTM, and CorroNet, respectively.

Both Corro-CNN and Corro-LSTM significantly reduced misclassification compared to regular CNN and LSTM models. Corro-CNN (Figure 4b) reduced the misclassification rate of C1 as C2 by 1.8% and the misclassification rate of C2 as C3 by 5.08%. Corro-LSTM (Figure 4d) reduced the misclassification rate of C1 as C2 by 7.65%. The improvement in the performance of CNN and LSTM after incorporating the CorroNet framework further demonstrated the effectiveness of the synergistic learning between AE and EN data. In contrast, all models performed exceptionally well on the stable corrosion stage (C3), nearly achieving 100% Accuracy. This indicated that C3 features are more distinct and easier to classify with minimal errors.

In addition to the 8:2 split ratio, we analyzed how different data split ratios affected the model's robustness in corrosion detection. The results and analysis are in Appendix A.3.

3.2. Ablation Experiments

To evaluate the key components in CorroNet, ablation experiments were conducted. CorroNet consists of the following critical components: (1) Using AE signals as anchors to enhance the learning of corrosion-related semantic features in EN signals. This is achieved through the multi-granularity synergistic learning of AE and EN, including feature alignment learning and probability distribution consistency learning; (2) introducing supervised contrastive learning to capture class-specific features of EN signals; (3) pre-training AE on the corrosion stage classification task to provide precise anchor points for optimizing feature mapping during training.

Therefore, the following seven ablation experiments were conducted:

(1) *w/o* FAL: the feature alignment learning (FAL) loss function L_{InfoNCE} was removed; (2) *w/o* DAL: the probability distribution alignment learning (DAL) loss function L_{KL} was removed; (3) *w/o* SCL: the supervised contrastive learning (SCL) L_{SC} of EN signals was removed; (4) *w/o* AE pre-training: the pre-training of the AE feature extractor in Stage 1 was removed, and the subsequent learning was conducted directly; (5) *w/o* AE: the AE signals were entirely excluded from the learning model, which means that only the supervised contrastive learning and the fine-tuning of EN signals were performed; (6) *w/o* pre-train: both pre-training stages (Stages 1 and 2) were excluded, and only the classification fine-tuning of EN signals was performed directly; (7) *w/o* EN: The EN signals was removed during model training, and only AE signals were used for model training and prediction. The full model is represented by CorroNet. The results are shown in Table 3.

Table 3. Results for ablation experiments. (The bolded values in the table indicate the optimal values).

	Ave. Acc.	Macro-F1	C1		C2		C3	
			Acc.	F1	Acc.	F1	Acc.	F1
w/o FAL	0.9440	0.9441	0.8794	0.9320	0.9667	0.9225	0.9860	0.9777
w/o FAL	0.9554	0.9552	0.8722	0.9315	**0.9977**	0.9398	0.9964	0.9944
w/o SCL	0.9470	0.9465	0.8479	0.9152	0.9946	0.9315	0.9986	0.9928
w/o AE pre-training	0.9586	0.9584	0.8870	0.9346	**0.9887**	0.9429	**1.0000**	**0.9978**
w/o AE	0.9373	0.9371	0.8636	0.9268	0.9482	0.9103	**1.0000**	0.9741
w/o Pre-train	0.9295	0.9287	0.8177	0.8872	0.9743	0.9104	0.9964	0.9884
w/o EN	0.8334	0.8311	0.7083	0.7907	0.9167	0.8627	0.8750	0.8400
CorroNet	**0.9701**	**0.9701**	**0.9203**	**0.9549**	0.9928	**0.9600**	0.9973	0.9953

w/o FAL (without feature alignment learning): Removing the feature alignment learning L_{InfoNCE} caused a 2.61% drop in Ave. Acc., with C1 Accuracy decreasing by 4.09%. This showed that feature alignment learning is essential for learning the corrosion features across stages.

w/o DAL (without distribution alignment learning): distribution alignment learning L_{KL} was particularly important for C1 classification, as its absence caused a noticeable 4.81% decline in C1 Accuracy and a 2.34% drop in F1 score.

w/o SCL (without supervised contrastive learning): Removing supervised contrastive learning resulted in a 7.24% drop in C1 Accuracy and a 3.97% decrease in F1 score. These showed that SCL is essential for learning class-specific features, especially for early-stage corrosion (C1).

w/o AE pre-training (without AE pre-training): Removing the pre-training of AE signals reduced Ave. Acc. by 1.05% and Macro-F1 score decreased by 1.17%. Additionally, the C1 Accuracy significantly dropped by 3.33%, highlighting the importance of AE pre-training for distinguishing early corrosion stages (C1).

w/o AE (without AE signals): Removing AE signals resulted in a significant performance degradation, especially in C1 (Accuracy drops by 5.67% and F1 score dropped by 2.81%) and C2 (Accuracy dropped by 4.46% and F1 score dropped by 4.97%). Additionally, the Ave. Acc. decreased by 3.28%, and the Macro-F1 score dropped by 3.3%. These highlight the importance of using AE signals as anchors to improve corrosion feature learning.

w/o pre-train (without pre-training in Stages 1 and 2): Removing both pre-training stages led to the worst performance, with Ave. Acc. decreasing by 4.06% and Macro-F1 by 4.14%. C1 Accuracy significantly dropped by 10.26%, with the F1 score decreasing by 6.77%. The F1 score for C2 also decreased by 4.96%, demonstrating that pre-training is critical for learning robust features. The sharp decline in C1 and C2 performance further underscored the importance of synergistic learning of AE and EN signals and the supervised contrastive learning for effective corrosion feature extraction.

w/o EN (without EN signals): Removing EN signals resulted in a significant performance drop, with the Ave. Acc. decreasing by 13.67% and the Macro-F1 score dropping by 13.9%. This highlights the importance of combining EN and AE signals for model learning.

Comparing the seven ablation experiments, it was found that removing EN signals caused the largest performance drop, which confirms their crucial role. Excluding AE signals or pre-training also led to considerable performance degradation, highlighting the importance of AE-EN multimodality and the pre-training stages. Overall, the ablation experiments strongly supported the design choices made in CorroNet, as each component contributed significantly to the model's ability to accurately monitor corrosion stages, especially in the early stages where feature differences are more subtle.

4. Conclusions

A novel deep learning framework, named CorroNet, is proposed in this paper, which provides a new approach to automated corrosion monitoring. The proposed CorroNet has the potential to detect corrosion processes on the surface of stainless steels in chloride solutions. The main conclusions are summarized as follows:

(1) CorroNet pioneers the integration of multi-scale learning from both AE and EN signals, effectively capturing the intricate characteristics of corrosion evolution at different stages. By combining the complementary strengths of these two signals, CorroNet offers a comprehensive approach to corrosion monitoring;
(2) To further improve the model's performance, CorroNet incorporates an adaptive learning mechanism that dynamically adjusts to the complexities of corrosion data. This mechanism not only optimizes feature extraction but also allows for more reliable detection of early-stage corrosion, which is often difficult to identify with traditional monitoring methods;
(3) A comprehensive metal corrosion monitoring task was conducted to validate CorroNet's real-world applicability. The results highlight its superior performance in predicting various corrosion stages. Ablation studies further underscore the significance of the key model components in achieving this performance.

Despite the promising capabilities of CorroNet, there are opportunities for further improvement. The current research centers on pitting corrosion, one of the most prevalent corrosion forms. However, distinct types of corrosion have different mechanisms. As different corrosion mechanisms may require tailored approaches, future work should explore diverse corrosion types and extend CorroNet's capabilities to handle multiple corrosion scenarios. Additionally, the development of a user-friendly graphical interface (GUI) is planned to facilitate better model interaction and accessibility for users in industrial settings.

Author Contributions: Conceptualization, R.W. and K.S.; methodology, R.W. and G.S.; software, R.W.; validation, K.W., X.M. and R.L.; resources, K.S.; data curation, G.S., F.Q. and L.Z.; writing—original draft preparation, R.W. and G.S.; writing—review and editing, K.S.; supervision, X.C.; project administration, K.S.; funding acquisition, K.S. All authors have read and agreed to the published version of the manuscript.

Funding: This research was funded by the National Key Research and Development Program of China (No. 2022YFC3004502), Ministry of Science and Technology of the People's Republic of China.

Data Availability Statement: The code is available upon request.

Conflicts of Interest: Authors Guangbin Shan, Feng Qiu and Linqi Zhu were employed by State Key Laboratory of Chemical Safety, SINOPEC Research Institute of Safety Engineering Co. Ltd. Authors Kang Wang, Xianglong Meng and Ruiqin Li were employed by Shanxi Beifang XingAn Chemical Industry Co., Ltd. The remaining authors declare that the research was conducted in the absence of any commercial or financial relationships that could be construed as a potential conflict of interest.

Abbreviations

AE	Acoustic emission
EN	Electrochemical noise
CNN	Convolutional neural network
LSTM	Long short-term memory network
PEM	Proton exchange membrane
CNN-TL	Transfer learning method based on convolutional neural networks
HDR	Hydrogen defect recognition
SOTA	State of the art
MLP	Multi-layer perceptron
InfoNCE	Information noise-contrastive estimation
KL	Kullback–Leibler divergence
CN	Cross-entropy
SCL	Supervised contrastive learning
FAL	Feature alignment learning
DAL	Distribution alignment learning
Ave. Acc.	Average Accuracy

Appendix A

Appendix A.1. The Transformer Encoder

This section provides the computational details of the Transformer encoder, as shown in Figure A1. The raw time-series input is represented as $x \in R^{H\times W}$, where H denotes the sequence length of the original sample matrix, and W denotes the number of features in the original sample. In the embedding module, x is first segmented into temporal slices and mapped into multiple one-dimensional token vectors $\left\{x_1,\ x_2,\ \ldots,x_p \in R^d\right\}$ through a linear projection. Consequently, the shape of x is transformed into $x \in R^{n\times d}$, where d is the dimension of the projected vectors, and n represents the number of tokens. A learnable position encoding $E_{pos} \in R^{n\times d}$ is then added to the token sequence x to retain positional information. The final output of the embedding module is denoted as $Z_0 \in R^{n\times d}$.

Next, Z_0 is processed through the multi-head attention computation. The attention mechanism extracts global features by computing the weighted sum of the values V, where the weights are derived from the interaction between queries Q and keys K across the tokens. Multi-head attention allows the model to simultaneously capture temporal features at multiple scales simultaneously. The computation of the multi-head attention mechanism is described using Equations (A1)–(A3):

$$\text{Attention}(Q,K,V) = \text{softmax}\left(\frac{QK^T}{\sqrt{d_k}}\right)V \quad \text{(A1)}$$

$$head_i = \text{Attention}\left(QW_i^Q, KW_i^K, VW_i^V\right) \quad \text{(A2)}$$

$$\text{MultiHead}(Q,K,V) = \text{Concat}(head_1,\ldots,head_h)W^O \quad \text{(A3)}$$

The dot-product (multiplicative) attention is computed according to Equation (A1). Here, Q, K, and V are obtained by applying normalization to the output sequence Z_0 from the embedding block, followed by multiplication with the transformation matrices W_q, W_k, and W_v, respectively. These are referred to as the query matrix $Q \in R^{n \times d_k}$, the key matrix $K \in R^{n \times d_k}$, and the value matrix $V \in R^{n \times d_v}$. The result of the i-th attention head is denoted as $head_i$, and all the heads are concatenated. The number of attention heads is denoted by h, and a linear transformation matrix W_o is applied to obtain the output $Z_1 \in R^{n \times d}$. Finally, Z_1 is fed into the MLP module, employing residual connections. To fully extract the features from the time-series data, the above attention computation and MLP module are repeated across multiple layers, with the model depth denoted by L. Therefore, the computation of the Transformer encoder module is summarized using Equations (A4) and (A5):

$$Z_l' = \text{MultiHead}(\text{LN}(Z_{l-1})) + Z_{l-1}, \qquad l = 1 \ldots L \tag{A4}$$

$$Z_l = \text{MLP}(\text{LN}(Z_l')) + Z_l', \qquad l = 1 \ldots L \tag{A5}$$

The output matrix Z_L from the final layer is subjected to average pooling to obtain the final feature vector, denoted as $z \in R^d$.

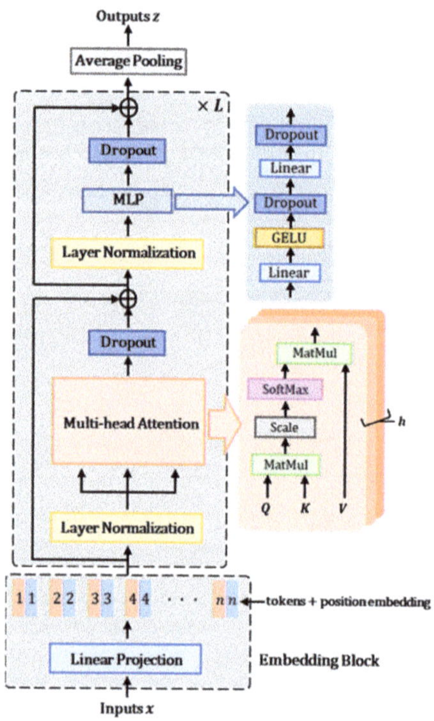

Figure A1. Detailed structure of the Transformer encoder.

Appendix A.2. Details of Data Collection

As shown in Figure 3, during the collection of AE signals, one side of the test steel sheet was in contact with the solution, while an AE sensor was placed on the other side. A PCI-2 acoustic emission detection system from Physical Acoustics Corporation (PAC) was used, with a 40 dB gain, and a WD-type broadband AE sensor was positioned above the specimen. The sampling frequency of the AE signal was set to 1024 Hz. The EN data were collected using a Gamry Reference 600 electrochemical workstation in ZRA mode, with a saturated calomel electrode (SCE) as the reference electrode, and the 304 steel specimen was used as the working electrode. The sampling frequency of the EN signal was 1 Hz. The corrosion process of the 304 steel specimen in 3.5% NaCl solution was simultaneously monitored using a combination of AE and EN methods. In addition to the experiments combining AE and EN, to expand the dataset and provide test data for evaluating the model's performance, additional corrosion experiments were conducted where only EN signals were collected. The experimental parameters for these tests were identical to those described above.

Appendix A.3. Detailed Results for Different Data Split Ratios

In addition to the aforementioned 8:2 split ratio for EN signals, we further considered the impact of different data split ratios on the model's performance to evaluate its robustness in corrosion detection under varying data partition scenarios. The results are shown in Table A1. Experimental results demonstrated that the proposed model maintained stable and robust corrosion monitoring capabilities despite changes in the data split ratio, further confirming the robustness of the proposed model.

Table A1. Results for different data split ratios.

	Ave. Acc.	Macro-F1	C1		C2		C3	
			Acc.	F1	Acc.	F1	Acc.	F1
5:5	0.9606	0.9605	0.8971	0.9425	0.9899	0.9480	0.9949	0.9910
6:4	0.9679	0.9654	0.9095	0.9473	0.9899	0.9661	0.9849	0.9830
7:3	0.9686	0.9645	0.9273	0.9477	0.9699	0.9639	0.9849	0.9820
8:2	0.9701	0.9701	0.9203	0.9549	0.9928	0.9600	0.9973	0.9953
9:1	0.9781	0.9513	0.9028	0.8930	0.9625	0.9677	0.9949	0.9932

References

1. Khalaf, A.H.; Xiao, Y.; Xu, N.; Wu, B.; Li, H.; Lin, B.; Nie, Z.; Tang, J. Emerging AI technologies for corrosion monitoring in oil and gas industry: A comprehensive review. *Eng. Fail. Anal.* **2024**, *155*, 107735. [CrossRef]
2. Shao, X.; Cai, B.; Ahmed, S.; Zhou, X.; Hu, Z.; Sui, Z.; Liu, X. Towards proactive corrosion management: A predictive modeling approach in pipeline industrial applications. *Process Saf. Environ. Prot.* **2024**, *190*, 1471–1480. [CrossRef]
3. Wu, X.; Han, E.-H. Acoustic emission during pitting corrosion of 304 stainless steel. *Corros. Sci.* **2011**, *53*, 1537–1546.
4. Wu, K.; Byeon, J.-W. Morphological estimation of pitting corrosion on vertically positioned 304 stainless steel using acoustic-emission duration parameter. *Corros. Sci.* **2019**, *148*, 331–337. [CrossRef]
5. Wu, K.; Jung, W.-S.; Byeon, J.-W. In-situ monitoring of pitting corrosion on vertically positioned 304 stainless steel by analyzing acoustic-emission energy parameter. *Corros. Sci.* **2016**, *105*, 8–16. [CrossRef]
6. Cottis, R. Simulation of electrochemical noise due to metastable pitting. *J. Corr. Sci. Eng.* **2000**, *3*. Available online: https://www.jcse.org/viewPaper/ID/42/QIgtNHePZr4acRFOy9UMYZd36DXHAj (accessed on 2 November 2020).
7. Homborg, A.; Tinga, T.; van Westing, E.; Zhang, X.; Ferrari, G.; de Wit, J.; Mol, J. A critical appraisal of the interpretation of electrochemical noise for corrosion studies. *Corrosion* **2014**, *70*, 971–987. [CrossRef] [PubMed]
8. Denissen, P.J.; Homborg, A.M.; Garcia, S.J. Interpreting electrochemical noise and monitoring local corrosion by means of highly resolved spatiotemporal real-time optics. *J. Electrochem. Soc.* **2019**, *166*, C3275. [CrossRef]
9. Homborg, A.; Olgiati, M.; Denissen, P.; Garcia, S. An integral non-intrusive electrochemical and in-situ optical technique for the study of the effectiveness of corrosion inhibition. *Electrochim. Acta* **2022**, *403*, 139619. [CrossRef]
10. Xia, D.-H.; Ji, Y.; Zhang, R.; Mao, Y.; Behnamian, Y.; Hu, W.; Birbilis, N. On the localized corrosion of AA5083 in a simulated dynamic seawater/air interface—Part 1: Corrosion initiation mechanism. *Corros. Sci.* **2023**, *213*, 110985. [CrossRef]
11. Recio-Hernández, J.A.; Landa, A.E.; Miguel, G.F.S.; Mejía-Sánchez, E.; Huerta, F.L.; Cruz, R.O.; Martínez, R.G. Electrochemical noise study of the passivation of AISI 1018 carbon steel as reinforcements embedded in ternary concretes during the setting process. *ECS Trans.* **2023**, *110*, 159. [CrossRef]
12. Recio-Hernández, J.A.; Landa, A.E.; Miguel, G.F.S.; Mejía-Sánchez, E.; Huerta, F.L.; Cruz, R.O.; Martínez, R.G. Electrochemical noise of SCC inhibition of a supermartensitic stainless steel in sour solution. *ECS Trans.* **2023**, *110*, 29. [CrossRef]
13. Homborg, A.; Mol, A.; Tinga, T. Corrosion classification through deep learning of electrochemical noise time-frequency transient information. *Eng. Appl. Artif. Intell.* **2024**, *133*, 108044. [CrossRef]
14. Rubio, M.; Sanchez, D.; Gazdzicki, P.; Friedrich, K.; Urquia, A. Failure mode diagnosis in proton exchange membrane fuel cells using local electrochemical noise. *J. Power Sources* **2022**, *541*, 231582. [CrossRef]

15. Guo, F.; Li, W.; Jiang, P.; Chen, F.; Liu, Y. Deep learning approach for damage classification based on acoustic emission data in composite materials. *Materials* **2022**, *15*, 4270. [CrossRef]
16. Liu, P.; Xu, C.; Xie, J.; Fu, M.; Chen, Y.; Liu, Z.; Zhang, Z. A CNN-based transfer learning method for leakage detection of pipeline under multiple working conditions with AE signals. *Process Saf. Environ. Prot.* **2023**, *170*, 1161–1172. [CrossRef]
17. Qiu, F.; Shen, Z.; Bai, Y.; Shan, G.; Qu, D.; Chen, W. Hydrogen defect acoustic emission recognition by deep learning neural network. *Int. J. Hydrog. Energy* **2024**, *54*, 878–893. [CrossRef]
18. Zhang, Z.; Zhao, Z.; Bai, P.; Li, X.; Liu, B.; Tan, J.; Wu, X. In-situ monitoring of pitting corrosion of AZ31 magnesium alloy by combining electrochemical noise and acoustic emission techniques. *J. Alloys Compd.* **2021**, *878*, 160334. [CrossRef]
19. Kawasaki, Y.; Fukui, S.; Fukuyama, T. Phenomenological process of rebar corrosion in reinforced concrete evaluated by acoustic emission and electrochemical noise. *Constr. Build. Mater.* **2022**, *352*, 128829. [CrossRef]
20. Kovac, J.; Alaux, C.; Marrow, T.J.; Govekar, E.; Legat, A. Correlations of electrochemical noise, acoustic emission and complementary monitoring techniques during intergranular stress-corrosion cracking of austenitic stainless steel. *Corros. Sci.* **2010**, *52*, 2015–2025. [CrossRef]
21. Calabrese, L.; Bonaccorsi, L.; Galeano, M.; Proverbio, E.; Di Pietro, D.; Cappuccini, F. Identification of damage evolution during SCC on 17-4 PH stainless steel by combining electrochemical noise and acoustic emission techniques. *Corros. Sci.* **2015**, *98*, 573–584. [CrossRef]
22. Kietov, V.; Mandel, M.; Krüger, L. Combination of electrochemical noise and acoustic emission for analysis of the pitting corrosion behavior of an austenitic stainless cast steel. *Adv. Eng. Mater.* **2019**, *21*, 1800682. [CrossRef]
23. Radford, A.; Kim, J.W.; Hallacy, C.; Ramesh, A.; Goh, G.; Agarwal, S.; Sastry, G.; Askell, A.; Mishkin, P.; Clark, J.; et al. Learning transferable visual models from natural language supervision. In Proceedings of the 38th International Conference on Machine Learning, Virtual, 18–24 July 2021.
24. Jia, C.; Yang, Y.; Xia, Y.; Chen, Y.-T.; Parekh, Z.; Pham, H.; Le, Q.V.; Sung, Y.-H.; Li, Z.; Duerig, T. Scaling up visual and vision-language representation learning with noisy text supervision. In Proceedings of the International Conference on Machine Learning, Virtual, 18–24 July 2021.
25. Yang, Y.; Pan, L.; Liu, L. Event camera data pre-training. In Proceedings of the IEEE/CVF International Conference on Computer Vision, Paris, France, 1–6 October 2023.
26. Van Den Oord, A.; Vinyals, O. Neural discrete representation learning. In Proceedings of the 31st Conference on Neural Information Processing Systems (NIPS 2017), Long Beach, CA, USA, 4–9 December 2017; Volume 30.
27. Girdhar, R.; El-Nouby, A.; Liu, Z.; Singh, M.; Alwala, K.V.; Joulin, A.; Misra, I. Imagebind: One embedding space to bind them all. In Proceedings of the IEEE/CVF Conference on Computer Vision and Pattern Recognition, Vancouver, BC, Canada, 17–24 June 2023.
28. Vaswani, A.; Shazeer, N.; Parmar, N.; Uszkoreit, J.; Jones, L.; Gomez, A.N.; Kaiser, Ł.; Polosukhin, I. Attention is all you need. In Proceedings of the 31st Conference on Neural Information Processing Systems (NIPS 2017), Long Beach, CA, USA, 4–9 December 2017; Volume 30.
29. Zhou, K.; Tong, Y.; Li, X.; Wei, X.; Huang, H.; Song, K.; Chen, X. Exploring global attention mechanism on fault detection and diagnosis for complex engineering processes. *Process Saf. Environ. Prot.* **2023**, *170*, 660–669. [CrossRef]
30. Huang, H.; Wang, R.; Zhou, K.; Ning, L.; Song, K. CausalViT: Domain generalization for chemical engineering process fault detection and diagnosis. *Process Saf. Environ. Prot.* **2023**, *176*, 155–165. [CrossRef]
31. Zhou, K.; Wang, R.; Tong, Y.; Wei, X.; Song, K.; Chen, X. Domain generalization of chemical process fault diagnosis by maximizing domain feature distribution alignment. *Process Saf. Environ. Prot.* **2024**, *185*, 817–830. [CrossRef]
32. Bi, X.; Wu, D.; Xie, D.; Ye, H.; Zhao, J. Large-scale chemical process causal discovery from big data with transformer-based deep learning. *Process Saf. Environ. Prot.* **2023**, *173*, 163–177. [CrossRef]
33. van den Oord, A.; Li, Y.; Vinyals, O. Representation learning with contrastive predictive coding. *arXiv* **2018**, arXiv:1807.03748.
34. Chen, X.; Xie, S.; He, K. An empirical study of training self-supervised vision transformers. In Proceedings of the IEEE/CVF International Conference on Computer Vision, Montreal, QC, Canada, 10–17 October 2021.
35. Hou, Y.; Aldrich, C.; Lepkova, K.; Machuca, L.; Kinsella, B. Analysis of electrochemical noise data by use of recurrence quantification analysis and machine learning methods. *Electrochim. Acta* **2017**, *256*, 337–347. [CrossRef]
36. Calabrese, L.; Galeano, M.; Proverbio, E.; Di Pietro, D.; Donato, A. Topological neural network of combined AE and EN signals for assessment of SCC damage. *Nondestruct. Test. Eval.* **2019**, *35*, 98–119. [CrossRef]
37. Jiang, S.; Zavala, V.M. Convolutional neural nets in chemical engineering: Foundations, computations, and applications. *AIChE J.* **2021**, *67*, e17282. [CrossRef]
38. Pham, T.D. Time-frequency time-space LSTM for robust classification of physiological signals. *Sci. Rep.* **2021**, *11*, 6936. [CrossRef] [PubMed]

processes

MDPI

Article

An Innovative NOx Emissions Prediction Model Based on Random Forest Feature Selection and Evolutionary Reformer

Xianyu Meng [1], Xi Li [2], Jialei Chen [2], Yongyan Fu [2], Chu Zhang [2], Muhammad Shahzad Nazir [2] and Tian Peng [2,3,*]

1 Nanjing Nanrui Able Engineering Technology Co., Ltd., Nanjing 211002, China; mengxy@nrec.com
2 Faculty of Automation, Huaiyin Institute of Technology, Huai'an 223003, China; lixi_1017@163.com (X.L.); chenjialei0408@163.com (J.C.); 17696933553@163.com (Y.F.); zhangchuhust@foxmail.com (C.Z.); msn_bhutta88@yahoo.com (M.S.N.)
3 Jiangsu Permanent Magnet Motor Engineering Research Center, Huaiyin Institute of Technology, Huai'an 223003, China
* Correspondence: husthydropt@126.com

Abstract: Developing more precise NOx emission prediction models is pivotal for effectively controlling NOx emissions from gas turbines. In this paper, a Reformer is combined with random forest (RF) feature selection and the chaos game optimization (CGO) algorithm to predict NOx in gas turbines. Firstly, RF evaluates the importance of data features and reduces the dimensionality of multidimensional data to improve the predictive performance of the model. Secondly, the Reformer model extracts the inherent pattern of different data and explores the intrinsic connection between gas turbine variables to establish a more accurate NOx emission prediction model. Thirdly, the CGO algorithm is a parameter-free meta-heuristic optimization algorithm used to find the best parameters for the prediction model. The CGO algorithm was improved using Chebyshev Chaos Mapping to improve the initial population quality of the CGO algorithm. To evaluate the efficiency of the proposed model, a dataset of gas turbines in north-western Turkey is studied, and the results obtained are compared with seven benchmark models. The final results of this paper show that RF can select appropriate input variables, and the Reformer can extract the intrinsic links of the data and build a more accurate NOx prediction model. At the same time, ICGO can optimize the parameters of the Reformer effectively.

Keywords: NOx prediction; reformer; chaos game optimization; random forest; Chebyshev chaotic mapping

1. Introduction

In response to the imperatives of energy conservation and emissions reduction, there is a growing focus on the utilization of renewable energy sources [1,2]. And carbon neutrality stands out as a pivotal strategy in advancing renewable energy [3]. Despite the increasing awareness of the importance of renewable energy, the consumption of power in the world today mainly relies on conventional fossil fuel power plants that use coal and natural gas due to the limited economic development of renewable energy sources [4]. According to statistics, 41% of carbon emissions come from thermal power generated [5,6]. However, conventional power generation methods generate significant amounts of carbon emissions and harmful gases. The combustion of gas turbines in power plants produces a large amount of nitrogen oxides (NOx), which are released into the atmosphere during the combustion process, leading to atmospheric damage and adverse effects on human health [7]. NOx is an umbrella term that includes nitrogen dioxide (NO2) and nitric oxide (NO) emissions, which are produced during combustion by the reaction of nitrogen and oxygen in the air [8,9] and is one of the most important hazardous air pollutants in the industry [10]. Improving the accuracy of NOx emission predictions is critical for improving gas turbine combustion efficiency and lowering pollutant and carbon emissions.

In the past decades, scholars have continued to explore the NOx emission prediction process by mining historical gas turbine operating data to find the relationship between real-time operating data and NOx emissions from gas turbines [11]. As machine learning (ML) methods continue to advance [12], researchers are increasingly focusing on constructing data-oriented models that accurately depict NOx emission processes. This is achieved by training these models using extensive sets of input and output process data. To forecast NOx emissions, ML techniques like extreme learning machines (ELMs) and support vector machines (SVMs) have gained significant popularity. These methods are extensively employed in the prediction of NOx emissions. Jia et al. [13] constructed a backpropagation (BP) neural network model to achieve NOx prediction. Zhou and Liang [14] developed a genetic algorithm and support vector machine (GA-SVM) model based on the traditional support vector machines for the study of NOx emission prediction. Liu et al. [15] developed a data-driven algorithm based on empirical mode decomposition (EMD) and multiple echo state networks (ESNs) for NOx emission prediction. Heysem et al. [16] used extreme learning machine (ELM) and random forest (RF) for predicting NOx emissions in a publicly available gas turbine dataset from Turkey. However, these models lack the consideration of time series data correlation, which leads to low prediction accuracy. For example, BP neural networks have limitations in dealing with nonlinear problems and are prone to falling into local minima; the generalization ability of ELM models is easily affected by the initialization of stochastic weights, leading to lower prediction accuracy, so the prediction methods can be further improved.

As a component of machine learning, deep learning has attracted great attention since it can extract deeper features between variable data automatically than shallow machine learning models [17], thus improving the accuracy of prediction. In recent years, deep neural network-based prediction methods have been applied to different fields, and NOx emission prediction has entered a new era. Among the various deep learning models, the transformer model has shown excellent performance in various prediction and classification problems. Wang et al. [18] proposed a hybrid CEEMDAN-AM-LSTM for NOx prediction, which has higher reliability and accuracy for non-stationary and nonlinear time series. However, the computational complexity of LSTM is high and often requires more parameters and computational effort, which is challenging when dealing with very long sequences. Ma et al. [19] proposed a hybrid deep learning framework combining Random Forest (RF) feature selection, Convolutional Neural Network (CNN), and Bidirectional Gated Recurrent Unit (BiGRU) for predicting NOx emissions from coal-fired boilers. Yousaf et al. [20] proposed a big data-based framework for feature selection through the ReliefF algorithm and training using Artificial Neural Network (ANN) and the Levenberg–Marquardt algorithm to predict the pressure loss coefficient of solar collectors. Li et al. [21] built a model to predict the temporal spread of seasonal influenza based on the transformer. Deng et al. [22] combined a random forest with a transformer to predict the trend of the gas and achieved the evaluation of the prediction performance by an appropriate evaluation index. The Transformer model, while having significant advantages in processing sequential data, may be affected by memory constraints and efficiency when processing long texts, leading to a decrease in processing efficiency. The reformer is a variant of a transformer that incorporates locality-sensitive hashing and reversible transformer networks to address attention and memory allocation issues. This paper adopts Reformer to investigate NOx emission prediction.

Since deep neural network models contain many hyperparameters, a manual enumeration method to adjust the optimization parameters will waste a lot of time and effort. To improve the accuracy of NOx emission prediction, this paper incorporates an improved optimization algorithm for the hyperparameters in Reformer. The introduction of algorithms to optimize the hyperparameters of the model is widely used in deep learning [23]. Li et al. [24] used Particle Swarm Optimization (PSO) to optimize the proposed model and proposed PSO-BP model for PV power forecasting. The model introduces the PSO algorithm for optimizing the threshold and initial weights of the BP, which compensates for the shortcomings of the BP model, such as randomly determined initial values and local optimization. Hu et al. [25] introduced the Improved Grasshopper Optimization Algorithm (IGOA) to optimize Long Short-Term Memory (LSTM) models for predicting short-term loads. The model uses IGOA to optimize the parameters of the LSTM model to improve the model prediction accuracy. The experiments in this paper will introduce the CGO algorithm for optimizing the Reformer model to find the optimal results. The CGO is a new meta-heuristic algorithm that outperforms other meta-heuristics in most cases [26].

Motivated by the aforementioned research techniques and applications, the Reformer model based on random forest and CGO algorithm is proposed to predict NOx emissions. The scientific contributions of this study can be outlined as follows:

(1) In order to construct a more stable NOx emission prediction model and fully exploit the intrinsic connection between gas turbine variables, this paper proposes to predict NOx emissions using Reformer. Reformer is based on the strong extraction ability and low computational complexity of the transformer and introduces locality-sensitive hashing and reversible residual network, which helps to reduce the memory footprint of the model. Experiments have proven that Reformer provides more robust and accurate prediction results for building NOx emission prediction models for gas turbines.

(2) To better utilize the correlation between data variables and identify and eliminate redundant data, several approaches can be employed; this paper uses RF to assess the importance of the data features and selects the variables of high importance to input into the model. RF can solve the multicollinearity problem by removing redundant features and thus reducing model computation or training time. Practical experiments have demonstrated that RF can effectively improve prediction performance.

(3) In order to find the optimal hyperparameters and improve model prediction accuracy, this paper introduces the CGO algorithm to optimize the model parameters. The CGO algorithm uses the basic concepts of fractal and chaotic games to build mathematical models. The position updating process of the CGO algorithm is carried out in four stages, each stage satisfying a more sensitive approach to global and local search of the search space, resulting in a better response. Empirical results indicate that the CGO algorithm effectively optimized the model hyperparameters, leading to a substantial improvement in model accuracy.

In this study, Section 2 introduces the principles of RF, Transformer, Reformer, and CGO. Section 3 describes the general model used in this research, including input, output, and network structure. Section 4 provides a detailed introduction and analysis of the dataset used in this study, including data source and preprocessing. Finally, Sections 5 and 6 compare and analyze the prediction results of the Reformer model and summarize the experimental results.

2. Methods

2.1. Transformer

Jin et al. [27] first proposed the transformer model in 2017. Compared to conventional deep learning models, the transformer model incorporates self-attention mechanisms and also introduces the concept of multi-head attention. The general organization of the transformer model is briefly shown in Figure 1, where 'Input Embedding' is the input embedding of the dimensional space of the hidden layer; 'Output Embedding' is the output embedding of the hidden layer dimensional space; 'Add&Norm' denotes residual concatenation and layer normalization operations. This section provides a further explanation of the transformer architecture.

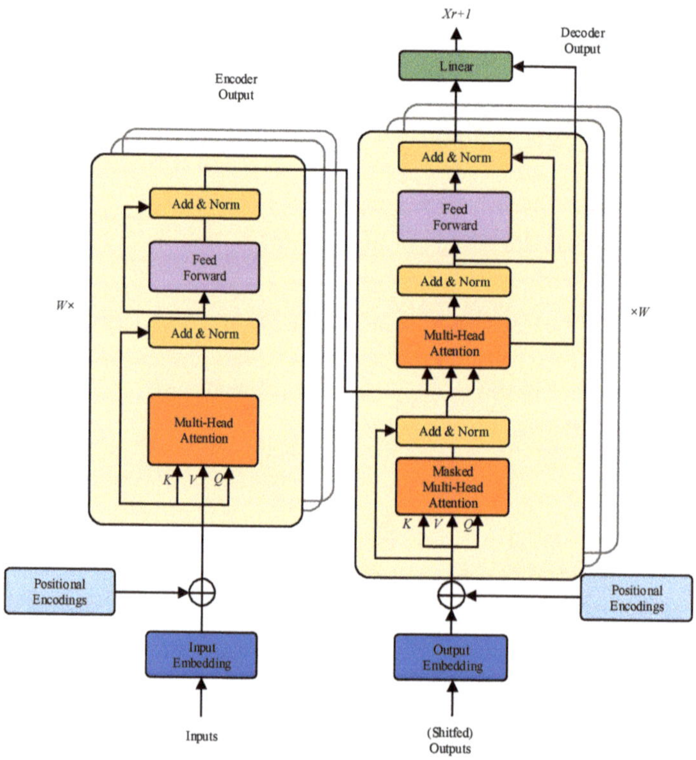

Figure 1. Architecture of transformer.

2.1.1. Positional Encoding

Transformer uses a self-attention encoder and eschews recurrent neural networks (RNNs) [28]. A self-attention encoder does not contain the sequential information inherent in RNNs. Therefore, the transformer proposes a trigonometric position encoding, which uses mainly sine and cosine functions.

$$PE_{(pos,i)} = \begin{cases} \sin(pos/10000^{\frac{i}{d_{\text{model}}}}, i = 2a \\ \cos(pos/10000^{\frac{i}{d_{\text{model}}}}, i = 2a+1 \end{cases} \quad (1)$$

where pos indicates the current position; a is a positive integer, i is the dimension, and d_{model} is the number of input features. When i = 2a, it means that the dimension is even, and PE

indicates the position information obtained at this point; when i = 2a + 1, the dimension is an odd number, and PE indicates the position information obtained at this point.

2.1.2. Encoder–Decoder

The encoder and decoder are the two main components of the transformer model [29]. The encoder's role is to translate the sequence of natural language into a hidden representation, and the decoder's role is to translate the hidden representation back into a sequence of natural language. There are $W = 6$ identical layers in the encoder, with each layer containing two sublayers. The two sub-layers are the fully connected feed-forward network and the multi-headed attention. Residual concatenation and layer normalization are applied to each of these two sub-layers. The decoder is also a stack of $W = 6$ identical layers. Like the encoder layer, the decoder also has the two sublayers of the encoder. At the same time, a third sublayer is added to the decoder, which provides multi-headed attention to the output of the encoder stack. To increase performance, the decoder, like the encoder, employs residual concatenation and normalization in each sublayer. Additionally, the self-attention sub-layers are adjusted to prevent the present position from taking up the position behind. Because of this masking, as well as the fact that the output embedding is offset by one place, forecasts for position j can only rely on known outputs at locations smaller than j. The encoder compresses the critical data from the input sequence into a fixed-length vector, which is then converted into an output by the decoder.

2.1.3. Scaled Dot-Product Attention

Scaled dot-product attention is the basis for the construction of a transformer. The attention mechanism here is self-attention. The formula for calculation is as follows:

$$\text{Attention}(Q, K, V) = \text{softmax}\left(\frac{QK^T}{\sqrt{d_k}}\right)V \tag{2}$$

where Q, K, and V represent the Query matrix, the Key matrix, and the Value matrix, respectively; the scaling factor dk represents the dimension of the matrices Q and K.

2.1.4. Multi-Head Attention

Multi-head attention is a variant of the attention mechanism commonly used in neural networks, whose main purpose is to improve the model's representation of input sequences by simultaneously considering different representation subspaces. [30]. Multi-head attention is a repetition of scaled dot-product attention for z times, and then the results obtained from z times of scaled dot-product attention are combined, and Q, K, and V are linearly transformed with a different parameter W each time, as shown in Equation (2). Then, another linear transformation is performed, at which point the value obtained is the value of multi-head attention, as shown in Equation (3). The multi-head attention derived from the aforementioned computation enables the model to acquire pertinent information across various representational dimensions. Figure 2 shows the framework for multi-headed attention.

$$\text{MultiHead}(Q, K, V) = \text{Coneat}(hd_1, \ldots, hd_h)M^{\text{o}} \tag{3}$$

$$\text{where } hd_i = \text{Attention}\left(QM_i^Q, KM_i^K, VM_i^V\right) \tag{4}$$

where M represents the linear transformation coefficient matrix. QM_i^Q, KM_i^K and VM_i^V represent the Query, Key, and Value matrices corresponding to the attention hd_i, respectively. $h = 8$ parallel attention layers are used in this study.

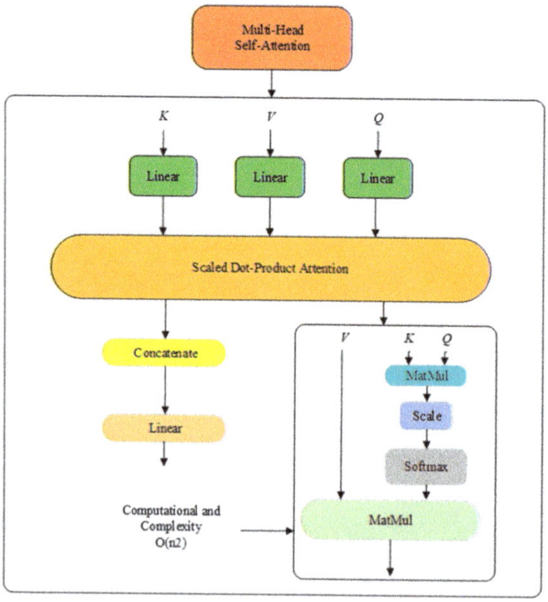

Figure 2. Framework of multi-head attention.

2.1.5. Feed-Forward Network

The feed-forward network (FNN) consists mainly of two linear transformations that use a relu as the activation function [31]. Transformer adds a feed-forward layer after the multi-head attention layer. The feed-forward serves to extract deeper features by mapping the data to a higher dimensional space and then to a lower dimensional space through a linear transformation [32]. The functional representation of *FNN* can be expressed as

$$FFN(x_{fnn}) = \max(0, x_{fnn}M_1 + a_1)M_2 + a_2 \tag{5}$$

where M_1 and M_2 denote the weights of the matrix; a_1 and a_2 denote the bias terms; x_{fnn} denotes the input to *FNN*; and max stands for relu activation function.

2.1.6. Advantages and Disadvantages of Transformer

Because RNN can only be computed serially, the inherent sequential property of the latter sequence, dependent on the former, prevents parallelization between training samples. Transformer addresses the limitation that RNN models cannot be computed in parallel. Then, self-attention can result in models that are easier to interpret. The model of attention distribution can be viewed. Every attention head may be taught to perform a variety of tasks.

Transformer cannot handle longer sequences (usually around 500) and is very GPU intensive. However, Reformer can take sequences of up to 64k in length, with much lower GPU resource consumption [33].

2.2. Reformer

In contrast to the transformer, Reformer incorporates a locality-sensitive hashing and reversible transformer network (revnet) to address the attention and memory allocation issues. Thus, when operating on a single GPU, the Reformer model can achieve extremely good behavior with less memory consumption than the transformer. The Reformer employs locality-sensitive hashing in place of dot product attention to alleviate the computational burden of attending to long sequences and utilizes revnets to conserve memory.

2.2.1. Locality Sensitive Hashing

Locality-sensitive hashing (LSH) can efficiently handle the nearest-neighbor problem for large amounts of high-dimensional data [34]. If the linear lookup is performed directly in high-dimensional space, the efficiency of the model will be reduced. LSH makes similar data in a higher dimensional space and maps to neighboring locations with higher probability. In this way, we can speed up data retrieval, improve search efficiency, and quickly find relevant data points in large-scale datasets. By adding LSH, the complexity of the model changes from $O(\mathrm{L}^2)$ to $O(\mathrm{LlogL})$.

2.2.2. Reversible Transformer Network

Reformer uses reversible residual layers instead of standard residuals and stores only one activation during training instead of M times, M being the number of layers [35]. The main idea of a revnet is to make the activation of any given layer be recovered from the activations of subsequent layers, using only the model parameters. During backpropagation from the output to the input of the network, layers can be reversed one after the other without the need to check the intermediate values used for backpropagation. The activation of each layer in revnet can be reconstructed exactly from the activation of the layer that follows and backpropagated without storing the activation in memory. By combining the attention and feed-forward layers inside the revnet block and applying the revnet idea to the Reformer, memory and GPU losses are effectively reduced.

2.3. Improved Chaos Game Optimization

2.3.1. Chaos Game Optimization

The chaos game optimization (CGO) algorithm uses a chaotic game approach to investigate fractal configurations and fractal self-similarity [36]. The overarching structure of the CGO algorithm is grounded in game theory, with chaotic game principles at its core. The CGO algorithm is a metaheuristic that contains no parameters and requires no internal parameter definition. One of the algorithm's most distinctive traits is its parameter-free architecture, where the main loop dynamically adjusts the exploration and exploitation stages.

Based on the essential ideas of CGO and the fundamental ideas of chaotic and fractal games, a mathematical model of the CGO algorithm is created.

$$X = \begin{bmatrix} X_1 \\ X_2 \\ \vdots \\ X_u \\ \vdots \\ X_m \end{bmatrix} = \begin{bmatrix} x_1^1 & x_1^2 & \cdots & x_1^v & \cdots & x_1^z \\ x_2^1 & x_2^2 & \cdots & x_2^v & \cdots & x_2^z \\ \vdots & \vdots & \vdots & \vdots & \ddots & \vdots \\ x_u^1 & x_u^2 & \cdots & x_u^v & \cdots & x_u^z \\ \vdots & \vdots & \vdots & \vdots & \ddots & \vdots \\ x_m^1 & x_m^2 & \cdots & x_m^v & \cdots & x_m^z \end{bmatrix}, \quad \begin{cases} u = 1, 2, \cdots m. \\ v = 1, 2 \cdots z. \end{cases} \tag{6}$$

where m denotes the number of candidate solutions in the search space, and z denotes the dimension of the seed.

The initial positions of Sierpinski triangle candidates are randomly determined as follows:

$$x_u^v(0) = x_{u,min}^v + \text{rand} \cdot (x_{u,max}^v - x_{u,min}^v), \begin{cases} u=1,2,\ldots,m \\ v=1,2,\ldots,z \end{cases} \tag{7}$$

where $x_{u,\min}^v$ and $x_{u,\max}^v$ denote the upper and lower bounds of the vth decision variable for the uth candidate solution; $x_u^v(0)$ is the initial value of the candidate solution; rand is a random number between 0 and 1.

Several eligible seeds produced in the search space are used to create the overall form of the Sierpinski triangle. For each qualified seed in the search space (X_u), a temporary triangle is chosen containing the following three seeds: the location of Global Best found so far (GB); the position of the Mean Group (MG_u); and the position of the uth solution candidate.

The primary goal of generating temporary triangles is to generate new candidate solutions in Sierpinski triangles. To achieve this goal, the algorithm establishes four seed generation methods as follows:

The method used to create the first seed is mathematically represented as follows:

$$Seed_u^1 = x_u + \eta_u \times (\varepsilon_u \times GB - \lambda_u \times MG_u), u = 1,2,\ldots,m \tag{8}$$

The second seed can move towards a point on the line joining X_u and MG_u, and the introduction of a random factor is used to constrain this movement. The mathematical representation of the process is as follows:

$$Seed_u^2 = GB + \eta_u \times (\varepsilon_u \times GB - \lambda_u \times MG_u), u = 1,2,\ldots,m \tag{9}$$

The third seed is randomly generated based on the position of $MG_{u,}$ and its mathematical model is shown below:

$$Seed_u^3 = MG_u + \eta_u \times (\varepsilon_u \times GB - \lambda_u \times GB), u = 1,2,\ldots,m \tag{10}$$

To execute the change phase of updating the qualified position in the search space, an additional procedure is employed to generate a fourth seed. The position of this seed is adjusted based on a random variation in a randomly selected decision variable. The mathematical representation of the fourth seed is provided as follows:

$$Seed_u^4 = X_u(x_u^k = x_u^k + R), k = [1,2,\ldots,z] \tag{11}$$

where X_u is the uth candidate solution, GB refers to the globally optimal candidate solution found so far, and MG_u refers to the average of some randomly selected candidate solutions. η_u is a randomly generated factorial used to model the seed movement restriction while ε_u and λ_u denote random integers of 1 or 2, respectively.

2.3.2. Chebyshev Chaotic Mapping

Individual distributions generated using random initialization are prone to problems such as coverage repetition and insufficient space traversal, which make the algorithms easily fall into local optimums and lead to slow convergence and low accuracy. Therefore, chaotic initialization is beginning to be widely used in algorithm optimization to improve the accuracy of the algorithm [37].

Chebyshev chaotic mapping [38] has good pseudo-random characteristics but is very sensitive to the initial parameters. In this paper, Chebyshev chaotic mapping is used instead of a random method to improve the initial population quality of the CGO algorithm. The use of the chaotic variable generation method can avoid the problem of duplicate coverage and insufficiency produced by randomization, thus improving the convergence speed and convergence accuracy of the algorithm. The iteration of Chebyshev chaotic mapping is

$$x_{n+1} = \cos(k\mathrm{arccos}x_n), x_n \in [-1, 1] \tag{12}$$

where K is the order, and 1 is the nth iteration chaotic value.

2.4. Random Forest

Breiman [39] introduced an algorithm called the Random Forest (RF) algorithm, a random forest-based importance measure in which random substitutions are made for each feature during the processing of data in a random forest. The random forest is an assessment of the importance of the feature variables and selects the variables with a high contribution based on the contribution of each category in the decision tree [36,37]. Typically, the magnitude of the contribution of each factor can be measured by the Gini index. The Gini index formula is as follows:

$$Gini_{index} = 1 - \sum_{c=1}^{|C|} \left(\mathrm{p}_{qc}^{(p)}\right)^2 \tag{13}$$

where p denotes that there are p decision trees, p_{qc} denotes the proportion of category c in q, and c denotes that there are c categories.

3. RF-ICGO-Reformer-Model

An evolutionary Reformer model based on RF and CGO is built in this paper in the following steps.

Step 1: Collect data from gas turbines in the same power plant.

Step 2: Pre-process the obtained dataset and use the random forest to analyze the data for selecting the input variables with high importance.

Step 3: The dataset is segmented into three components: a training set, a test set, and a validation set, which are all input into the Reformer model for analysis and training.

Step 4: By utilizing the Chebyshev chaotic mapping to enhance the CGO algorithm, improved performance can be achieved in terms of global search capability and convergence speed, thereby making the algorithm more suitable for complex optimization problems.

Step 5: The hyperparameters in the Reformer were optimized using CGO to continuously improve the performance of the NOx emission prediction model.

Step 6: The resulting metrics for NOx prediction were calculated: MAE, RMSE, MAE, MAPE, and R^2.

Figure 3 shows the overall flowchart of Reformer built by CGO to predict NOx emissions based on random forest.

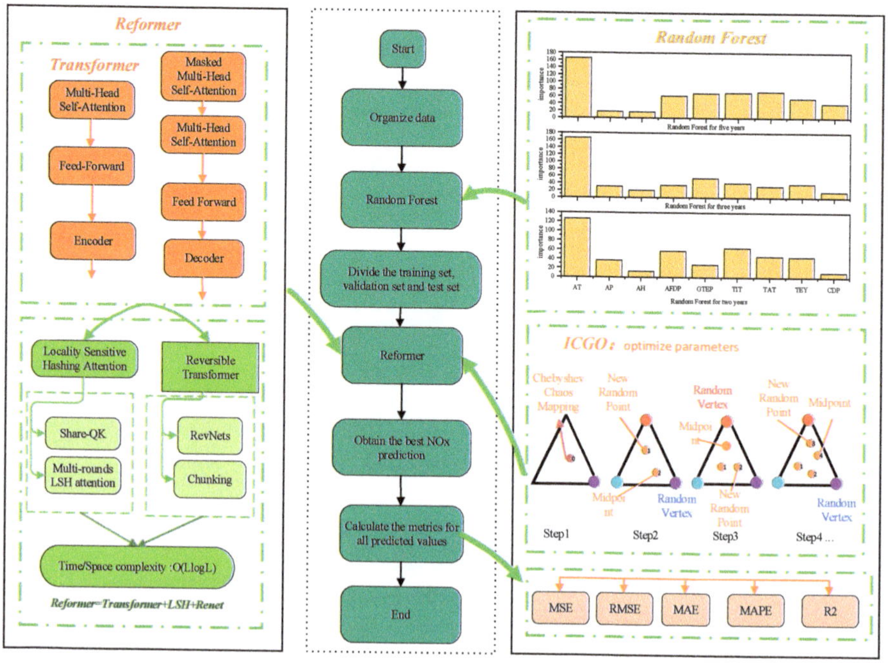

Figure 3. The overall flowchart of the proposed model.

4. Case Study

4.1. Data Source and Preprocessing

The data are made up of 9 independent variables and 2 target variables, averaged from hourly sensor measurements. Data are collected for a total of five years, with the number of data values for 2011 being 7411; for 2012, 7628; for 2013 and 2014, 7152; and for 2015, 7158. The nine independent variables consist of ambient temperature, differential air filter pressure, atmospheric pressure, humidity levels, exhaust pressure from the gas turbine, temperature at the turbine inlet, temperature after the turbine, the turbine's energy output, and the turbine itself. The two independent variables are CO and NOx, and this study focuses on the prediction of NOx. Table 1 shows the names, abbreviations, and basic data information of the nine independent variables.

Heysem et al. [16] used the first three years of data for training and validation and the last two years for testing to ensure repeatability and comparability of the work. To demonstrate the predictive effect of the prediction model more clearly, this paper considers whether to extract some of the data for prediction presentation. Kochueva and Nikolskii [40] examine the possibility of treating the entire five-year dataset as a unified training set to develop a single model or determining whether it is essential to partition the data based on either the year or the season. In the paper, the authors mention that the median NOx emissions in 2014 and 2015 were 10% lower than in 2011–2013. To validate the viability of the proposed model in this study, the data can be split into two subsets. One subset includes the 2011–2013 dataset, and the second subset includes the 2014–2015 dataset.

Table 1. Basic information of datasets from 11 sensor measures over 5 years.

Variable	Abbr	Unit	Min	Max	Mean
Ambient temperature	AT	°C	−6.23	37.10	17.71
Ambient pressure	AP	mbar	985.85	1036.56	1013.07
Ambient humidity	AH	(%)	24.08	100.20	77.87
Air filter difference pressure	AFDP	mbar	2.09	7.61	3.93
Gas turbine exhaust pressure	GTEP	mbar	17.70	40.72	25.56
Turbine inlet temperature	TIT	°C	1000.85	1100.89	1081.43
Turbine after temperature	TAT	°C	511.04	550.61	546.16
Compressor discharge pressure	CDP	mbar	9.85	15.16	12.06
Turbine energy yield	TEY	MWH	100.02	179.50	133.51
Carbon monoxide	CO	mg/m^3	0.00	44.10	2.37
Nitrogen oxides	NOx	mg/m^3	25.90	119.91	65.29

To sum up, the first experiment in this paper will be discussed using five years of research data, where three years of data from 2011, 2012, and 2013 will be used as the training and validation set, and data from 2014 and 2015 will be used as the test set. The second experiment in this paper will be discussed using data from 2011, 2012, and 2013; the third experiment in this paper will be discussed using data from 2014 and 2015. For the models applied to the last two datasets, 80% of them are utilized as the training and validation set, with the remaining 20% serving as the test set. The division of the three datasets is illustrated in Figure 4, while Table 2 presents the data analysis for the five-year dataset.

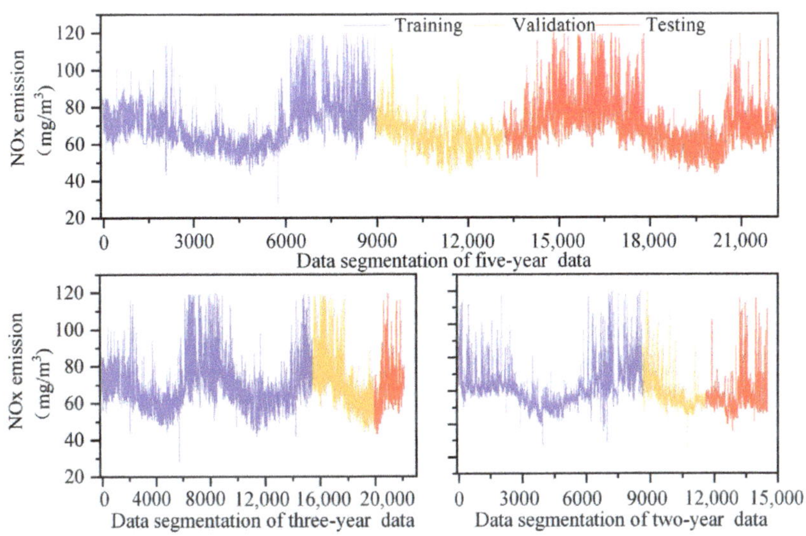

Figure 4. Data segmentation of the three datasets.

Table 2. Basic information on the NOx emission datasets.

Years	Dataset	Data Length	Max	Min	Mean	Kurtosis	Std-Dev	Skewness
	All(m/s)	7000	119.3	27.77	66.96	5.689	10.50	1.164
2011	Training(m/s)	4900	116.0	43.27	66.46	3.345	9.487	0.635
	Testing(m/s)	2100	119.3	27.77	68.15	6.380	12.46	1.544
	All(m/s)	7000	119.8	41.78	68.10	6.254	9.962	1.245
2012	Training(m/s)	4900	119.8	42.09	69.36	5.345	10.84	1.018
	Testing(m/s)	2100	115.7	41.78	65.16	9.216	9.216	1.534
	All(m/s)	7000	119.9	43.20	69.95	4.905	12.16	1.030
2013	Training(m/s)	4900	119.9	44.22	70.97	4.783	12.61	1.106
	Testing(m/s)	2100	119.9	43.20	67.58	4.227	10.68	0.571
	All(m/s)	7000	118.5	27.18	59.87	6.417	9.791	1.279
2014	Training(m/s)	4900	118.5	27.18	59.49	6.446	10.06	1.269
	Testing(m/s)	2100	116.1	35.66	60.76	6.327	9.065	1.392
	All(m/s)	7000	119.7	25.91	59.95	6.870	11.15	1.630
2015	Training(m/s)	4900	119.7	25.91	61.58	6.280	11.23	1.490
	Testing(m/s)	2100	115.8	35.60	56.16	11.186	10.00	2.363

In terms of data pre-processing, this paper uses RF to reduce the dimensionality of the data; see Section 2.4 for more information on the RF method. Figure 5 shows a histogram of the importance of the variables in the dataset of this paper. As can be seen from the graph, the importance of different variables is different. Variables with higher importance are selected and entered into the model to enhance its predictive accuracy. In this paper, the first six sets of parameters with higher importance were finally selected as input variables.

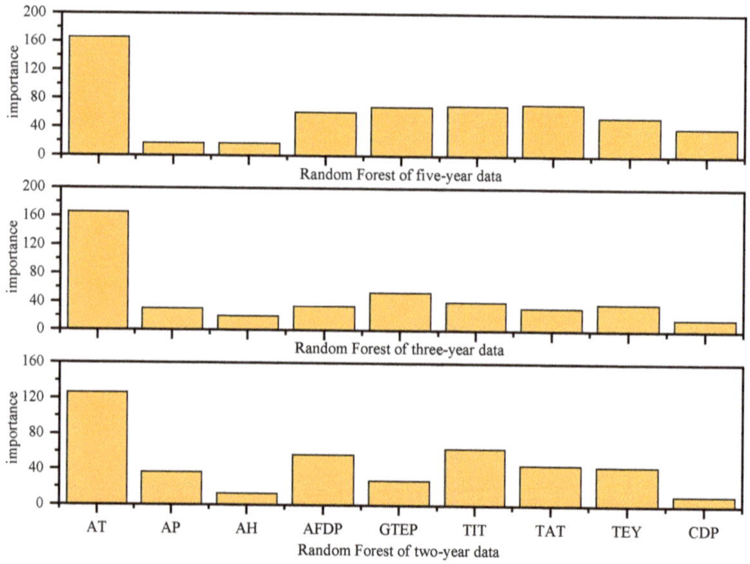

Figure 5. Random forest importance distributions for three datasets.

The ICGO algorithm is evaluated on four test functions and compared with Particle Swarm Optimization (PSO), Harris Hawk Optimization (HHO), Atomic Search Optimization (ASO), Grasshopper Optimization Algorithm (GOA), and Chaotic Game Optimization (CGO). To ensure a fair comparison, all algorithms are set to the same number of iterations (200) and population size (30), with other parameters kept at their default values. Figure 6 presents the convergence curves of the six optimization algorithms on several benchmark test functions, providing a clear visualization of their optimization performance. The results clearly demonstrate that the ICGO algorithm achieves the fastest convergence speed.

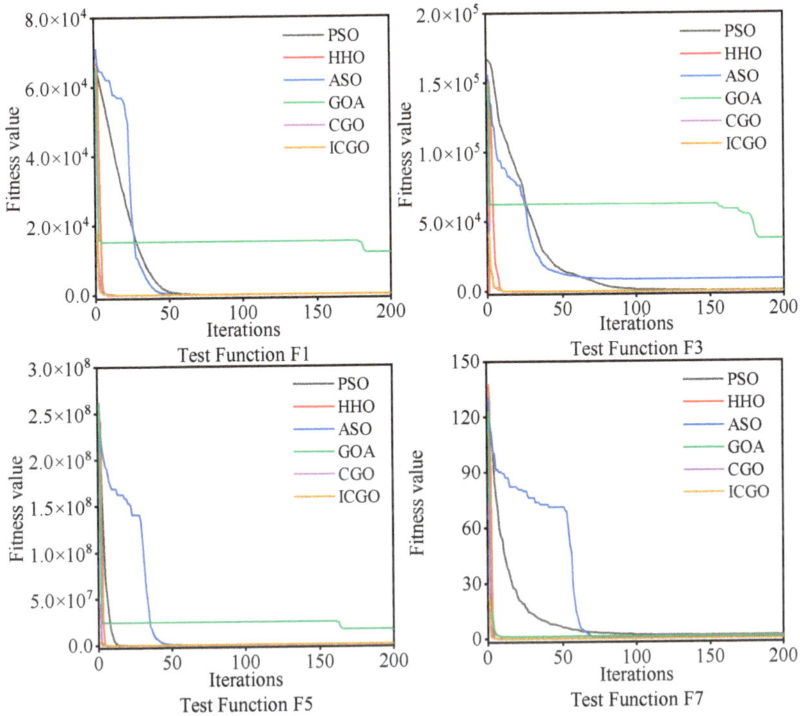

Figure 6. Convergence curves of six algorithms under partial benchmarking functions.

4.2. Performance Metrics

There are five error metrics, including Mean Square Error (MSE), Mean Absolute Error (MAE), Mean Square Error (RMSE), Mean Absolute Percentage Error (MAPE), and Coefficient of Determination (R^2). Among them, MSE can reflect the overall level of prediction error; the smaller the value, the higher the prediction accuracy of the model; RMSE is similar to MSE, but it can be a more intuitive numerical range of error size; again, the smaller the value, the better; MAE provides the average absolute size of prediction error; the smaller the value means that the model predicts more accurately. MAPE reflects the percentage of prediction error relative to the true value, the smaller the value means that the model predicts more accurately; the smaller the value means the model predicts more accurately; the smaller the value means the model predicts more accurately. MAPE can reflect the percentage of prediction error relative to the true value, and the smaller the value is, the higher the prediction accuracy of the model is the value of R^2 is between 0 and 1, and the closer the value is to 1, the better the prediction effect of the model is. Therefore, in the context of NOx emission prediction, these indicators can fully reflect the prediction performance of the model.

$$\text{MSE} = \frac{1}{r}\sum_{i=1}^{r}(b_i - a_i)^2 \tag{14}$$

$$\mathrm{RMSE} = \sqrt{\frac{1}{r}\sum_{i=1}^{r}(b_i - a_i)^2} \tag{15}$$

$$\mathrm{MAE} = \frac{1}{r}\sum_{i=1}^{r}|b_i - a_i| \tag{16}$$

$$\mathrm{MAPE} = \sum_{i=1}^{r}|\frac{b_i - a_i}{a_i}| \times \frac{100\%}{r} \tag{17}$$

$$\mathrm{R}^2 = 1 - \frac{\sum_{i=1}^{r}(a_i - b_i)^2}{\sum_{i=1}^{r}(b_i - \overline{a_i})^2} \tag{18}$$

where a_i is the predicted value, b_i is the true value, $\overline{a}_i$ is the mean of a_i, and n is the total number of samples.

5. Comparisons Results

Based on the purpose of improving the accuracy of NOx emission continuously, this paper will establish a NOx emission prediction based on RF, CGO, and Reformer. To verify the superiority of the Reformer-RF-CGO model and the effectiveness of the CGO algorithm proposed in the paper and to show that the constructed model has an improved effect on the NOx prediction accuracy, a total of eight prediction models are used in this paper. These models are applied to the same dataset. The single and combined models compared with the models in this paper include BP, ELM, LSTM, transformer, Reformer, Reformer-CGO, Reformer-RF, and Reformer-RF-CGO.

5.1. Comparison of Predicted Results of 2011–2015

Five widely used error metrics are examined in order to assess all of the experimental models. These error indicators, which are defined in the paper, include MAE, MSE, RMSE, MAPE, and R^2. These metrics can be useful in revealing deviations between predicted and actual values. A comparison of the prediction results for the five-year data test set is given in Table 3. Figure 7 displays graphs of the prediction outcomes for the suggested model and the comparison models over a five-year period with linear fitting curves. Figure 8 compares the metrics for the eight models using five radar plots, a probability density plot, a line chart, and two histograms. The hyperparameters of the model are set as follows: d_ff denotes the dimension of the intermediate vector representation of the FFN after self-attention, the range of d_ff is set from 64 to 1500, the range of dropout is set from 0.000005 to 0.5, and the range of learning rate is set from 0.000001 to 0.1.

Table 3. Metrics for the prediction performance of NOx for the five-year dataset.

Model	MSE(mg/m^3)	MAE(mg/m^3)	RMSE(mg/m^3)	MAPE(%)	R^2
BP	35.9098	4.3543	5.9925	0.0736	0.6790
ELM	32.6497	4.1340	5.7140	0.0696	0.7073
LSTM	33.9815	4.3268	5.8294	0.0726	0.6953
Transformer	25.3384	3.1771	5.0337	0.0520	0.7728
Reformer	24.1673	3.0616	4.9160	0.0501	0.7833
Reformer-CGO	23.5225	2.8415	4.8500	0.0457	0.7891
Reformer-RF	23.0061	2.8013	4.7965	0.0452	0.7937
Reformer-RF-CGO	22.7358	2.7745	4.7682	0.0448	0.7962
Reformer-RF-ICGO	22.3882	2.5817	4.7316	0.0410	0.7993

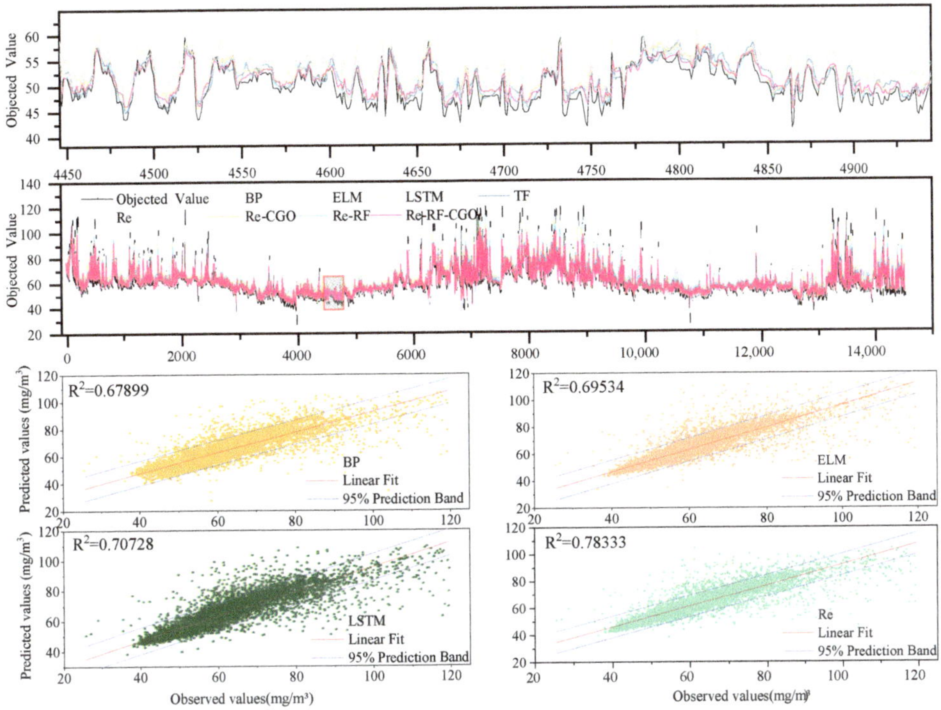

Figure 7. Plot of prediction results for the first dataset and scatterplot of predictors.

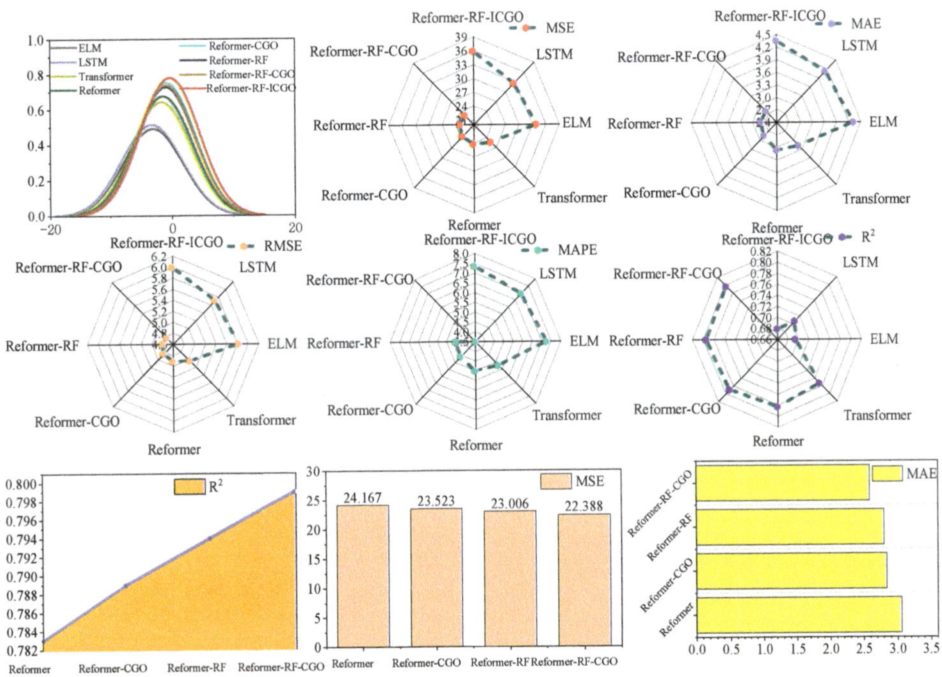

Figure 8. Comparison of science indicators results for eight models in a five-year dataset.

Firstly, a detailed analysis of Table 3 leads to the following conclusions: The Reformer -RF-ICGO model has the highest R^2 and the lowest MAE, MSE, RMSE, and MAPE, with the unoptimized Reformer performing worse than the optimized model. The MSE of the Reformer-RF-ICGO proposed in this study is 22.388 mg/m^3, MAE is 2.582 mg/m^3, RMSE is 4.732 mg/m^3, MAPE is 4.105% and R^2 is 0.799. while the MSE of the Reformer is 24.167 mg/m^3, MAE of 3.062 mg/m^3, RMSE of 4.916 mg/m^3, MAPE of 5.008%, and R^2 of 0.783.

Secondly, by trying to compare the metrics of the BP, ELM, transformer, and Reformer, it can be seen that the transformer and Reformer have superior prediction performance to more conventional ML models like BP, ELM, and LSTM.

Next, by comparing Reformer with Reformer-CGO and Reformer-RF, it is found that for MSE, Reformer is 0.645 mg/m^3 higher than Reformer-CGO and 1.161 mg/m^3 higher than Reformer-RF. Reformer is 1.779 mg/m^3 higher than Reformer-RF-ICGO. This shows that ICGO can optimize the hyperparameters of the model and also demonstrates that RF can effectively improve the accuracy of the model.

Next, by comparing Reformer, Reformer-CGO, Reformer-RF, Reformer-RF-CGO, and Reformer-RF-ICGO, the Reformer-RF-ICGO outperformed Reformer, Reformer-CGO, Reformer-RF, and Reformer-RF-CGO in terms of MSE, RMSE, and MAPE. While the R^2 of Reformer-RF-ICGO is the largest and the best fit.

Then, through the detailed analysis in Figures 6 and 7, the following conclusions can be drawn. Firstly, it can be seen from the waveform plots that the Reformer-RF-ICGO model proposed in this paper works best, and it can also be seen that the results of Reformer-RF-CGO are most similar to the true values. Comparison with BP, ELM, and LSTM shows that Reformer has the largest R^2 and the best fit. Comparing the four models Reformer, Reformer-CGO, Reformer-RF, and Reformer-RF-CGO, it can be observed from the five radar plots, a probability density plot, a line chart, and two histograms that CGO and RF can improve the Reformer accuracy. From the probability density plot, it can be concluded that the model proposed in this study is closest to the zero point and has the best prediction. Moreover, through the radar plot, it is obvious that the model proposed in this paper has the largest R2 and the smallest MSE, MAE, RMSE, and MAPE, and all indicators are the optimal results. In order to be able to show the curve fitting effect of the predicted results of the proposed model and the comparison model more clearly, the fitted curves in the red box are locally enlarged in this paper, and the enlarged plots are placed on the top of the red box.

Finally, the Reformer-RF-ICGO model of this study demonstrated the best performance in terms of error metrics and fitting effect.

5.2. Comparison of Predicted Results of 2011–2013 and 2013–2015

In order to further validate the feasibility of the model proposed in this paper, the five-year data are divided into two datasets: 2011–2013 as the second dataset, and 2014–2015 as the third dataset model to further validate the model feasibility.

Tables 4 and 5 compare the indicators for the predicted results for the second and third datasets, respectively. Figure 9 presents the graphical predictions of both the proposed model and the comparative model over a five-year period, complemented by a linear fit curve for the second dataset. Figure 10 illustrates the performance metrics of the eight models across five radar plots, a probability density plot, a line graph, and two histograms, all specific to the second dataset. Figure 11 displays the predicted outcomes for the proposed and comparative models over five years, including a linear fit curve for the third dataset. Figure 12, in turn, presents a comparative analysis of the eight models through five radar plots, a probability density plot, a line graph, and two histograms, focusing on the third dataset.

Table 4. Metrics for the prediction performance of NOx for the three-year dataset.

Model	MSE(mg/m³)	MAE(mg/m³)	RMSE(mg/m³)	MAPE(%)	R^2
BP	26.5917	3.2650	5.1567	0.0498	0.6937
ELM	25.9591	3.4431	5.0950	0.0524	0.7010
LSTM	25.5996	3.4689	5.0596	0.0536	0.7051
Transformer	24.0896	3.2956	4.9081	0.0501	0.7225
Reformer	23.7782	3.2236	4.8763	0.0490	0.7261
Reformer-CGO	22.3726	3.0694	4.7300	0.0470	0.7423
Reformer-RF	22.8676	3.1344	4.7820	0.0480	0.7366
Reformer-RF-CGO	22.3573	3.0942	4.7283	0.0473	0.7425
Reformer-RF-ICGO	22.1270	3.0432	4.7039	0.0465	0.7451

Table 5. Metrics for the prediction performance of NOx for the two-year dataset.

Model	MSE(mg/m³)	MAE(mg/m³)	RMSE(mg/m³)	MAPE(%)	R^2
BP	32.5860	3.0171	5.7084	0.0504	0.6347
ELM	30.0297	3.1950	5.4799	0.0543	0.6634
LSTM	31.2122	3.2710	5.5868	0.0544	0.6501
Transformer	29.3286	3.1977	5.4156	0.0551	0.6712
Reformer	28.3223	2.9464	5.3219	0.0494	0.6825
Reformer-CGO	27.6839	2.8181	5.2615	0.0471	0.6897
Reformer-RF	27.7995	2.7877	5.2725	0.0464	0.6884
Reformer-RF-CGO	27.2303	2.8292	5.2183	0.0477	0.6948
Reformer-RF-ICGO	26.8834	2.7408	5.1849	0.0457	0.6987

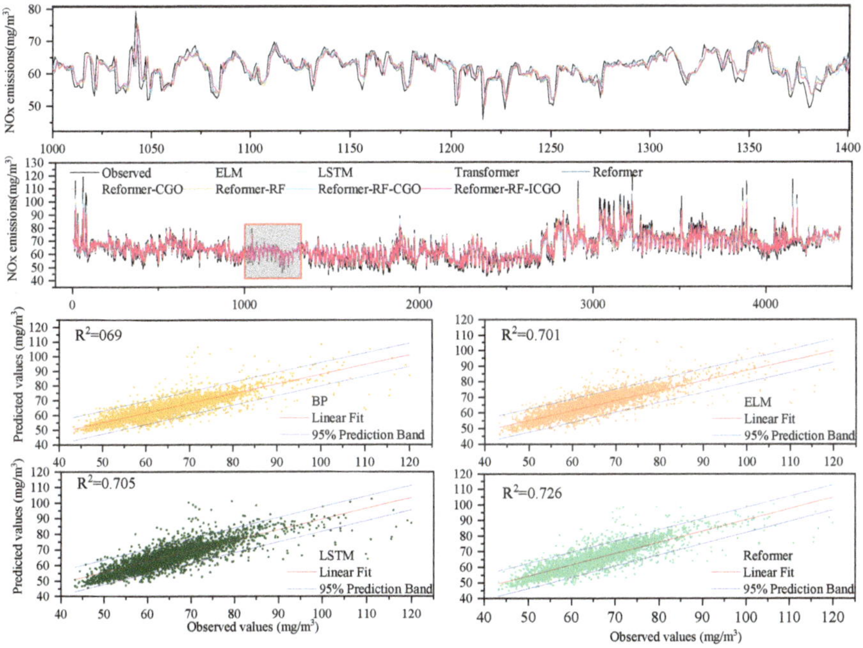

Figure 9. Plot of predicted results for the second dataset and scatterplot of predicted indicators.

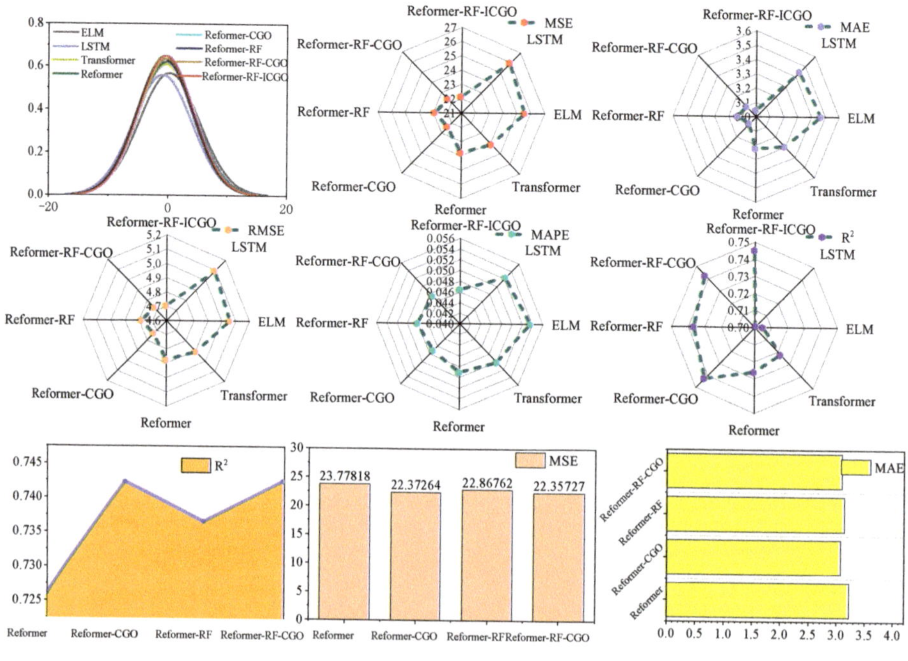

Figure 10. Comparison of science indicators results for eight models from a three-year dataset.

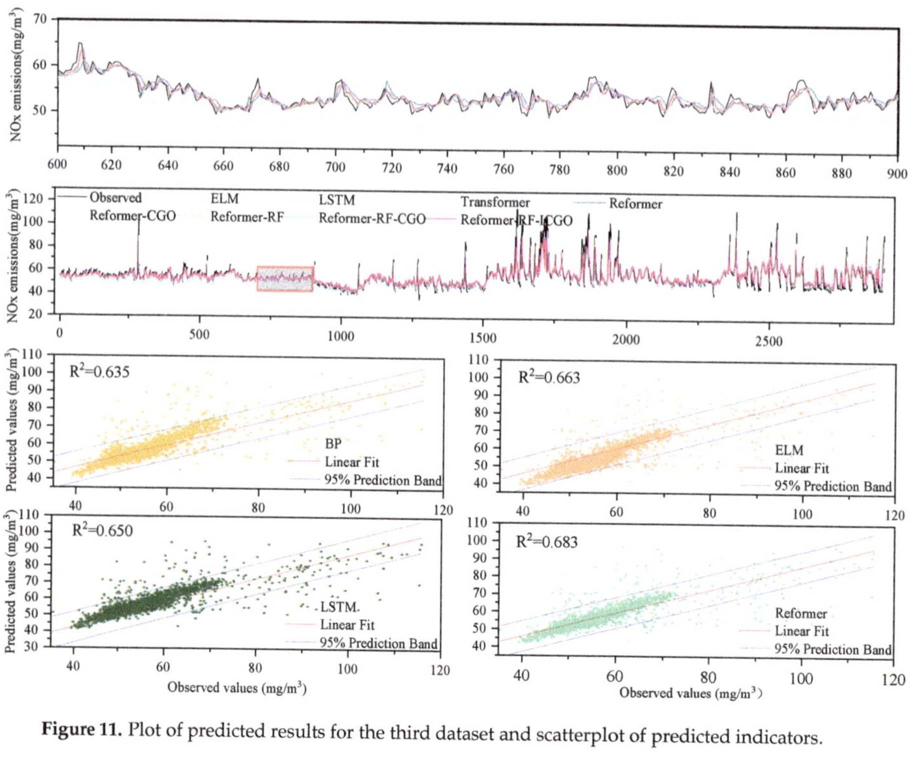

Figure 11. Plot of predicted results for the third dataset and scatterplot of predicted indicators.

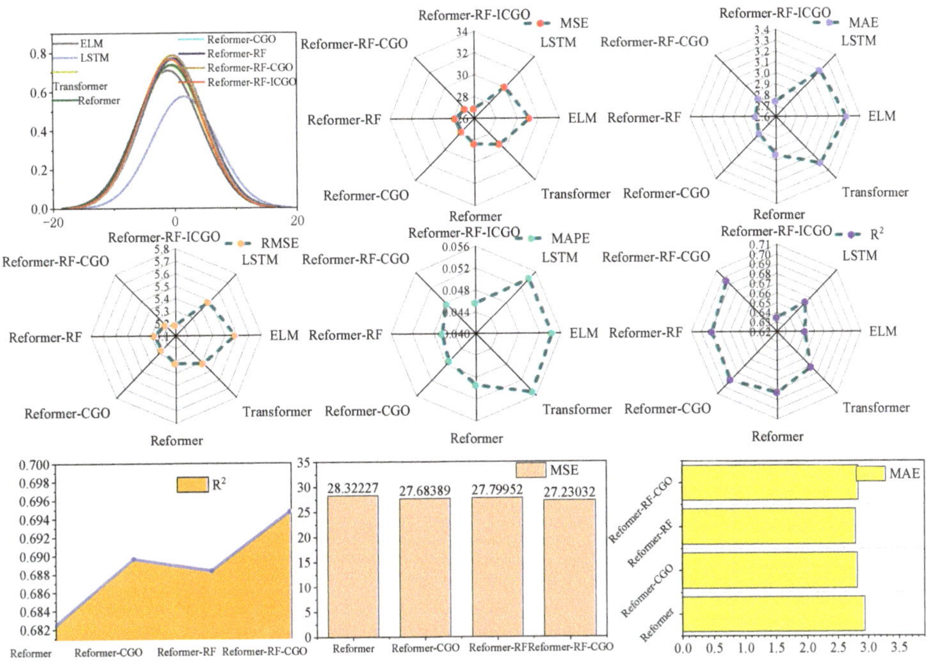

Figure 12. Comparison of science indicators results for eight models from a two-year dataset.

A detailed analysis of Tables 4 and 5 and Figures 8–11 is as follows.

Firstly, for the single model comparison, in the second and third datasets, MSE, RMSE, MAE, and MAPE of Reformer are all lower than BP, ELM, LSTM, and transformer, while R^2 of Reformer is the highest. Also, the plots of prediction results and linear chart for both datasets in Figures 8 and 10 show that Reformer has a better fit. This further indicates that Reformer predicts better performance than the base model.

Secondly, for the second and third datasets, Reformer, Reformer-CGO, Reformer-RF-CGO, and Reformer-RF-ICGO were compared. In the second dataset, compared to Reformer, MSE, RMSE, and MAE, MAPE for Reformer-CGO decreased by 1.4056 mg/m^3, 0.1542 mg/m^3, 0.1463 mg/m^3, 0.2% and R^2 increased by 0.0162, respectively. Compared to Reformer-RF-CGO, MSE, RMSE, MAE, MAPE for Reformer-RF-ICGO decreased by 0.2303 mg/m^3, 0.051 mg/m^3, 0.0244 mg/m^3, 0.08%, and R^2 increased by 0.0023, respectively. In the third dataset, compared to Reformer, MSE, RMSE, MAE, MAPE for Reformer-CGO decreased by 0.6384 mg/m^3, 0.1283 mg/m^3, 0.0604 mg/m^3, 0.23%, respectively, and R^2 increased by 0.0072. Compared to Reformer-RF-CGO MSE, RMSE, MAE, MAPE for Reformer-RF-ICGO decreased by 0.3469 mg/m^3, 0.0884 mg/m^3, 0.0334 mg/m^3, 0.2%, and R^2 increased by 0.039, respectively. Meanwhile, radar plots, line plots, and histograms of the predictors in Figures 10 and 11 clearly show that the algorithm-optimized model significantly outperforms the unoptimized model in predicting NOx, which indicates that ICGO can effectively find the optimal hyper-parameters.

The Reformer-RF and Reformer were then compared for the second and third datasets. In the second dataset, the MSE, RMSE, MAE, and MAPE of Reformer-RF decreased by 0.9106 mg/m^3, 0.0892 mg/m^3, 0.0943 mg/m^3, and 0.1%, respectively, compared to Reformer, while R^2 increased by 0.0105. In the third dataset, the MSE, RMSE, MAE, and MAPE of the Reformer-RF decreased by 0.5228 mg/m^3, 0.1587 mg/m^3, 0.0494 mg/m^3, and

0.3 mg/m^{3}, respectively, compared to the Reformer, while R^2 increased by 0.0059. In addition, in Figures 10 and 11, the radar plots, line plots, and histograms of the predictors show that the model incorporating RF is significantly better at predicting NOx than a single one, indicating that RF can effectively improve the prediction of the model.

Finally, this result further demonstrates the superiority and stability of the Reformer model in predicting NOx. It also demonstrates that ICGO can effectively optimize the parameters, and RF can effectively optimize the model.

This study uses three datasets to validate the feasibility of the model. Experiments on the three datasets show that the model is highly predictive and stable for NOx emissions. Further, as can be seen from the prediction result metrics in Tables 3–5, the prediction effect of the smaller 2-year dataset, 2014–2015, is slightly inferior to that of the 5-year and 3-year datasets. This is because the smaller number of training samples leads to a decrease in sample diversity and feature coverage, which increases the risk of overfitting the model and reduces its generalization ability, ultimately reducing prediction accuracy.

5.3. Diebold–Mariano Test Results

In order to further verify the validity of the proposed model, the DM test was conducted between the Reformer-RF-ICGO model proposed in this paper and other comparison models. The DM test results of the three datasets are shown in Table 6. Based on the data in the table, it can be seen that the DM test values between the Reformer-RF-ICGO model and the other comparative models are all greater than 0. The statistical results show that the Reformer-RF-ICGO model has the best prediction effect.

Table 6. DM test results for the three datasets.

Model	Five-Year Dataset	Three-Year Dataset	Two-Year Dataset
BP	16.00	10.28	9.48
ELM	15.61	9.46	8.18
LSTM	15.20	9.23	7.90
Transformer	9.06	5.83	7.79
Reformer	10.20	5.30	4.15
Reformer-CGO	6.51	6.59	4.98
Reformer-RF	6.98	5.94	4.01
Reformer-RF-CGO	6.79	4.52	3.36

All cases were carried out on a device with RTX3070 GPU (NVIDIA Corporation, Santa Clara, CA, USA), Pytorch 1.11.0 (USA), Python 3.8.0 (PSF, Wilmington, DE, USA), and Cuda 11.3 (NVIDIA Corporation, Santa Clara, CA, USA).

6. Conclusions

This paper aims to develop an evolutionary Reformer model based on RF and ICGO for predicting NOx emissions. The model first uses RF to downscale the sample features and then feeds the downscaled data into the Reformer model. Then, the hyperparameters of Reformer are optimized using ICGO to improve the NOx prediction accuracy. To investigate the prediction performance of the models, eight models are constructed in this study, namely, BP, ELM, LSTM, transformer, Reformer, Reformer-CGO, Reformer-RF, Reformer-RF-CGO and Reformer-RF-ICGO. and MSE, RMSE, MAE, R^2 and MAPE statistical metrics are used to compare the developed comparison models. The main findings of this study are as follows:

(1) It is clear from the highest R^2 and the lowest RMSE, MAE, and MSPE that the Reformer-RF-ICGO has higher predictive performance than the other models.
(2) Comparing the Reformer with the base models, it is clear that the Reformer has better predictive capability.
(3) Comparing the RF-treated model with the non-RF-treated model, it was concluded that using RF to analyze the importance of gas turbine variables yielded better prediction results, indicating that RF can effectively process the data.
(4) The introduction of the ICGO algorithm to optimize the hyperparameters of the Reformer model gives better prediction results than the unoptimized model, suggesting that ICGO can effectively find the optimal hyperparameters.
(5) Improvement of the CGO algorithm using Chebyshev Chaos Mapping showed increased prediction accuracy compared to the original CGO, indicating that the improved CGO algorithm is effective.

Overall, the Reformer-RF-ICGO model proposed in this paper is promising for NOx emission prediction from gas turbines.

7. Further Research

In this paper, we effectively predicted NOx emissions by constructing the Reformer-RF-ICGO model and achieved satisfactory results. Nevertheless, in order to further enhance the performance of the model, future research can deeply explore and improve the following key areas:

(1) The accuracy of the prediction can be significantly improved by introducing advanced error correction strategies. The implementation of error correction mechanisms should be considered in the modeling and prediction of NOx emissions to reduce the deviation between the output of the prediction model and the actual measured values.
(2) The presence of high-dimensional input variables not only increases the complexity of the model but also may lead to a decrease in computational efficiency. Therefore, careful selection of input variables becomes the key to improving the performance of NOx emission prediction models. Future research can explore more comprehensive feature selection methods to avoid incomplete feature selection that may be caused by a single method and then improve the prediction accuracy of the model.
(3) Given the significant time lag characteristic of the SCR system, there may be a delay of more than 45 s in the effect of input variables on NOx concentration, and the measurement of NOx concentration itself has a lag. All of these time lag factors may have an impact on the accuracy of the prediction model. Therefore, future modeling efforts should fully consider the time-lag effect to ensure that the model can more accurately reflect the true dynamics of the system.

Author Contributions: X.M., conceptualization, data curation, methodology. X.L., methodology, investigation, software, writing—original draft. J.C., writing—review and editing. Y.F., software, visualization. C.Z., visualization, writing—review and editing. M.S.N., writing—review and editing. T.P., conceptualization, writing—review and editing. All authors have read and agreed to the published version of the manuscript.

Funding: This work is supported by the National Natural Science Foundation of China (NSFC) (No. 62306123, No. 62303191), the Postgraduate Research & Practice Innovation Program of Jiangsu Province (SJCX24_2140, SJCX24_2141), the Postgraduate Science & Technology Innovation Program of Huaiyin Institute of Technology (HGYK202412, HGYK202413), the Natural Science Foundation of the Jiangsu Higher Education Institutions of China (No. 23KJD480001), and the Double-innovation

Doctor Program of Jiangsu Province (No. JSSCBS20201033 and No. JSSCBS20201037). Special thanks are given to the "Qinglan Project" of Jiangsu Province.

Data Availability Statement: The data presented in this study are available on request from the corresponding author.

Conflicts of Interest: Author Xianyu Meng was employed by Nanjing Nanrui Able Engineering Technology Co., Ltd. The remaining authors declare that the research was conducted in the absence of any commercial or financial relationships that could be construed as a potential conflict of interest.

Abbreviations

RF	Random Forest	CGO	Chaos Game Optimization
ICGO	Improved Chaos Game Optimization	PSO	Particle Swarm Optimization
BiGRU	Bidirectional Gated Recurrent Unit	NOx	Nitrogen Oxides
NO	Nitric Oxide	GOA	Grasshopper Optimization Algorithm
ML	Machine Learning	IGOA	Improved Grasshopper Optimization Algorithm
ELM	Extreme Learning Machines	HHO	Harris Hawk Optimization
BP	Backpropagation	ASO	Atomic Search Optimization
SVM	Support Vector Machines	GA	Genetic Algorithm
NO_2	Nitrogen Dioxide	ESNs	Echo State Networks
RMSE	Mean Square Error	MSE	Mean Squared Error
CNN	Convolutional Neural Network	MAE	Root Mean Absolute Error
ANN	Artificial Neural Network	MAPE	Mean Absolute Percentage Error
LSTM	Long Short-Term Memory	EMD	Empirical Mode Decomposition
RNN	Recurrent Neural Network	R^2	Coefficient Of Determination
		LSH	Locality Sensitive Hashing
		FNN	Feed-Forward Network

References

1. Mcingani, I.; Meyer, E.L.; Overen, O.K. The Impact of Ambient Weather Conditions and Energy Usage Patterns on the Performance of a Domestic Off-Grid Photovoltaic System. *Energies* **2024**, *17*, 5013. [CrossRef]
2. Markowski, J.; Leszczyński, J.; Ferreira, P.F.V.; Dranka, G.G.; Gryboś, D. Analysis of Electricity Supply and Demand Balance in Residential Microgrids Integrated with Micro-CAES in Northern Portugal. *Energies* **2024**, *17*, 5005. [CrossRef]
3. Chong, P.; Li, J.; Zhu, X.; Jing, D.; Deng, L. Numerical Study on Combustion Characteristics of a 600 MW Boiler Under Low-Load Conditions. *Processes* **2024**, *12*, 2496. [CrossRef]
4. Tang, Z.; Wang, S.; Chai, X.; Cao, S.; Ouyang, T.; Li, Y. Auto-encoder-extreme learning machine model for boiler NOx emission concentration prediction. *Energies* **2022**, *256*, 124552. [CrossRef]
5. Gu, J.; Song, Y.; Wang, Y.; Meng, W.; Shi, L.; Deng, K. Prediction of heat release and NOx emissions for direct-injection diesel engines using an innovative zero-dimensional multi-phase combustion model. *Fuel* **2022**, *329*, 125438. [CrossRef]
6. Li, F.; Zhang, D.; Zhang, J.; Kou, G. Measuring the energy production and utilization efficiency of Chinese thermal power industry with the fixed-sum carbon emission constraint. *Int. J. Prod. Econ.* **2022**, *252*, 108571. [CrossRef]
7. Wang, H.; Yang, X.; Li, Z.; Zhang, C.; Zhu, X.; Zhang, R.; Du, J.; Zhang, S. Numerical Investigation on the Effect of Air Humidification and Oxygen Enrichment on Combustion and Emission Characteristics of Gas Boiler. *Processes* **2024**, *12*, 2282. [CrossRef]
8. Huselstein, E.; Garnier, H.; Richard, A.; Guernion, P.Y.; Control, I. *Experimental Modeling of NOx Emissions in Municipal Solid Waste Incinerator*; IFAC World Congress: Barcelona, Spain, 2003.
9. Xie, P.; Gao, M.; Zhang, H.; Niu, Y.; Wang, X. Dynamic modeling for NOx emission sequence prediction of SCR system outlet based on sequence to sequence long short-term memory network. *Energies* **2020**, *190*, 116482. [CrossRef]
10. Lyu, T.; Gan, Y.; Zhang, R.; Wang, S.; Li, D.; Zhuo, Y. Development of a Real-Time NOx Prediction Soft Sensor Algorithm for Power Plants Based on a Hybrid Boost Integration Model. *Energies* **2024**, *17*, 4926. [CrossRef]
11. Zhang, Y.; Cen, Y.; Luo, G. Causal direction inference for air pollutants data. *Comput. Electr. Eng.* **2018**, *68*, 404–411. [CrossRef]

12. Navada, B.R.; Sravani, V.; Venkata, S.K. Enhancing Industrial Valve Diagnostics: Comparison of Two Preprocessing Methods on the Performance of a Stiction Detection Method Using an Artificial Neural Network. *Appl. Syst. Innov.* **2024**, *7*, 104. [CrossRef]
13. Jia, M.; Zhang, W.; Song, Q.; Qian, B.; Ye, Y.; Xu, K.; Wang, X. Prediction of CO/NOx emissions and the smoldering characteristic of sewage sludge based on back propagation neural network. *Environ. Pollut.* **2024**, *342*, 123049. [CrossRef]
14. Wang, J.; Kang, J.; Liang, H. *Prediction of the NOx Emissions from Thermal Power Plant Based on Support Vector Machine Optimized by Genetic Algorithm*; Springer: Berlin/Heidelberg, Germany, 2011.
15. Fanjun, L.; Ying, L.; Xiaohong, W.; Zhao, L.; Wenting, W. Prediction Method for NOx Emissions in Cement Arotary Kiln Based on EMD and ESNs. In Proceedings of the 2021 China Automation Congress (CAC), Beijing, China, 22–24 October 2021; pp. 5128–5132.
16. Kaya, H.; Tufekci, P.; Uzun, E. Predicting CO and NOxemissions from gas turbines: Novel data and abenchmark PEMS. *Turk. J. Electr. Eng. Comput. Sci.* **2019**, *27*, 4783–4796. [CrossRef]
17. Gifalli, A.; Amaral, H.L.M.D.; Bonini Neto, A.; de Souza, A.N.; Frühauf Hublard, A.V.; Carneiro, J.C.; Neto, F.T. Forecasting Electricity Consumption Using Function Fitting Artificial Neural Networks and Regression Methods. *Appl. Syst. Innov.* **2024**, *7*, 100. [CrossRef]
18. Wang, X.; Liu, W.; Wang, Y.; Yang, G. A hybrid NOx emission prediction model based on CEEMDAN and AM-LSTM. *Fuel* **2022**, *310*, 122486. [CrossRef]
19. Ma, H.; Peng, T.; Zhang, C.; Ji, C.; Li, Y.; Nazir, M.S. Developing an evolutionary deep learning framework with random forest feature selection and improved flow direction algorithm for NOx concentration prediction. *Eng. Appl. Artif. Intell.* **2023**, *123*, 106367. [CrossRef]
20. Yousaf, S.; Shafi, I.; Din, S.; Paul, A.; Ahmad, J. A big data analytical framework for analyzing solar energy receptors using evolutionary computing approach. *J. Ambient. Intell. Humaniz. Comput.* **2019**, *10*, 4071–4083. [CrossRef]
21. Li, L.; Jiang, Y.; Huang, B. Long-term prediction for temporal propagation of seasonal influenza using Transformer-based model. *J. Biomed. Inform.* **2021**, *122*, 103894. [CrossRef] [PubMed]
22. Deng, K.; Xiong, W.; Zhu, L.; Zhang, H.; Li, Z. Prediction of dissolved gas in power transformer oil based on random forests algorithm. In Proceedings of the 2015 5th International Conference on Electric Utility Deregulation and Restructuring and Power Technologies (DRPT), Changsha, China, 26–29 November 2015; pp. 1531–1534.
23. Ai, C.; He, S.; Hu, H.; Fan, X.; Wang, W. Chaotic time series wind power interval prediction based on quadratic decomposition and intelligent optimization algorithm. *Chaos Solitons Fractals* **2023**, *177*, 114222. [CrossRef]
24. Li, Y.; Wan, Y.; Xiao, J.; Zhu, Y. Prediction of Photovoltaic Power Generation Based on POS-BP Neural Network. In *Bio-Inspired Computing: Theories and Applications*; Pan, L., Liang, J., Qu, B., Eds.; Springer: Singapore, 2020; pp. 443–453.
25. Hu, H.; Xia, X.; Luo, Y.; Zhang, C.; Nazir, M.S.; Peng, T. Development and application of an evolutionary deep learning framework of LSTM based on improved grasshopper optimization algorithm for short-term load forecasting. *J. Build. Eng.* **2022**, *57*, 104975. [CrossRef]
26. Talatahari, S.; Azizi, M. Chaos Game Optimization: A novel metaheuristic algorithm. *Artif. Intell. Rev.* **2021**, *54*, 917–1004. [CrossRef]
27. Jin, T.; Zhao, Z.; Wang, P.; Yu, J.; Wu, F. Interaction augmented transformer with decoupled decoding for video captioning. *Neurocomputing* **2022**, *492*, 496–507. [CrossRef]
28. Vaswani, A.; Shazeer, N.; Parmar, N.; Uszkoreit, J.; Jones, L.; Gomez, A.N.; Kaiser, L.; Polosukhin, I.J. Attention Is All You Need. *arXiv* **2017**, arXiv:1706.03762.
29. Wu, L.; Wang, Y.-Q.; Liu, J.-B.; Shan, D.-H. Developing a time-series speed prediction model using Transformer networks for freeway interchange areas. *Comput. Electr. Eng.* **2023**, *110*, 108860. [CrossRef]
30. Yu, G.; Luo, Y.; Deng, R. Automatic segmentation of golden pomfret based on fusion of multi-head self-attention and channel-attention mechanism. *Comput. Electron. Agric.* **2022**, *202*, 107369. [CrossRef]
31. Li, C.; Huang, X.; Song, R.; Qian, R.; Liu, X.; Chen, X. EEG-based seizure prediction via Transformer guided CNN. *Measurement* **2022**, *203*, 111948. [CrossRef]
32. Srivastava, N.; Hinton, G.; Krizhevsky, A.; Sutskever, I.; Salakhutdinov, R. Dropout: A Simple Way to Prevent Neural Networks from Overfitting. *J. Mach. Learn. Res.* **2014**, *15*, 1929–1958.
33. Kitaev, N.; Kaiser, Ł.; Levskaya, A. Reformer: The Efficient Transformer. *arXiv* **2020**, arXiv:2001.04451.
34. Lu, K.; Wang, H.; Xiao, Y.; Song, H. Why locality sensitive hashing works: A practical perspective. *Inf. Process. Lett.* **2018**, *136*, 49–58. [CrossRef]
35. Gomez, A.N.; Ren, M.; Urtasun, R.; Grosse, R.B. *The Reversible Residual Network: Backpropagation Without Storing Activations*; NIPS: Okazaki, Japan, 2017.

36. Azizi, M.; Aickelin, U.; Khorshidi, H.A.; Shishehgarkhaneh, M.B. Shape and size optimization of truss structures by Chaos game optimization considering frequency constraints. *J. Adv. Res.* **2022**, *41*, 89–100. [CrossRef]
37. Peng, T.; Fu, Y.; Wang, Y.; Xiong, J.; Suo, L.; Nazir, M.S.; Zhang, C. An intelligent hybrid approach for photovoltaic power forecasting using enhanced chaos game optimization algorithm and Locality sensitive hashing based Informer model. *J. Build. Eng.* **2023**, *78*, 107635. [CrossRef]
38. Shakiba, A. A novel randomized one-dimensional chaotic Chebyshev mapping for chosen plaintext attack secure image encryption with a novel chaotic breadth first traversal. *Multimed. Tools Appl.* **2019**, *78*, 34773–34799. [CrossRef]
39. Breiman, L. Random Forests. *Mach. Learn.* **2001**, *45*, 5–32. [CrossRef]
40. Kochueva, O.; Nikolskii, K. Data Analysis and Symbolic Regression Models for Predicting CO and NOx Emissions from Gas Turbines. *Computation* **2021**, *9*, 139. [CrossRef]

processes

MDPI

Article

A Fault Judgment Method of Catalyst Loss in FCC Disengager Based on Fault Tree Analysis and CFD Simulation

Yuhui Li [1], Yunpeng Zhao [1], Zeng Li [1], Nan Liu [1], Chunmeng Zhu [1], Shouzhuang Li [2,*], Xiaogang Shi [1], Chengxiu Wang [1] and Xingying Lan [1,*]

1 State Key Laboratory of Heavy Oil Processing, China University of Petroleum (Beijing), Beijing 102249, China
2 Laboratory for Chemical Technology (LCT), Department of Materials, Textiles and Chemical Engineering, Faculty of Engineering & Architecture, Ghent University, 9000 Ghent, Belgium
* Correspondence: shouzhuang.li@ugent.be (S.L.); lanxy@cup.edu.cn (X.L.)

Abstract: Catalyst loss is a typical fault that impacts the long-term operation of the fluidized catalytic cracking (FCC) in the oil refining process. The FCC disengager is a critical place for separating the catalyst from oil gas. A fast and precise fault-cause judgment of catalyst loss is vital for avoiding catalyst loss failures. In this study, a novel fault judgment method of catalyst loss failures with quantitative criteria was established via the fault tree analysis (FTA) method, based on the relationship model between flow field signals and faults in the FCC disengager investigated by computational fluid dynamics (CFD). The FTA method defines three intermediate events: catalyst fragmentation, process fault and mechanical fault. In CFD results, it was found that the detailed fault reason can be inferred based on the changes in the characteristic parameters within the disengager. For example, when the catalyst loss rate of the FCC disengager may rapidly increase by a factor of around 200. Furthermore, the pressure drop of the cyclone separator decreases by around 35%, which indicates that the dipleg has fractured. The new fault judgment method has been applied in cases of catalyst loss in two industrial disengagers. The method accurately pinpointed the sudden reduction in inlet velocity and blockage fault at the cyclone separator as the main factors leading to catalyst loss faults, respectively. The judgment results are consistent with actual reasons, demonstrating the reliability of the method. This study could contribute to providing theoretical support and enhancing the accuracy for the diagnosis of catalyst loss faults, thereby ensuring the safe and stable operation of the FCC unit.

Keywords: disengager; catalyst loss; computational fluid dynamics (CFD); fault tree analysis (FTA)

1. Introduction

As an important means for crude oil secondary processing, fluidized catalytic cracking (FCC) is the main approach to convert heavier feedstock into lighter products in petroleum refineries, such as gasoline and middle distillates [1–4]. In the FCC unit, the catalyst should be separated from the oil gas in the disengager after catalytic cracking reactions. Cyclone separators are the core component in the separation system of the FCC disengager. Since the separation efficiency of the cyclone separator for catalyst particles is less than 100%, a small amount of catalyst loss is unavoidable. During normal operation, a catalyst loss rate below 0.6 kg/t is acceptable, which is considered as a normal loss in the process of FCC operation [5,6]. When the catalyst loss failure occurs (loss rate exceeds 0.6 kg/t), the primary hazard is the fluidization performance of the catalyst circulating inside the FCC disengager deteriorating due to the escape of fine catalyst particles [7,8]. Secondly, due to the decreased catalyst–oil ratio, the reaction depth of the feedstock oil in the reactor will be affected, resulting in the product distribution and quality failing to meet requirements. Lastly, the failure of catalyst loss can lead to equipment deterioration. The increased solids content in the slurry exacerbates the wear on pumps, valves and pipelines within the oil slurry system [9–11]. The severe catalyst loss may ultimately result in an unplanned shutdown of the FCCU, affecting the long-term stable operation of the FCC and posing serious safety hazards and economic losses. Based on the extensive cooperation experience between the author's research team and numerous petroleum refineries, in recent years, the increasing severity of heavy and poor-quality feeding oil has led to a rapid increase in the frequency of catalyst loss faults. Therefore, it is urgent for oil refineries to establish an effective judgment method to identify the cause of the catalyst loss failure and thus ensure the long-term stable operation of the FCC unit in addition to enhancing economic benefits.

The process of catalyst loss in the FCC disengager involves numerous process parameters and equipment parameters. The interaction between the different equipment makes it very difficult to determine the cause of catalyst loss failure. Currently, the diagnosis of catalyst loss failure in FCC units is primarily conducted by analyzing the distributed control system (DCS) operation data, such as pressure, temperature and catalyst particle spectrum data and refinery LIMS (Laboratory Information Management System) data on-site [12–14]. Song et al. [12] extracted the characteristics of the catalyst loss faults and the main factors in an industrial cyclone separator. Based on this, a stepwise diagnosis approach was proposed to determine the causes and location of catalyst loss faults of the cyclone separator. However, compared to the single cyclone separator, the flow field in the FCC disengager is more complex. It is a challenge to apply the derived rules directly to large-scale industrial FCC disengagers. The fault tree analysis (FTA) was employed to facilitate the identification of the interconnections among various factors within the failure system. FTA is a graphical deductive method, which is widely used in industrial process fault diagnosis. The normalized fault tree consists of three types of events ("top event", "intermediate event" and "bottom event") and three types of logic gates ("with", "or" and "without"). The FTA method considers the least expected incident of the system as the "top event" of the fault tree. Starting from the top event, it analyzes the causes of the faults that led to the top event layer by layer and then progresses to the various "bottom events" where the fault mechanism is clear. Through qualitative analysis, all failure modes leading to the top event can be identified. The method has been used in industrial fault cause analysis widely [15–21]. Vasić et al. [16] employed the fault tree analysis (FTA) method to investigate the reasons behind the failure of the belt conveyor idler roller bearing. Through this analysis, they identified potential causes of damage within the fault tree and suggested measures aimed at eliminating or significantly reducing the damage. Wang et al. [21] applied the FTA method to construct a fault tree model and diagnostic process for catalyst loss in FCC unit. It was found that there are many causes leading to catalyst loss; these factors are both independent and interrelated. Nonetheless, the existing fault diagnosis methods require a high level of experience from operators and fall short in precisely pinpointing the type and location of faults. Furthermore, many fault sources in industrial FCC disengagers are located inside the reactors, where they are highly concealed and difficult to detect. As a result, the causes of catalyst loss in industrial FCC disengagers can only be pinpointed during the equipment shutdown.

The design of fault diagnosis methods for catalyst loss faults of the FCC disengager relies heavily on a vast amount of fault data. However, the time and material costs required to obtain these fault data are both extremely high [4]. In recent years, computational fluid dynamics (CFD) has rapidly developed and is now being utilized in numerous simulations of flow and heat transfer behaviors in the FCC disengager [22–27], as well as the internal cyclone separator [28–30]. Song et al. [26] investigated the gas flow behaviors and gas residence time distribution in an industrial FCC disengager space. The simulation results indicate that, in a disengager where multiple primary and secondary cyclones are arranged, the flow distribution at the inlets of the secondary cyclones tends to be uneven. Therefore, it is crucial to address this concern during the design phase to ensure the enhanced operational efficiency of the secondary cyclones. Zhang et al. [27] simulated an industrial FCC disengager to establish a hybrid modeling framework for the safety monitoring of coking rates. It is evident that the movement of oil gas is predominantly confined to the internal cyclone separator and the dilute phase area of the disengager. Upon entering the cyclone separator, a prominent vortex effect is triggered, resulting in the majority of the oil gas being expelled through the gas phase outlet of the primary cyclone separator and subsequently entering the dilute phase area of the disengager. The above literature shows that the separation process of catalyst particles is closely related to the movement trajectories of oil gas and the pressure balance depends on the normal fluidization of the particles between the different units. However, almost all current flow field simulations related to the FCC disengager are conducted under normal operating conditions. The flow field during abnormal operations has not been taken into account and flow field simulation analyses for catalyst loss faults have not been performed.

It can be seen that the FTA can be used to identify the causes of catalyst loss, but it cannot provide quantitative criteria to pinpoint the specific reason behind it. The CFD simulation is capable of capturing the flow and heat transfer dynamics within the entire disengager space and can assist in providing a theoretical basis for diagnosing catalyst loss faults in the FCC disengager. However, it is deficient in real-time assessment capabilities. Therefore, the FTA and CFD simulations were combined for the first time in this paper to propose a fault judgment of catalyst loss in FCC disengagers. In this study, FTA was applied to determine the potential causes resulting in catalyst loss faults. A comparison and summary were conducted on the variations in the pressure drop, temperature, catalyst particle size distribution and other signals. Then, the CFD simulation was used to analyze the flow field during normal conditions and abnormal conditions. The changes in the pressure drop, temperature, particle size distribution of the catalyst and other signals were compared and summarized. Finally, based on the mechanistic analysis and simulation result, a new fault judgment method with quantitative criteria for catalyst loss in the FCC disengager has been proposed. The accuracy of this method was verified through an industrial example. This study is expected to provide theoretical guidance for the prevention and diagnosis of catalyst loss issues in disengagers. In addition, it also can reduce reliance on the operational experience of workers and enhance operational stability and production efficiency of FCC unit.

2. Fault Tree Analysis for Catalyst Loss Fault

The catalyst loss fault of the FCC disengager is a complex industrial issue; it is necessary to adopt the FTA method to analyze the corresponding causes and hazards of the fault. Currently, the catalyst loss fault in industrial FCC disengagers is primarily addressed through shutdown maintenance, adjusting catalyst mixture ratios and modifying process parameters. However, fault cases are too random and scattered, lacking a systematic classification and organization of the causes of catalyst loss. In order to enhance the reliability of the results when analyzing catalyst loss issues in FCC disengagers using the fault tree analysis (FTA) method, the research and summary of the literature on catalyst loss faults reported in Chinese refineries have been conducted in the present work. The catalyst loss fault in the disengager was identified as the top event of the fault tree. By investigating the causes and mechanisms of industrial disengager catalyst loss over the past 10 years, the catalyst fragmentation, the mechanical fault and the process fault were identified as intermediate events. A similar approach was employed to conduct in-depth analyses until the basic events, which led to the catalyst loss fault, were fully identified. As shown in Table 1, twelve basic events were considered in this fault tree.

Table 1. Basic events of catalyst loss in FCC disengager.

No	Basic Event	Typical Industrial Cases
X1	Catalyst striking	2.2 Mt/a FCCU [31]
X2	Catalyst abrasion	2.2 Mt/a FCCU [32]
X3	High inlet velocity of cyclone	3.6 Mt/a FCCU [33]
X4	Low inlet velocity of cyclone	3.6 Mt/a FCCU [33]
X5	High operation pressure	1.0 Mt/a FCCU [34]
X6	Low operation pressure	0.8 Mt/a FCCU [35]
X7	Stripper blockage	1.4 Mt/a FCCU [36]
X8	Isolated shell blockage	2.9 Mt/a FCCU [37]
X9	Cyclone blockage	1.3 Mt/a FCCU [38]
X10	Vortex arm hole	2.9 Mt/a FCCU [39]
X11	Cyclone hole	1.1 Mt/a FCCU [40]
X12	Dipleg fracture	1.8 Mt/a FCCU [41]

The historical data from the LIMS in oil refineries include the catalyst attrition rate, replacement frequency and results from analyzing samples of spent catalyst and escaped catalyst. Catalyst fragmentation is identified as one of the intermediate events causing catalyst loss in FCC disengagers, which could increase the content of catalyst fines. Due to the low separation capacity of the separation system for catalyst fines, when a catalyst fragmentation fault occurs, the catalyst loss rate will increase. In the FCC disengager, catalyst particles are subjected to fragmentation from striking and abrasion during conveying and separation operations, leading to alterations in their physical properties [42]. However, the resultant particle size distributions in the catalysts differ significantly between the two mechanisms. The former is due to the lack of mechanical strength of the carrier of the catalyst, whereas the latter results from the friction wear between particles, particle groups, wall surfaces and mixture phases [43–45].

The analysis of process failures relies on historical monitoring data of process parameters and quality control reports. If the process fault occurs during the operation of the FCC disengager, the inlet conditions to the cyclone separator are affected, which results in the catalyst loss rate increasing. The unstable operation of the reaction–regeneration system of the FCC unit will directly impact the separation performance of the cyclone separator. Abruptions of the inlet velocity and the operating pressure can lead to a sharp deterioration in separation performance, resulting in decreased separation efficiency. These are evident through a considerable increase in catalyst concentration at the cyclone outlet and catalyst loss fault.

Mechanical faults were analyzed based on equipment maintenance records and failure logs. By statistically analyzing the type, frequency and impact of mechanical failures over the past 10 years, the mechanical factor was identified as an intermediate event causing catalyst loss in FCC disengagers [31–41]. Mechanical faults can significantly result in abnormal gas–solid flow patterns within the unit and degrade the separation performance of the disengager separation system, causing abnormal catalyst loss [46]. The internals of the FCC disengager are exposed to erosive wear from flowing particles and induced vibration from the unstable two-phase flow. Due to the prolonged high temperature, high velocity and high concentration of particles, frequent mechanical failures like blockage faults and leakage faults have occurred. The typical mechanical faults are shown in Figure 1.

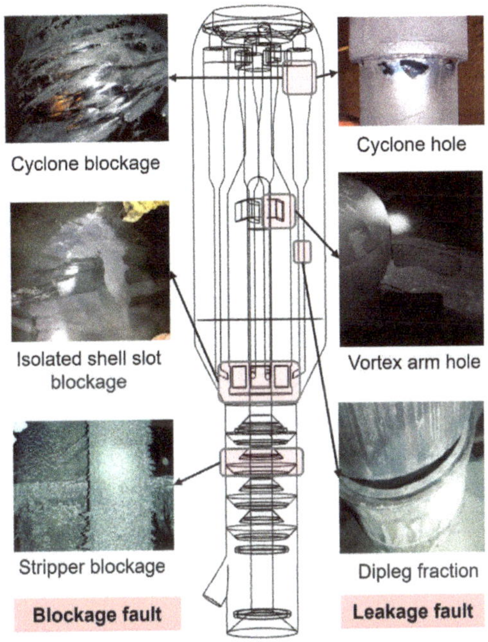

Figure 1. Mechanical damage in the disengager.

A fault tree model of catalyst loss was developed based on the results of the fault tree analysis for the catalyst loss fault, which is shown in Figure 2. According to the definitions of the two logic gates, an "or" gate has an output as long as one event occurs, while an "and" gate requires all events to occur for an output. From the fault tree, it is clear that the fault tree contains all "or" gates and no "and" gates. There are numerous fault events individually leading to the catalyst loss fault. Therefore, it is necessary to enhance preventive measures in the weak link of the FCC disengager to prevent catalyst loss. In the fault tree, the group of several bottom events causing the top event is known as the cut set. The minimum cut set is the irreducible set of bottom events that led to the top event. A minimum cut set represents a failure mode resulting in the top event. Based on the analysis, it is concluded that the number of minimum cut sets for the catalyst loss fault tree is 12. A lower order indicates a higher risk of catalyst loss. The occurrence of any one of these minimum cut sets can lead to catalyst loss in the FCC disengager, bringing a high risk of failure.

The contribution made by each bottom event of a fault tree to the occurrence of the top event is called the importance of the bottom event. The study of the importance of the bottom events plays an important role in the development of the catalyst loss judgment process for the FCC disengager. It can reduce the troubleshooting time and improve the credibility of the entire system effectively. Based on the principle of importance analysis [18], the structural importance of each bottom event mentioned in this study is the same. Each bottom event contributes the same to the catalyst loss of the FCC disengager.

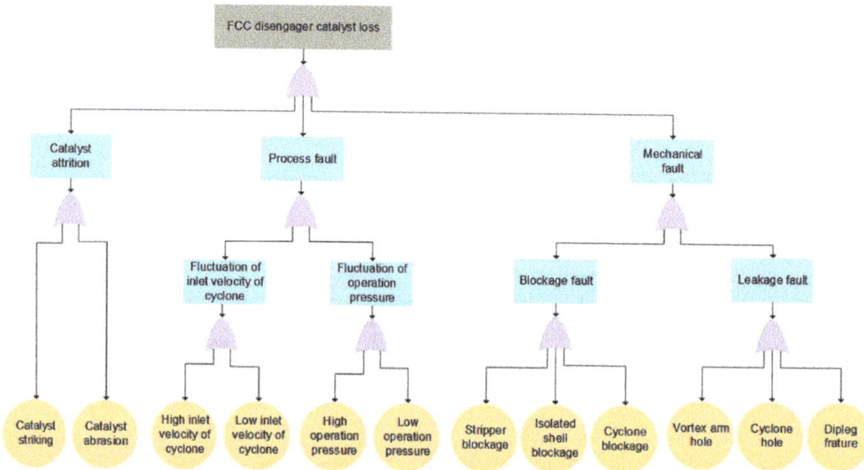

Figure 2. Fault tree analysis of catalyst loss in FCC disengager.

3. CFD Methods

Upon the detection of a catalyst loss malfunction in the FCC disengager, the immediate priority is to acquire relevant operational data from the FCC unit and meticulously compare them against historical normal operation data. Based on the results of the fault tree analysis, the fluid flow and heat transfer conditions inside the FCC disengager during various bottom events occurrences were simulated by utilizing the CFD simulation method, with boundary conditions or internal parameters being altered. This enables the rapid extraction of fault data, facilitating the establishment of typical fault data characteristics.

3.1. Simulation Object

The research object of this paper is the disengager of an 800,000 t/a catalytic cracking unit of a refining company in North China. During a recent operational phase, it was discovered that there was a notably high rate of catalyst loss. Figure 3a shows the geometric model of the FCC disengager, which shows that its inner diameter and height are 6.2 m and 28.7 m, respectively. The disengager mainly consists of a vortex quick separator (VQS) system, a secondary separation system with four top cyclone separators, a dilute phase space and a stripping section. At the top of the disengager, there is an anti-coking steam ring. Upon entering the disengager, the oil–gas and catalyst mixture first undergoes preliminary separation in the rough cyclone. The catalyst and oil gas not captured by the VQS system proceeds directly into the overhead cyclone separators via a guide tube. Catalyst particles separated by the system descend into the dipleg and accumulate in the dense phase zone, whereas unseparated catalyst particles exit the disengager alongside the oil gas. In the bottom stripping section, the vapor extraction section has multi-stage vapor extraction steam to enhance the vapor extraction effect and maximize the recovery of the oil gas entrained in the catalyst.

The geometric structures were discretized using unstructured grids, with a volume range of 9.00×10^{-10} m^3~4.73×10^{-2} m^3. Considering the variations in the oil–gas velocity gradients and wall effects, local grid refinement was applied to critical areas such as the steam rings and the inlet and outlet of the cyclone separators. Polyhedral grid conversion technology was employed to optimize the grid quality and quantity, significantly enhancing computational efficiency while ensuring calculation accuracy. Through the grid independence study, 1.2 million polyhedral grids are adopted (Figure 3b).

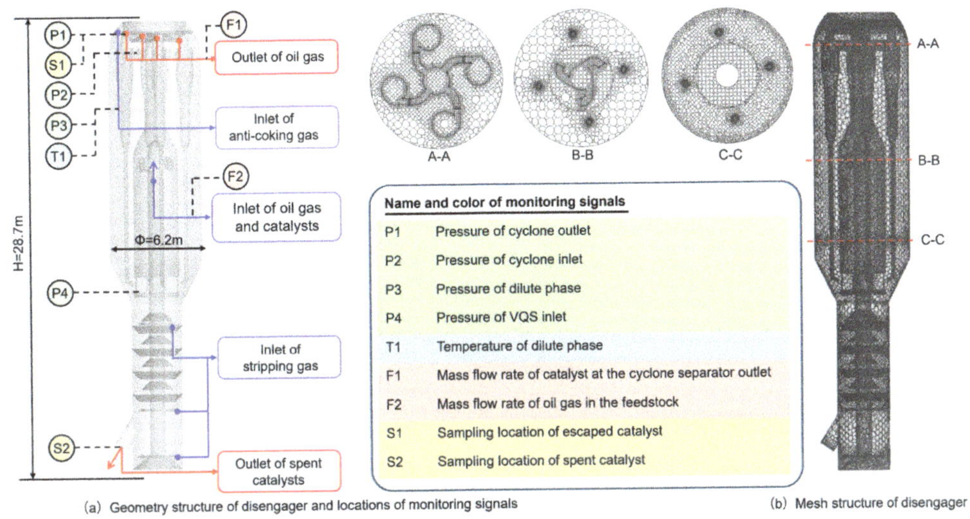

Figure 3. Geometry and mesh structure diagram of industrial disengager.

Considering the actual industrial composition of dry gas, liquefied gas, gasoline, diesel and oil slurry, the physical properties of the oil–gas mixture were determined through the application of the Universal Group Activity Coefficient (UNIFAC) model. The properties of the oil–gas mixture are shown in Table 2. The steam utilized in the system was superheated at 350 °C, with a density of 2.54 kg/m^3 and a viscosity of 2.858×10^{-5} Pa·s. The solid phase was industrial FCC catalyst particles with an average particle size of 85 μm. Its density and specific heat capacity are 1500 kg/m^3 and 1100 $J \cdot kg^{-1} \cdot K^{-1}$.

Table 2. Properties of oil–gas mixture (operation conditions: T = 514 °C, P = 0.146 MPa).

Properties	Mass Fraction, %	Density, kg/m^3	Viscosity, $\times 10^{-5}$ Pa·s	Relative Molecular Mass, g/mol
Dry gas	2.72	0.91	2.56	18.21
Liquefied gas	16.33	2.56	1.91	49.81
Gasoline	40.47	4.71	1.62	91.65
Diesel	20.05	9.86	1.23	192.15
Oil slurry	3.63	727.13	-	395.48
Mixture oil–gas	100.00	3.71	1.68	93.42

3.2. Governing Equations

Extensive numerical simulations have been conducted to investigate the fluidization state of the gas phase and solid phase in the FCC disengager. The selection of gas–solid flow models is vitally important for the accuracy of the CFD simulation. By combining calculation accuracy and efficiency, the Eulerian–Eulerian model was applied in this paper to describe the behavior of two oil gas and catalyst particle flows in the FCC disengager. For enhanced precision in simulating the turbulence process of oil and gas, the Reynolds Stress

Model (RSM) was utilized. The modified Gidaspow drag force model [47] was applied to improve the reduction in the drag force due to the presence of particle clusters. The basic equations used in this work are shown in Table 3.

Table 3. Basic equations used for simulation.

Name	Basic Equations
Continuity equations of gas and solid phases	$\frac{\partial}{\partial t}(\alpha_m\rho_m) + \frac{\partial}{\partial x_j}(\alpha_m\rho_m u_m) = 0$
Momentum equations of gas and solid phases	$\frac{\partial}{\partial t}(\alpha_m\rho_m u_m) + \frac{\partial}{\partial x_j}(\alpha_m\rho_m u_m u_m) = -\alpha_m\frac{\partial p}{\partial x_j} + \frac{\partial \tau_{m,ji}}{\partial x_j} + \alpha_m\rho_m g_i - \beta(u_{mi} - u_{mi})$
Reynolds stress transport equation	$\frac{\partial}{\partial t}(\rho\overline{u'_i u'_j}) + \frac{\partial}{\partial x_k}(\rho u_k\overline{u'_i u'_j}) = \frac{\partial}{\partial x_k}\left(\frac{\mu_t}{\sigma k}\frac{\partial\overline{u'_i u'_j}}{\partial x_k}\right) + \frac{\partial}{\partial x_k}\left[\mu\frac{\partial}{\partial x_k}\left(\overline{u'_i u'_j}\right)\right] + P_{ij} + G_{ij} + \varphi_{ij} + \varepsilon_{ij}$
Turbulent kinetic energy equation	$\frac{\partial}{\partial t}(\rho k) + \frac{\partial}{\partial x_j}(\rho k u_i) = \frac{\partial}{\partial x_j}\left[\left(\mu + \frac{\mu_t}{\sigma_k}\right)\frac{\partial}{\partial x_k}\right] + \frac{1}{2}(P_{ii} + G_{ii}) - \rho\varepsilon$
Turbulent energy consumption rate equation	$\frac{\partial}{\partial t}(\rho\varepsilon) + \frac{\partial}{\partial x_j}(\rho u_i\varepsilon) = \frac{\partial}{\partial x_j}\left[\left(\mu + \frac{\mu_t}{\sigma_\varepsilon}\right)\frac{\partial\varepsilon}{\partial x_j}\right]C_{\varepsilon 1}\frac{1}{2}(P_{ii} + C_{\varepsilon 3}G_{ii})\frac{\varepsilon}{k} - C_{\varepsilon 2}\rho\frac{\varepsilon^2}{k}$
Energy conservation equation	$\frac{\partial}{\partial t}(\rho E) + \frac{\partial}{\partial x_i}[\overline{u_i}(\rho E + p)] = \frac{\partial}{\partial x_i}\left[\left(k + \frac{c_p\mu_t}{\mathrm{Pr}_t}\right)\frac{\partial T}{\partial x_j}\right]$
Modified Gidaspow drag force model	$\begin{cases}\beta = \frac{3}{4}C_D\frac{\alpha_s\alpha_g\rho_g\left\lvert\vec{u}_p - \vec{u}_g\right\rvert}{d_p^*}\alpha_g^{-2.65} & \alpha_g > 0.8\\ \beta = 150\frac{\alpha_s(1-a_g)\mu_g}{\alpha_g\left(d_p^*\right)^2} + 1.75\frac{\rho_g\alpha_s\left\lvert\vec{u}_p - \vec{u}_g\right\rvert}{d_p^*} & \alpha_g \le 0.8\end{cases}$

3.2.1. Boundary Conditions

The types of boundary conditions involved in the simulation calculations under normal operation conditions were determined according to the process design manual for the disengager, as shown in Figure 3a. The mass flow inlet boundary condition is used for the inlet of catalyst and oil–gas. The same mass flow inlet boundary condition is used for the inlet of anti-coke steam, stripping steam and fluidized steam. The velocity inlet boundary condition is used for the catalyst outlet at the bottom of the stripper section. The gas phase outlet in the second-stage cyclone is the gas phase outlet in the system and the pressure outlet boundary condition was used. The no-slip boundary condition is used for all walls in the unit. The lining material of the disengager is of a hex metal mesh structure, with a calculated heat transfer coefficient of 0.75 W/(m^2·K). The detailed main operating conditions in the disengager under normal conditions are listed in Table 4.

Table 4. The operating conditions in the disengager under normal conditions.

Name	Value
Total mass flow rate, kg/s	33.16
Feedstock temperature, °C	514
Mass ratio of catalyst to oil, -	6
Outlet pressure of secondary cyclone, kPa	146
Fluidized gas mass flow rate, kg/s	1.22
Anti-coking gas mass flow rate, kg/s	0.4
Anti-coking gas temperature, °C	425
Striping gas (top) mass flow rate, kg/s	0.11
Stripping gas (middle) mass flow rate, kg/s	0.49
Stripping gas (bottom) mass flow rate, kg/s	0.14
Stripping gas temperature, °C	254

The results of the fault tree analysis indicate that the causes of the catalyst loss fault include many factors such as catalyst fragmentation, mechanical fault and process fault. Since each bottom event contributing to the catalyst loss fault has an equal contribution, the characteristics of the flow field inside the FCC disengager for the twelve types of bottom events listed in Table 1 were simulated in this section. The setting of fault conditions is crucial to the impact on changes in signal characteristics. When the catalyst striking caused by high-velocity gas flow occurs, the average particle size decreases, resulting in a unimodal distribution in the particle size distribution (PSD) of the catalyst. Conversely, when catalyst abrasion occurs at a low speed, the PSD of the catalysts exhibits a bimodal distribution [42,48]. Considering the variation in catalyst fragmentation conditions, the average particle diameter of the FCC catalyst was assumed to be 80 μm, with an alternative mixture containing 95% (vol) at 80 μm and 5% (vol) at 30 μm, respectively. The influence of the gas in-leakage rate (the ratio between the gas in-leakage volumetric flowrate and the inlet gas volumetric flowrate) on signals such as the pressure drop is minimal when it is less than 8% [49]. Therefore, the gas in-leakage rate was set to 10%. In the case of a partial blockage, only a portion of the separation capacity was weakened, and the resulting signal change was not significant. Therefore, the complete blockage was selected for consideration. The operating manual specifies that the operating flexibility for the processing capacity of this FCC unit is within a variation range of ±20%. The inlet velocity was determined within −30% and 30% of the normal value to simulate the conditions at the abruption of inlet velocity. The operation pressure of the disengager is constrained by the pressure balance of the reactor–regenerator system. The analysis of historical data reveals that, even in cases of catalyst loss faults, the operation pressure of the disengager differs minimally from the outlet pressure under normal operating conditions. Therefore, the operation pressure was selected with a variation range of ±5% around the normal pressure value. The specific setting approaches for different fault operation conditions were summarized in Table 5. The stability of the simulation can be ascertained by monitoring the flow rate of the cyclone separator outlet. Considering the blockage process as a gradual accumulation, a significant amount of computational resources would be required to calculate until the separation system completely loses functionality. Therefore, a fault time of 250 s was selected for the blockage fault. Meanwhile, a fault time of 150 s was chosen to acquire the stable flow fields under other fault conditions.

Table 5. Settings for catalyst fragmentation fault condition.

Fault Type	Fault Condition	Setting Approach	Data	Fault Time (s)
Catalyst fragmentation	Catalyst striking	Reduce the average particle diameter of the FCC catalyst	80 μm	150
	Catalyst abrasion		80 μm (95 vol%) + 30 μm (5 vol%)	150
Mechanical fault	Blockage fault	Cyclone/isolated shell slot	Degree of blockage = 100%	250
		Stripper section	Degree of blockage = 50%	250
	Leakage fault	Cyclone/Vortex arm/Dipleg fraction	Degree of leakage rate = 10%	150
Process fault	Abruption of the inlet velocity	Low/high velocity of inlet	12.08 m/s, 22.33 m/	150
	Abruption of the operation pressure	Low/high operation pressure	138.7 kPa, 153.3 kPa	150

The finite volume method was utilized to discretize the governing equations. The phase-coupled SIMPLE scheme was adopted for the pressure–velocity coupling. The gradients were computed using the least squares cell-based method. The calculation time step was set to 0.01 s. Convergence was assumed for each time step when all the residuals fell below 0.001 and a maximum of 50 iterations per time step was adopted. The key indicators for monitoring catalyst loss encompass the catalyst loss rate and PSD of the catalyst, as well as the variation rate of the temperature of the dilute phase and the variation in the pressure drop of the cyclone, VQS and dilute phase. The sampling locations of these signals in the disengager are shown Figure 1. The aforementioned parameters under faulty conditions, which can be calculated by the method listed in Table 6, were analyzed through CFD simulation in the present work.

Table 6. Calculation methods of key indicators for monitoring catalyst loss.

Indicators	Meaning	Calculation Method
Catalyst loss rate	-	$F1/F2$
PSD of escaped catalyst	-	PSD of S1
PSD of spent catalyst	-	PSD of S2
$T_{dilute,t}/T_{dilute,normal}$	Temperature variation rate of dilute phase	$T1_{catalyst\ loss\ fault\ condition}/T1_{normal\ condition}$
$\Delta P_{CYC,t}/\Delta P_{CYC,normal}$	Pressure drop variation rate of cyclone separator	$P2 - P1_{catalyst\ loss\ fault\ condition}/P2 - P1_{normal\ condition}$
$\Delta P_{VQS,t}/\Delta P_{VQS,normal}$	Pressure drop variation rate of VQS	$P4 - P2_{catalyst\ loss\ fault\ condition}/P4 - P2_{normal\ condition}$
$\Delta P_{dilute,t}/\Delta P_{dilute,normal}$	Pressure drop variation rate of dilute phase	$P3 - P1_{catalyst\ loss\ fault\ condition}/P3 - P1_{normal\ condition}$

3.2.2. Model Validation

The normal operating conditions of the disengager were simulated. The simulated data were compared with the industrial and DCS data. The results are shown in Table 7. $P_{\text{cyclone-outlet}}$, P_{dilute}, T_{dilute}, $T_{\text{gas-out}}$ and $u_{\text{cyclone-inlet}}$ are important and measurable performance parameters of the cyclone separator, which usually influence the catalyst separation process in the disengager. The pressure at the inlet of the cyclone and the dilute phase space results were 146.1 kPa and 150.0 kPa, which are very close to the industrial data of 146.0 kPa and 148.3 kPa, respectively. This indicates that the pressure balance inside the disengager was accurate. The temperature of the dilute phase was also closely aligned with the industrial data. In summary, the model developed in this paper can accurately describe the gas–solid flow field in the disengager under normal conditions.

Table 7. Comparison between simulation results and DCS data under normal conditions.

Item	$P_{\text{cyclone-outlet}}$	P_{dilute}	T_{dilute}	$T_{\text{gas-out}}$	$v_{\text{cyclone-inlet}}$
Simulation	146.1 kPa	150.0 kPa	418.9 °C	511.2 °C	16.9 m/s
Industrial	146.0 kPa	148.3 kPa	424.0 °C	507.7 °C	16.2 m/s
Relative error (%)	0.07	1.15	1.20	0.69	4.32

4. CFD Results

4.1. Catalyst Fragmentation

Figure 4a depicts the catalyst loss rate at catalyst fragmentation. In the figure, the horizontal axis ranging from −50 s to 0 s represents normal operation conditions, while the period after 0 s indicates the occurrence of a fault, which was highlighted in the diagram. It can be seen that, regardless of whether catalyst striking or catalyst abrasion occurs, the catalyst particle loss rate rapidly increased to reach the shutdown zone (1.5 kg/t). The final catalyst loss rate fluctuated around 12 kg/t and 16 kg/t, respectively. The reduction in the catalyst particle diameter leads to a decrease in the centrifugal force on the particles and an increase in the drag force. Due to these effects, when the particles entered the cyclone, the smaller particles entered the short-circuit flow and moved directly upward into the top cyclone. Consequently, the catalyst concentration at the top cyclone inlet increased, which resulted in a decrease in the separation efficiency of the top cyclone.

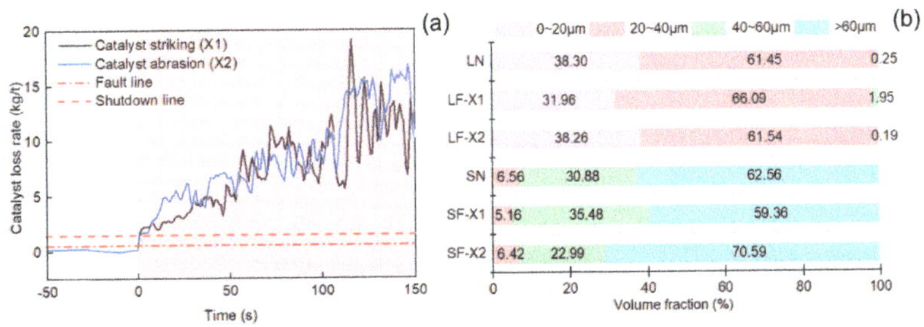

Figure 4. Catalyst loss rate (**a**) and the content of catalyst samples (**b**) at catalyst fragmentation.

Figure 4b displays the content of catalyst samples at catalyst fragmentation. The content of 40–60 μm particles in the escaped catalyst (LF-X1) after the catalyst striking increased by 1.7%. Conversely, the content of particles with diameters of 40–60 μm in the escaped catalyst particles (LF-X2) is 0.19%, which is 0.06% less than that in the normal conditions. It indicates that the catalyst striking is mainly caused by the escape of larger diameter particles, while the catalyst abrasion is mainly caused by the escape of smaller diameter particles. Additionally, the content of particles larger than 60 μm in the spent catalyst (SF-X1) decreases by 3.2% compared to the value of the normal conditions (LN), whereas the content of particles with diameters larger than 60 μm in the spent catalyst (SF-X2) is 8.03% more than that in the normal conditions (SN). This indicates that when X2 occurs, more catalyst particles with smaller diameters accumulate at the variable diameter section.

Figure 5a–c illustrate the patterns of pressure drop variation in cyclone separators, VQS and dilute phase during the catalyst fragment. The increase in the solid content within the separation system leads to an increase in the pressure drop of the cyclone, VQS and dilute phase. When the fault duration time is around 15 s (Figure 4a), the catalyst loss rate rapidly increases, resulting in abrupt changes in parameters such as $\Delta P_{VQS,t}/\Delta P_{VQS,normal}$, $\Delta P_{dilute,t}/\Delta P_{dilute,normal}$ and $T_{dilute,t}/T_{dilute,normal}$, which are closely related to the solid holdup in the dilute phase at this time. However, since the inlet of the cyclone separator is not directly connected to the dilute phase space, the change trend of $\Delta P_{CYC,t}/\Delta P_{CYC,normal}$ remains relatively continuous. When X1 occurs, the separation efficiency is slightly higher than when X2 occurs (Figure 4a), suggesting that most of the particles are separated. Therefore, the particle concentration at the inlet of the cyclone separator decreases, but the solid holdup at the variable diameter section of the disengager increases. Consequently, the increase in the pressure drop of the cyclone separator is around 1.3, which is smaller than that during X2. The pressure drop of the VQS and the dilute phase are about 2.3 and 6.2, which are greater than those during X2. As depicted in Figure 5d, during catalyst striking and the catalyst abrasion fault, the temperature in the dilute phase of the disengager decreases by 6.2% and 20.9%, respectively. This is because, as the particle size decreases, the thickness of the gas film covering the particle surface becomes thinner. When the particle diameter decreased, the heat carried by individual particles decreased, resulting in a larger specific surface area and a larger relative heat transfer area of particles. Meanwhile, the number of particles per unit volume increases, enhancing collision effects, which can severely disrupt the heat transfer boundary layer of the particles and reduce the heat transfer resistance between the gas and solid. This explains why the decrease in the dilute-phase temperature is more significant when X2 occurs.

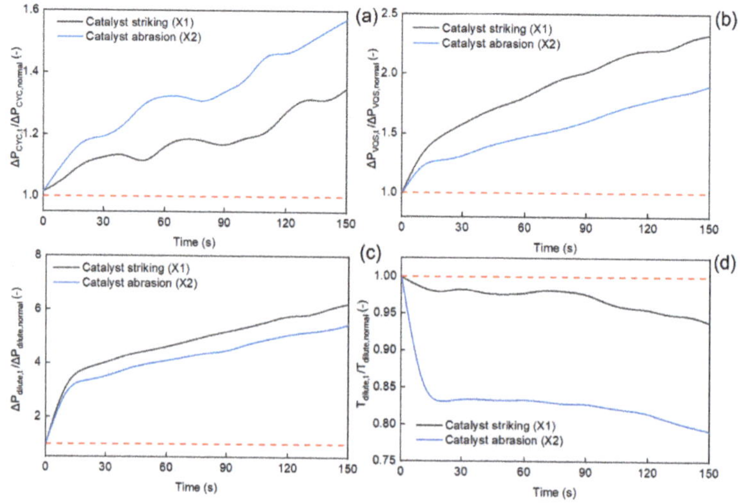

Figure 5. Flow field characteristic variation at catalyst fragmentation: (**a**) pressure drop variation in cyclone separators, (**b**) pressure drop variation in VQS, (**c**) pressure drop variation in dilute phase and (**d**) temperature in dilute phase.

4.2. Process Fault

Figure 6a illustrates the catalyst loss rate when the process fault occurred. It can be observed that high inlet velocities result in a rapid increase in the catalyst loss rate above the shutdown line (X3). This is because, at high inlet velocities, the turbulence intensity of the internally rotating gas flow increases, leading to the intense diffusion and rebound of the particles [50]. Small particles are more likely to escape as a result, causing a decrease in separation efficiency. When the low inlet velocity of a cyclone process fault occurs (X4), the catalyst loss rate initially fluctuates between the fault line and the shutdown line. After stabilization, it then fluctuates predominantly around the fault line. This is because the low inlet velocity reduces the centrifugal force on the particles, leading to a weaker separation capacity and lower separation efficiency. The operating pressure of the FCC unit directly affects the inlet velocity of the cyclone and the discharge process at the dipleg. When the operating pressure changes suddenly, the catalyst loss rate increases rapidly and fluctuates around the shutdown line (X5~X6). More particles were carried to the dilute phase, due to the fact that the higher pressure led to a higher gas density, which increased the interaction between the gas phase and the solid phase. With the increase in the gas density, the tangential velocity decreases, which enhances the separation efficiency of the cyclone separator.

Figure 6b displays the content of catalyst samples at the process fault. It can be seen that the particle content in different size ranges within the spent catalyst changes consistently (SF-X3~SF-X6). On the contrary, the particle content varies differently among different particle sizes in the escaping catalyst. The content of escaped particles with a diameter of 40–60 μm at a low inlet velocity decreases by 0.04% compared to the value at normal operation (LF-X4). In contrast, the content of particles with a diameter of 40–60 μm, which escaped from the cyclone separator, increases by 6.44% at a high inlet velocity (LF-X3). When the operation pressure fluctuates suddenly, the variation in the content of particles with a diameter of 20–40 μm in the escaped catalyst is opposite. Specifically, as the operation pressure increases (LF-X5), the content of particles within this size range rises; conversely, when the pressure operation decreases, the content of particles within this size range reduces (LF-X6).

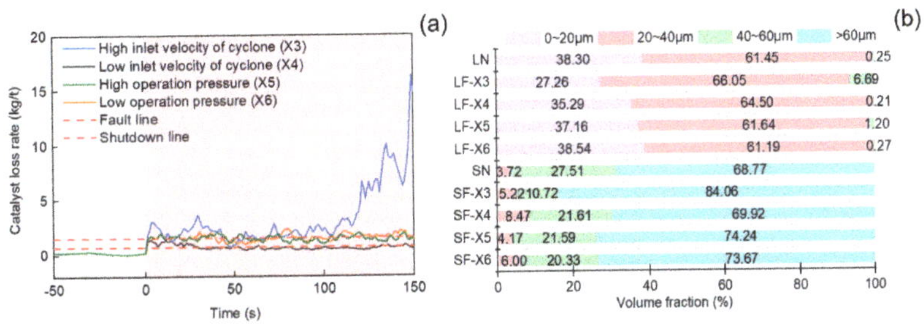

Figure 6. Catalyst loss rate (**a**) and the content of catalyst samples (**b**) at the process fault.

Due to the sudden increase in the catalyst loss rate when the fault duration lasts for approximately 15 s, this results in abrupt changes in all the parameters shown in Figure 7a–d around this time. Figure 7a illustrates the patterns of pressure drop variation in cyclone separators. It can be seen that the pressure drop of the cyclone increases by a factor of 1.8 at a high inlet velocity (X3). Conversely, at a low inlet velocity, the pressure drop of the cyclone decreases by 45.5% (X4). Moreover, when there is an abruption of operation pressure, the pressure drop of the cyclone will fluctuate around the normal value (X5-X6). As shown in Figure 7b, apart from the process fault of the low inlet velocity of the cyclone separator (X4), when other process faults occur, the pressure drop in VQS increases, with an increase factor ranging from 1.2 to 2.7. Figure 7c,d show the pattern of pressure drop variation in the dilute phase and the pattern of temperature in the dilute phase. Whenever a process fault occurs, the former increases while the latter decreases.

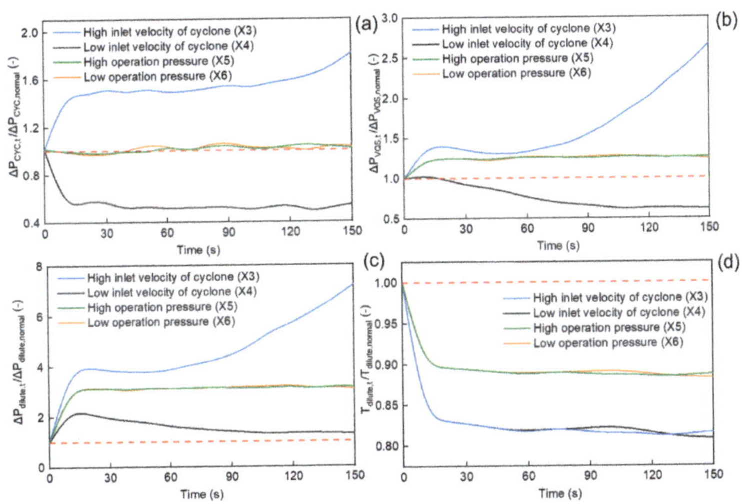

Figure 7. Flow field characteristic variation at process fault: (**a**) pressure drop variation in cyclone separators, (**b**) pressure drop variation in VQS, (**c**) pressure drop variation in dilute phase and (**d**) temperature in dilute phase.

4.3. Mechanical Fault

The mechanism of the catalyst loss caused by the blockage fault and leakage fault varies. The blockage fault typically occurs due to the falling off of internal parts or coke forming in the disengager system, which can obstruct the normal flow of the catalyst. When the cyclone is blocked, the catalyst accumulates in the cyclone separation space, reducing the gas flow rotation space and creating a shortcut channel that bypasses the cyclone separator vortex finder, resulting in a rapid decline in separation capacity [51]. The inherent issue of blockage at the isolated shell slot and the stripper section leads to a reduction in the catalyst flow space. When a leakage fault appears, the external high-pressure oil gas will flow back into the faulty part, disrupting the trajectory of the catalyst and bringing a significant reduction in separation efficiency [52].

Figure 8a shows the catalyst loss rate in the disengager corresponding to blockages in the cyclone (X9), the isolated shell slot (X8) and the stripper section (X7), respectively. It can be seen that the catalyst loss rate in the disengager immediately increases to more than 1.5 kg/t. The maximum values of the catalyst loss rate are 137.2, 7.8 and 123.8 kg/t, respectively. Figure 8b shows the catalyst loss rate in the disengager during the leakage fault. If the oil gas backflow occurs at the cyclone separator wall (X11) or the wall of the vortex arm (X10), the curve of the catalyst loss rate fluctuates above the shutdown line. When the dipleg is fractured (X12), the curve of the catalyst loss rate fluctuates below the fault line and above the shutdown line. The final value of the catalyst loss rate varies from 3.83~20.98 kg/t.

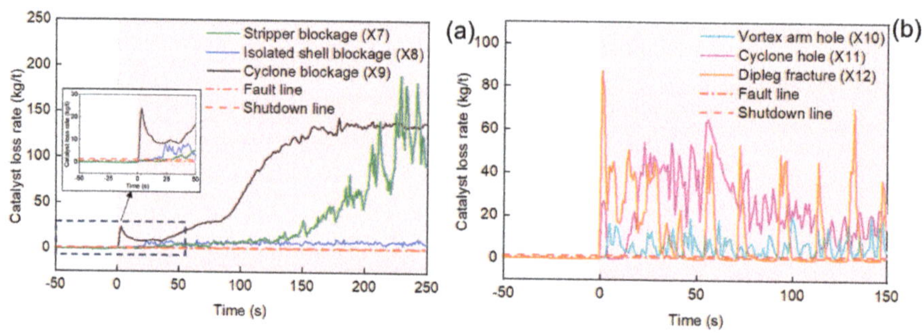

Figure 8. Catalyst loss rate at blockage fault (**a**) and leakage fault (**b**).

Figure 9a reveals the content of catalyst samples at the blockage fault. When the stripper section is blocked, the content of particles with a diameter of 40–60 μm in the escaped catalyst increases by an average of 7.20% (LF-X7). The content of particles with diameters larger than 60 μm in the spent catalyst decreases by 15.44% (SF-X7). When the isolated shell slot was obstructed, the content of catalyst particles with a diameter of 40–60 μm, which escaped from the cyclone separator, increased by an average of 2.66% (LF-X8). The content of particles with diameters larger than 60 μm in the spent catalyst decreases by 6.93% compared to the normal condition (SF-X8). In the case of blockage at the cyclone, the content of particles with a diameter of 0–20 μm, which escaped from the faulty cyclone (LF1-X9), is reduced by 18.58%. Figure 9b shows the content of catalyst samples at the leakage fault. When the gas leakage fault occurs at the above three positions, the patterns of change in PSD of spent catalyst particles are consistent. Specifically, the content of particles with a diameter smaller than 60 μm increases and the content of particles with a diameter larger than 60 μm decreases (SF-X10~SF-X12).

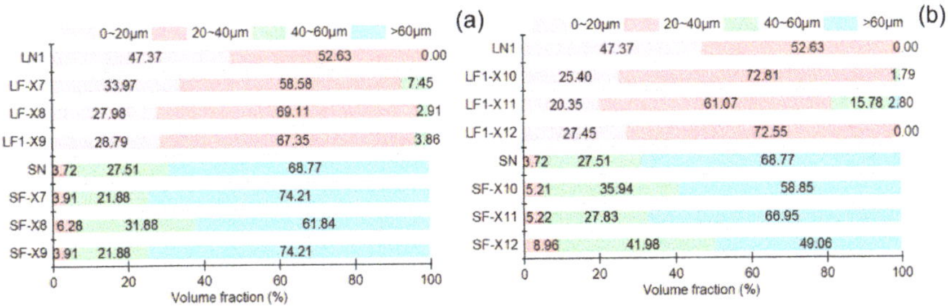

Figure 9. The content of catalyst samples at blockage fault (**a**) and leakage fault (**b**).

Figure 10a depicts the variation in the pressure drop in the cyclone separator during the mechanical happens. Due to the accumulation of catalysts in the isolated shell and cyclone, the pressure at the inlet of the VQS and the cyclone increased. When the blockage fault at the stripper section (X7) and isolated shell (X8) lasted for 250 s, the factors of increase were 3.2 and 1.4, respectively. Moreover, when the vortex arm was holed (X10), the trajectory of the oil gas and fluidized gas were disturbed. The pressure drop of the cyclone separator increased by a factor of 1.3. Upon the occurrence of a mechanical fault in the cyclone separator, a reduction in the pressure drop of the cyclone separator was observed. When the blockage fault at the cyclone separator occurs (X9), the pressure drop of oil gas passing through the cyclone is the dominant pressure drop. Hence, the pressure drop of the fault cyclone gradually decreases by 49.8%. When the cyclone separator wall hole (X11) and cyclone dipleg fraction (X12) occurred, the pressure drop of the cyclone separator decreased by 54% and 35%, respectively.

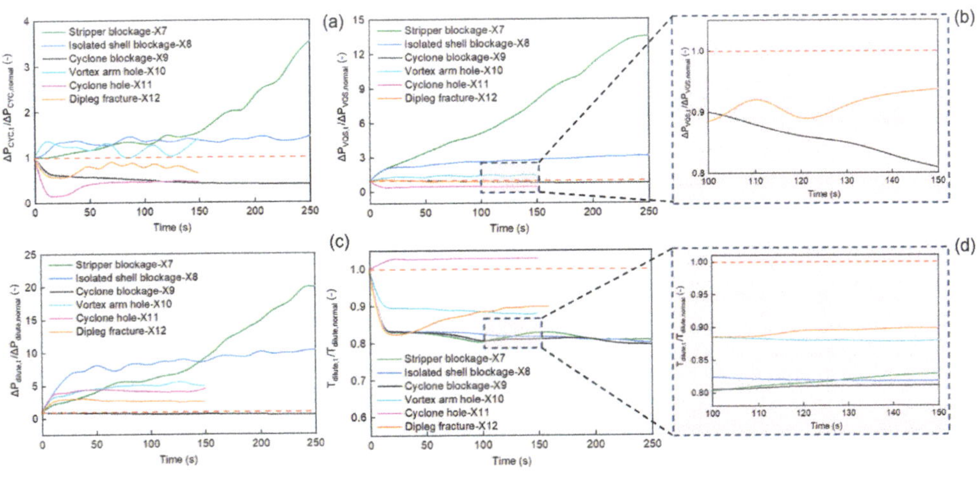

Figure 10. Flow field characteristic variation at mechanical fault: (**a**) pressure drop variation in cyclone separators, (**b**) pressure drop variation in VQS and local enlargement, (**c**) pressure drop variation in dilute phase and (**d**) temperature in dilute phase and local enlargement.

Figure 10b demonstrates the variation in the pressure drop in the VQS during the mechanical happens. The variation pattern of the pressure drop in VQS is consistent with that of the cyclone separator shown in Figure 5a. Figure 10c,d depict the variation in the pressure drop and temperature in the dilute phase during the mechanical happens, respectively. Except for the occurrence of a blockage fault in the cyclone separator, the pressure drop increases by a factor of 2.6~19.9 when the other mechanical fault occurs (X8–X12). It can be observed from Figure 10d that, when most mechanical faults occur, the temperature in the dilute phase space decreases by 10.1~20.5% compared to the normal value. Conversely, when the cyclone wall is holed (X11), the temperature of the dilute phase increases. This is mainly because the low-temperature anti-coke steam flows directly into the cyclone without entering the dilute phase, which leads to a decrease in the content of the water steam in the dilute phase.

5. Fault Judgment Method of Catalyst Loss

5.1. Establishment of Fault Diagnosis Methods

Based on the characteristics of catalyst loss faults, when the catalyst loss fault occurs in the FCC unit, several online operational data points and PSD of particles undergo significant impact and fluctuation. The following are typical flow characteristics under various bottom events leading to the catalyst loss of FCC disengagers.

(1) When the particle size decreases due to catalyst striking, the content of particles with a diameter of 40–60 μm in the escaped catalyst particles increases by 1.7%. Conversely, the content of particles with a diameter of 40–60 μm in the spent catalyst particles decreased by 0.06%, which is opposite to the condition of catalyst striking;

(2) When a mechanical failure occurs in the FCC disengager, the basic parameters, such as the pressure drop, temperature and PSD of particles, all vary in different degrees. When the location of the blockage fault is the cyclone separator, the pressure drops of the cyclone separator, the VQS and the dilute phase decrease 31.1~49.8%, 23.9% and 47.0%, respectively. When the blockage fault occurs at the isolated shell slot and the stripper section, the above pressure drop increases by a factor of 1.4, 3.0 and 10.1 compared to normal conditions. When the location of the leakage fault is the cyclone separator, the pressure drop increases by a factor of 3.2, 12.2 and 19.8 for the cyclone, VQS and dilute phase, respectively;

(3) When the inlet velocity at the cyclone separator increases rapidly, the pressure drop of the cyclone separator and the VQS increase by a factor of 1.7 and 2.6. On the contrary, when the inlet velocity at the cyclone separator decreases rapidly, the pressure drop of the cyclone separator, VQS and dilute phase decrease by 44.2%, 38.4% and 38.4%, respectively. When the operation pressure increases rapidly, the content of particles with a diameter of 20–40 μm in the escaped catalyst particles decreases by 0.26%. When the operation pressure decreases rapidly, the content of particles with a diameter of 20–40 μm in the escaped catalyst particles increases by 0.19%.

By obtaining the variation patterns of characteristic parameters through CFD simulation results, it is possible to retroactively analyze and distinguish various abnormal operating conditions related to catalyst loss of FCC disengagers. Based on the analyses regarding the change patterns during the catalyst loss, a judgment process (Figure 11) was proposed to identify the cause of the catalyst loss in the disengager. The meanings of the bottom events corresponding to X1 to X12 in Figure 11 have been listed in Table 1.

As can be seen from Figure 11, when the increase in the catalyst loss rate is less than a factor of 30, it may indicate the occurrence of process faults or catalyst fragmentation. At this point, emphasis should be placed on checking specific signals such as the pressure drop of the cyclone separator, dilute and the PSD in LF and SF, to identify the specific fault type among X1–X6 and X8. When the catalyst loss rate increases by a factor of more than 30 compared to the normal consumption, it is likely due to blockage faults (X7 and X9) or gas leakage faults (X10 and X12) in the mechanical equipment. At this point, emphasis should be placed on checking specific signals such as the pressure drop of the cyclone separator, VQS and the temperature of the dilute, to determine the specific type of mechanical fault. The following are the specific criteria for judgment.

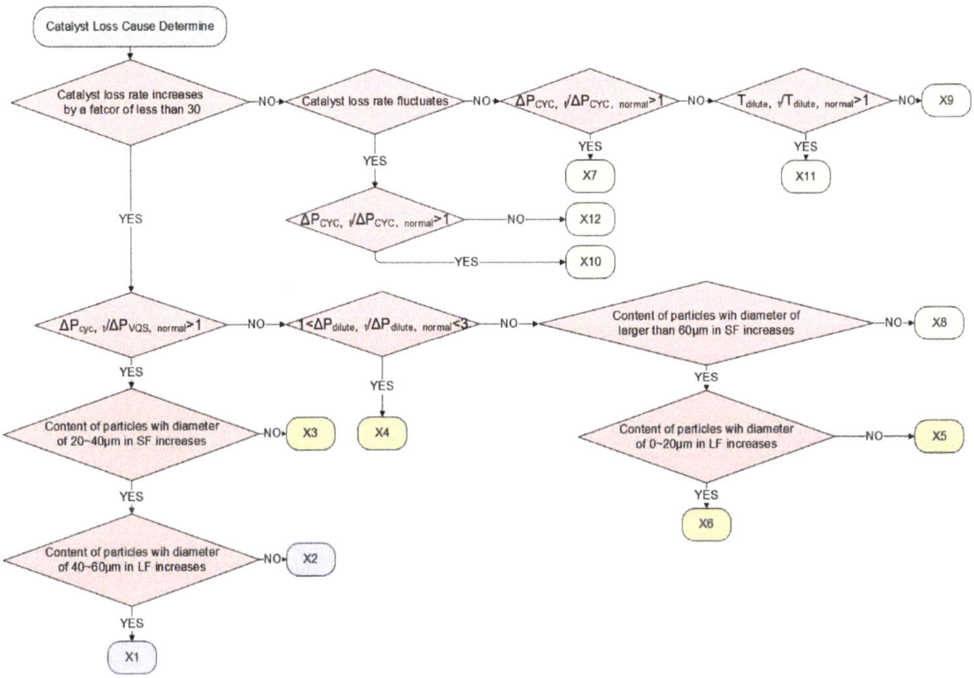

Figure 11. Fault judgment method of catalyst loss in FCC disengager.

(1) When the catalyst loss rate increases by a factor of less than 30, the $\Delta P_{CYC,t}/\Delta P_{CYC,normal} > 1$ and the content of particles with a diameter of 20~40 μm in SF increases, it can be excluded that the catalyst loss is caused by catalyst fragmentation. Building upon this, if the content of particles with a diameter of 40~60 μm in LF decreases, the presence of catalyst striking can be excluded. Conversely, it suggests the presence of catalyst abrasion;

(2) When the catalyst loss rate increases by a factor of less than 30 and the $\Delta P_{CYC,t}/\Delta P_{CYC,normal} < 1$, it can be excluded that the catalyst loss is caused by the process fault. If the $1 < \Delta P_{dilute,t}/\Delta P_{dilute,normal} < 3$, it can be excluded that the catalyst loss is caused by the low inlet velocity of the cyclone. If the $\Delta P_{dilute,t}/\Delta P_{dilute,normal} > 1$, it can be excluded that the catalyst loss is caused by the abruption of operation pressure or the blockage at the isolated shell;

(3) When the catalyst loss rate increases by a factor of larger than 30 and fluctuates around the shutdown line (catalyst loss rate is 1.5 kg/t), it may suggest the presence of a blockage at the stripper, cyclone or cyclone hole. If the $\Delta P_{CYC,t}/\Delta P_{CYC,normal} > 1$, the blockage at the stripper can be excluded. If it the $\Delta P_{CYC,t}/\Delta P_{CYC,normal} < 1$ and the $\Delta T_{dilute,t}/\Delta T_{dilute,normal} < 1$, it suggests that the cyclone separator is blocked. Conversely, if the $\Delta P_{CYC,t}/\Delta P_{CYC,normal} < 1$ and the $\Delta T_{dilute,t}/\Delta T_{dilute,normal} > 1$, it indicates that gas leakage has occurred in the cyclone separator. When the catalyst loss rate increases rapidly by a factor of larger than 30, it is necessary to consider the changes in the cyclone pressure drop along with other factors to make an accurate judgment. If the $\Delta P_{CYC,t}/\Delta P_{CYC,normal} < 1$, a dipleg fracture may occur. Conversely, it suggests the presence of a vortex arm hole.

5.2. Cases of Catalyst Loss Fault in Two Commercial FCC Disengagers

To validate the diagnosis method proposed in this paper, the causes of catalyst loss in two different commercial FCC disengagers were analyzed. The first industrial FCC disengager was introduced in Section 3.1, with an 800,000 t/a processing capacity. The detailed structure was shown in Figure 3a. During a certain operating period, the average slurry solid content (mass concentration of solid particles in the slurry oil) is 6.5 g/L and the catalyst loss rate is 0.73 kg/t, which is higher than the normal data (0.6 kg/t). It was initially determined that the catalyst loss fault may have occurred in the disengager. According to the judgment process proposed in Figure 11, the cause of the catalyst loss was investigated and the key measurement characters in the disengager were analyzed. By analyzing the DCS data, it can be found that, during the catalyst loss period, the temperature of the dilute phase decreased by about 36.4%. The pressure drop of the dilute phase and cyclone increased by about 0.5% and 8.3%, respectively (Figure 12). Therefore, it is presumed that the low inlet velocity of the cyclone is the main cause of the catalyst loss, which corresponds to fault type X4. After calculation, the current inlet velocity of the cyclone is found to be 16.2 m/s, which resides at the lowest boundary of the cyclon operational conditions. By redesigning the structure of the cyclone separators, the inlet area of cyclone separators was decreased to adjust the inlet velocity of the cyclone. The cyclone inlet line velocity was adjusted within a suitable range, within 19.9 m/s, the cyclone efficiency was improved. The phenomenon of catalyst loss in the FCC disengager was improved effectively.

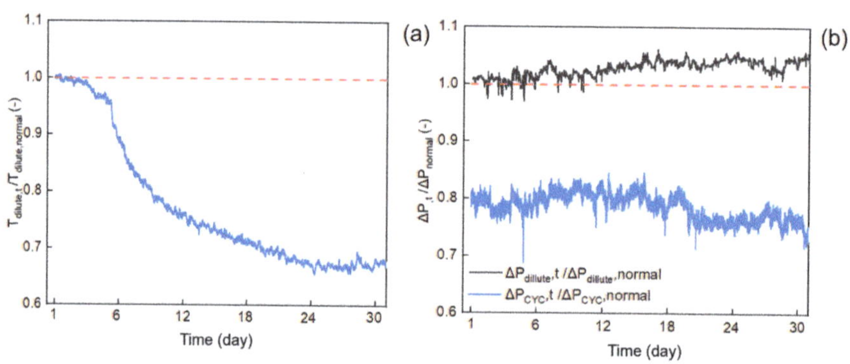

Figure 12. Flow characteristics of the first FCC disengager: (**a**) temperature in dilute phase; (**b**) pressure drop variation in the cyclone separator and dilute phase.

The second industrial disengager is the component of the FCC unit in an oil refinery, located in southern China, with a processing capacity of 1.7 Mt/a [38]. The connection structure of the rough cyclone and the cyclone separators inside the disengager differs from the first industrial unit, which is called the "soft-connection type"; there is a certain gap between the outlet of the rough cyclone and the inlet of the top cyclone (Figure 13a). During the upward flow process within the riser, the feedstock oil gas undergoes a cracking reaction with the catalyst particles. Then, the mixture is divided into three streams and enters three quick cyclones for the first-stage separation of oil gas from the catalyst particles. After most of the catalyst particles are separated, the mixture enters six top cyclone separators for the second-stage separation, further separating the catalyst from the oil gas. Subsequently, the oil gas enters the gas chamber, ultimately entering the subsequent unit. The separation process within this disengager is similar to that of the first one, thus the judgment method established in this work is expected to be applied in diagnosing the reason for the catalyst loss fault in this disengager.

Due to the impact of a typhoon, a fluctuation occurred in the operation of the FCC unit, resulting in a sudden decrease in the total catalyst inventory. The solid content in the slurry oil increased from the normal range of 2–3 g/L to 28.41 g/L, and the peak point even reached 92.38 g/L. During this catalyst loss fault period, adjustments were made multiple times to the process parameters of the unit, but the solid content in the slurry oil remained at a high value. Through an analysis of the catalyst loss rate, it was discovered that during the fault period, the catalyst loss rate surged to over 30 times the normal rate, thereby eliminating the possibilities of issues related to catalyst fragmentation and process faults. The volume content of the escaped catalyst particles with a diameter of more than 40 μm increased. Further investigation into the pressure drop of the top cyclone separator revealed that it had decreased from 10.19 kPa to 8.98 kPa. In other words, the $\Delta P_{CYC,t}/\Delta P_{CYC,normal} < 1$, which indicates that the catalyst loss fault may be caused by a blockage or leakage fault within the cyclone separator. Since the temperature in the dilute phase did not increase, it is speculated that the main reason for the catalyst loss is the blockage at the cyclone separator. Upon conducting a shutdown inspection, it was found that there was severe coking inside the FCC disengager (Figure 13b,c), which led to a blockage fault at the cyclone separator, preventing the discharge of catalyst particles. This resulted in the reduction in the separation capacity of the top cyclone separator, thereby further substantiating the reliability of the fault diagnosis procedure outlined in this paper. By utilizing techniques such as installing V-shaped pins and deflector vanes, as shown in Figure 13d, the coke blocks were stabilized and segmented successfully. After operating the FCC unit for a production period, the formation and detachment of coke blocks were effectively prevented, solving the issue of coke detachment from the outer wall of the outer wall of cyclone separator vortex finder. This ensured the long-term operation of the catalytic cracking unit.

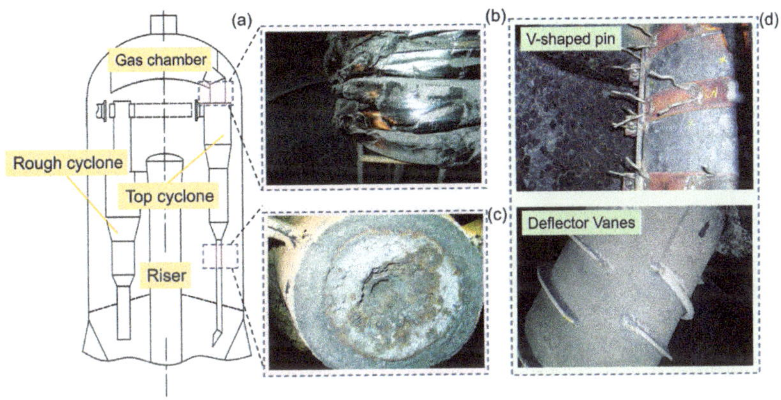

Figure 13. The structure of the second FCC disengager (**a**) and the condition during the catalyst loss fault (**b–d**).

Through the analysis of catalyst loss faults in two industrial FCC disengagers, we found that the diagnostic method proposed in this paper can make relatively accurate judgments on the type and location of fault reasons. It demonstrates good universality for different types of FCC disengagers. However, the current research still has certain limitations. One limitation is the lack of capability to quantitatively assess the degree of the fault reason. Another challenge lies in precisely isolating the individual causes when the catalyst loss fault results from the interactive effect of two or more factors. This is because of the unpredictability surrounding catalyst loss fault symptoms and their causes. A fault symptom can stem from various underlying issues related to different faults, or a single fault can manifest in multiple symptoms. In the future, the foundation of our previous work [53–55] will be built upon. The Bayesian Network (SRMBN) method will be adopted to continue the research on catalyst loss failures in FCC units. It is expected to quantify the severity of faults and identify catalyst loss caused by different types of faults. In addition, it is necessary to further integrate simulation data obtained from the CFD method with industrial data. In this process, in order to achieve the monitoring of catalyst loss faults in the FCC disengagers, it is planned to refer to our previous experience in monitoring the coking rate in FCCUs using CFD simulations and machine learning methods [27]. More accurate fault diagnosis methods for catalyst loss faults need to be established. A real-time monitoring platform should be built to reduce the diagnosis time for staff working in oil refinery and improve the accuracy of the fault judgment method.

6. Conclusions

This research studies the behaviors of gas–solid flow, heat transfer and separation within an 800,000 t/a FCC disengager under different abnormal operating conditions. Based on the FTA method and CFD simulation, the changes in the pressure drop, temperature, particle size distribution of catalyst and other signals were analyzed. A new fault tree-based judgment method for identifying the cause of the catalyst loss in the FCC disengager was proposed. The main conclusions could be summarized as follows:

(1) There are three intermediate events resulting in the catalyst loss fault, including catalyst fragmentation, the process fault and the mechanical fault. In addition, there are twelve factors, such as catalyst striking, abrupt changes in operating pressure and a cyclone separator blockage, which can lead to abnormal catalyst loss faults in the FCC disengager. Each of these factors contributes equally to the catalyst loss fault;

(2) The type of cause and the location of the catalyst loss fault can be inferred based on changes in the characteristic parameters within the disengager. It can be found that an increase in the catalyst loss rate by a factor of larger than 30 indicates the presence of a mechanical fault. When the catalyst loss rate increases by a factor of less than 30, it can be excluded that the catalyst fragmentation, process fault or blockage at the isolated shell has occurred;

(3) The fault judgment method in this study was applied to identify the cause of catalyst loss in two industrial disengager cases. In the first case, the catalyst loss rate increased by 4%, the $\Delta P_{CYC,t}/\Delta P_{CYC,normal} < 1$ and the $\Delta T_{dilute,t}/\Delta T_{dilute,normal} > 1$, suggesting the abnormal catalyst loss was caused by the low inlet velocity. In the second case, the catalyst loss rate surged by over 30, with a reduction in pressure drop in the cyclone and more catalyst particles larger than 40 μm escaping. A blockage at the cyclone separator was identified as the main cause. Both inferences align with actual identification results.

The present work can provide a theoretical basis for the rapid diagnosis of the causes of catalyst loss of FCC disengagers, thereby minimizing the reliance on the operational experience of personnel. In the future, research is expected to focus on integrating CFD data with industrial data to further enhance the reliability and accuracy of the judgment method.

Author Contributions: Conceptualization, N.L. and C.Z.; Methodology, Y.L., Y.Z. and Z.L.; Software, Y.L.; Validation, Y.L.; Formal analysis, S.L.; Investigation, Y.L., Z.L., N.L. and C.Z.; Resources, X.L.; Data curation, Y.Z. and X.S.; Writing—original draft, Y.L.; Writing—review & editing, Y.Z., S.L., C.W.

and X.L.; Visualization, Y.L.; Supervision, S.L., X.S., C.W. and X.L.; Project administration, X.L. All authors have read and agreed to the published version of the manuscript.

Funding: This research was funded by National Natural Science Foundation of China (No. 22021004).

Data Availability Statement: Data are contained within the article.

Conflicts of Interest: The authors declare no conflicts of interest.

Abbreviations

C_p	Specific heat capacity: J·kg^{-1}·K^{-1}
$C_{\varepsilon 1}$, $C_{\varepsilon 2}$, $C_{\varepsilon 3}$	Model number
LF	Escaped catalyst from cyclone under fault
LF1	Escaped catalyst from CYC1 under fault
LN	Escaped catalyst from cyclone under normal operation
SF	Spent catalyst under fault
SF1	Spent catalyst at CYC1 under fault
SN	Spent catalyst under normal operation
d_p^*	Equivalent diameter of catalyst, μm
E	Energy, J
g	Gravitational acceleration, m·s^{-2}
k	Turbulent kinetic energy, m^2·s^{-2}
P_r	Planck number
P	Pressure, kPa
Re	Reynolds number
T	Temperature, K
t	Time, s
u	Velocity, m·s^{-1}
x	x-direction coordinate
α	Volume fraction
β	Coefficient of friction
ε	Turbulent kinetic energy dissipation rate, m^2·s^{-3}
μ	Viscosity, Pa·s
ε	Turbulent kinetic energy dissipation rate, m^2·s^{-3}
ρ	Density, kg·m^{-3}
δ	gas in-leakage rate, %
σ_k	Density, kg·m^{-3}
τ	Divergence of gas stress tensor, Pa
$\Delta P_{CYC,t}/\Delta P_{CYC,normal}$	Pressure drop of cyclone separator, -
$\Delta P_{VQS,t}/\Delta P_{VQS,normal}$	Pressure drop of VQS, -
$\Delta P_{dilute,t}/\Delta P_{dilute,normal}$	Pressure drop of dilute phase, -
$T_{dilute,t}/T_{dilute,normal}$	Temperature of dilute phase, -
Subscript	
i, j, k	Coordinate direction
m	Solid or gas phase
g	Gas phase
p	Solid phase

References

1. Gao, J.; Shi, X.; Lan, X.; Xu, C. A Technical Roadmap for China's Petrochemical Industry Upgrading to Achieve Carbon Neutrality. *Engineering* **2023**, *29*, 55–58. [CrossRef]
2. Bai, P.; Etim, U.J.; Yan, Z.; Mintova, S.; Zhang, Z.; Zhong, Z.; Gao, X. Fluid catalytic cracking technology: Current status and recent discoveries on catalyst contamination. *Catal. Rev.* **2019**, *61*, 333–405. [CrossRef]

3. Rodríguez-Fragoso, M.; Zavala-Salazar, S.; Moreno-Montiel, M.; Bouchot, C.; Elizalde-Solis, O.; Ramírez-Jiménez, E. A kinetic model for the FCC process as function of the feedstock composition. *Chem. Eng. J.* **2023**, *474*, 145489. [CrossRef]
4. He, J.; Peng, W.; Liu, M.; Huang, X.; Han, S. The flow characteristics of gas–solid two-phase flow in an inclined pipe. *Adv. Powder Technol.* **2025**, *36*, 104725. [CrossRef]
5. Tian, W.; Wang, S.; Sun, S.; Li, C.; Lin, Y. Intelligent prediction and early warning of abnormal conditions for fluid catalytic cracking process. *Chem. Eng. Res. Des.* **2022**, *181*, 304–320. [CrossRef]
6. Li, P.; Ren, Y.; Chen, X. Analysis of operation status of catalytic cracking unit in SINOPEC. *Pet. Process. Petrochem.* **2022**, *53*, 53–59.
7. Mclean, J.B. FCC catalyst properties can affect cyclone erosion. *Oil Gas J.* **2000**, *98*, 33–36.
8. Shaw, D.F.; Walter, R.E. How FCCU trickle valves affect catalyst losses. *Hydrocarb. Process.* **2007**, *171*, 75–84.
9. Peng, W.; Wei, Y.; Liu, Y.; Han, S.; Bao, B. Effect of flow patterns in the recirculation catalyst standpipe on the performance of FCC high-efficiency regenerator. *Chem. Eng. Sci.* **2022**, *248*, 117154. [CrossRef]
10. Zheng, M.; Chen, G.; Han, J. Failure analysis on two austenitic stainless steels applied in cyclone separators of catalytic cracking unit. *Eng. Fail. Anal.* **2011**, *18*, 88–96. [CrossRef]
11. Jianjun, W.; Yajun, L.; Weiwei, X.; Qiang, L.; Dandekar, A. Effects of blade roughness on particle deposition in flue gas turbines. *Powder Technol.* **2019**, *353*, 426–432. [CrossRef]
12. Song, J.; Wang, D.; He, Y.; Lei, P.; Peng, W.; Wei, Y. A Stepwise Diagnosis Method for the Catalyst Loss Fault of the Cyclone Separator in FCC Units. *Separations* **2023**, *10*, 28. [CrossRef]
13. Peng, W.; Xu, B.; Yang, L.; Xu, Y.; He, J. Analysis of the cyclone separator wear and catalyst loss in FCC unit regenerator. *Pet. Refin. Eng.* **2023**, *53*, 37–40.
14. Niccum, P. 20 questions: Identify probable causes for high FCC catalyst loss: Here is a list to troubleshoot your catalyst problems: Refining Developments. *Hydrocarb. Process.* **2010**, *89*, 29–38.
15. Yang, F.; Xu, M.; Lei, W.; Lv, J. Artificial intelligence methods applied to catalytic cracking processes. *Big Data Min. Anal.* **2023**, *6*, 361–380. [CrossRef]
16. Yazdi, M.; Mohammadpour, J.; Li, H.; Huang, H.; Zarei, E.; Pirbalouti, R.G.; Adumene, S. Fault tree analysis improvements: A bibliometric analysis and literature review. *Qual. Reliab. Eng. Int.* **2023**, *39*, 1639–1659. [CrossRef]
17. Zhu, C.; Zhang, T. A review on the realization methods of dynamic fault tree. *Qual. Reliab. Eng. Int.* **2022**, *38*, 3233–3251. [CrossRef]
18. Pan, K.; Liu, H.; Gou, X.; Huang, R.; Ye, D.; Wang, H.; Glowacz, A.; Kong, J. Towards a systematic description of fault tree analysis studies using informetric mapping. *Sustainability* **2022**, *14*, 11430. [CrossRef]
19. Song, Z.; Feng, Y.; Lu, C. Superimposable neural network for health monitoring of aircraft hydraulic system. *Eng. Fail. Anal.* **2024**, *160*, 108063. [CrossRef]
20. Zhao, J.; Zhu, Y.; Liu, X.; Jiang, R.; Ding, B.; Chen, Y. Root cause analysis of a cracked primary heat exchanger in a gas wall-mounted boiler. *Eng. Fail. Anal.* **2023**, *153*, 107583. [CrossRef]
21. Wang, D.; Sun, L.Q.; Yan, C.Y.; Wei, Y.D. Mechanisms and fault tree analysis of catalyst loss in fluid catalytic cracking (FCC) unit. *Chem. Ind. Eng. Prog.* **2019**, *38*, 3534–3539.
22. Lan, X.; Xu, C.; Wang, H.; Gao, J. Simulation on flow rules of liquid droplets in FCC disengager. *J. Chem. Eng. Chin. Univ.* **2007**, *21*, 423–428.
23. Yan, P.; Lan, X.; Xu, C.; Gao, J. Preliminary study on the vaporization ratio of the slurry in the residue fluid catalytic cracking disengager. *Chin. J. Chem. Eng.* **2007**, *15*, 315–319. [CrossRef]
24. Lan, X.; Gao, J.; Xu, C. Analysis of coking mechanisms in residue fluid catalytic cracking disengagers. *Mod. Chem. Ind.* **2007**, *27*, 46–48.
25. Song, J.; Sun, G.; Chao, Z.; Wei, Y.; Shi, M. Gas flow behavior and residence time distribution in a FCC disengager vessel with different coupling configurations between two-stage separators. *Powder Technol.* **2010**, *201*, 258–265. [CrossRef]
26. Song, J.; Sun, G.; Chao, Z.; Wei, Y. Gas flow behavior in industrial FCC disengager vessels with different coupling configurations between two-stage separators. *Powder Technol.* **2011**, *207*, 444–453. [CrossRef]
27. Zhang, M.; Yang, Z.; Zhao, Y.; Lv, M.; Lan, X.; Shi, X.; Gao, J.; Li, C.; Yuan, Z.; Lin, Y. A hybrid safety monitoring framework for industrial FCC disengager coking rate based on FPM, CFD, and ML. *Process Saf. Environ. Prot.* **2023**, *175*, 17–33. [CrossRef]
28. Zhang, Z.; Yan, S.; Dong, S.; Dong, K.; Zhang, Y.; Wang, B. Study of the short-circuit flow and circulation flow's impact on separation performance of cyclone separator with volute-helical inlet. *Adv. Powder Technol.* **2024**, *35*, 104281. [CrossRef]
29. Yang, L.; Chen, C.; Hsu, W.; Fukui, K.; Fukasawa, T.; Huang, A.; Kuo, H. Effect of the operation temperature on the hydrodynamics and performances of a cyclone separator. *Adv. Powder Technol.* **2022**, *33*, 103791. [CrossRef]
30. Brar, L.S.; Rahmani, F.; Wasilewski, M. Performance analysis of multi-inlet cyclone separators considering different shapes and locations of the inlet ducts. *Adv. Powder Technol.* **2024**, *35*, 104325. [CrossRef]

31. Yang, Q.; Wang, F.; Han, S.; Zhan, Q.; He, J.; Fang, L.; Peng, W. Changing law of catalyst average particle size during catalyst loss in fluid catalytic cracking units. *Chem. Ind. Eng. Prog.* **2025**, *1*, 1–10. [CrossRef]
32. Liu, M.; Yu, H.; Guo, J. Characterization of iron-poisoning catalyst and its influence on fluid catalytic cracking unit. *Petrochem. Technol. Appl.* **2024**, *42*, 436–440.
33. Li, C.; Piao, J.; Qin, J.; Ji, Z.; Xu, Z. Comprehensive Application of Quantitative Control over the Catalyst Loss in Catalytic Cracking Unit's Start-up Process. *Chem. Eng. Mach.* **2024**, *51*, 317–322.
34. Ji, Y.; Liu, J.; Li, P.; Han, S.; Yang, Q.; Peng, W. Diagnosis and analysis of catalyst loss caused by crushing in FCCU. *Pet. Refin. Eng.* **2024**, *54*, 56–59+64.
35. Yang, Z.; Wang, J.; Cai, X.; Wu, J.; Wei, Y. Cause analysis of process fault of cyclone separator in FCCU. *China Powder Sci. Technol.* **2020**, *26*, 75–80.
36. Dng, J.; Zhou, Z.; Chai, X. Analysis of blocking in stripping section of settler and countermeasures. *Pet. Refin. Eng.* **2014**, *44*, 11–14.
37. Hao, Y.; Cheng, J. Application and improvement of FCC VQS. *Petrochem. Ind. Appl.* **2013**, *32*, 82–86.
38. Wen, P.; Yan, C.; Wei, Y. Cause analysis and countermeasures for the agent runout of the disengager of FCCU. *Pet. Refin. Eng.* **2023**, *53*, 32–35.
39. Xia, M.; Zhang, Y.; Cheng, Q.; Hao, Y.; Wang, J.; Lu, C. Analysis and countermeasure of crack and wear in reaction system of FCC Unit. *Pet. Refin. Eng.* **2020**, *50*, 42–45.
40. Sun, L.; Xie, M.; Yang, L.; Peng, W. Analysis of Wear Holes on Cone Wall of Cyclone Separator in FCC Unit. *J. Chengde Pet. Coll.* **2024**, *54*, 56–59+64.
41. Bai, X. Cause Analysis and Countermeasures of Regenerator Runner in Heavy Oil Catalytic Crack Device. *Shandong Chem. Ind.* **2019**, *48*, 132–133.
42. Peng, W.; Yang, Q.; Huang, X.; Fang, Q.; Han, S.; He, J.; Gao, Z. Properties analysis of catalyst particles during stripper distributor fault in industrial fluid catalytic cracking unit. *Fuel* **2025**, *381*, 133279. [CrossRef]
43. Maurer, S.; Durán, S.R.; Künstle, M.; Biollaz, S.M. Influence of interparticle forces on attrition and elutriation in bubbling fluidized beds. *Powder Technol.* **2016**, *291*, 473–486. [CrossRef]
44. Ferreira, J.M.; Sousa-Aguiar, E.F.; Aranda, D.A.G.J.C. FCC catalyst accessibility—A review. *Catalysts* **2023**, *13*, 784. [CrossRef]
45. Kukade, S.; Kumar, P.; Rao, P.V.; Choudary, N.V. Comparative study of attrition measurements of commercial FCC catalysts by ASTM fluidized bed and jet cup test methods. *Powder Technol.* **2016**, *301*, 472–477. [CrossRef]
46. Zhou, F.; Guo, Y.; Qian, H.; Fu, S.; Yuan, H. Effect of wall roughness on particle deposition on the outer wall of the vortex finder in a cyclone separator. *Adv. Powder Technol.* **2024**, *35*, 104428. [CrossRef]
47. Gao, J.; Chang, J.; Lan, X.; Yang, Y.; Lu, C.; Xu, C. CFD modeling of mass transfer and stripping efficiency in FCCU strippers. *AICHE J.* **2010**, *54*, 1164–1177. [CrossRef]
48. Portnikov, D.; Santo, N.; Kalman, H. Simplified model for particle collision related to attrition in pneumatic conveying. *Adv. Powder Technol.* **2020**, *31*, 359–369. [CrossRef]
49. Song, J.; He, Y.; Meng, F.; Li, S.; Wei, Y. Effects of gas in-leakage on separation performance and gas-phase flow field of a FCC cyclone. *Powder Technol.* **2023**, *415*, 118201. [CrossRef]
50. Kępa, A. The efficiency improvement of a large-diameter cyclone–the CFD calculations. *Sep. Purif. Technol.* **2013**, *118*, 105–111. [CrossRef]
51. Mao, H. Analysis on blockage failure of secondary cyclone separator in a FCC unit. *Pet. Refin. Eng.* **2023**, *53*, 49–52.
52. Liu, F.; Li, C.; Zeng, X.; Chen, J.; Guan, J.; Yang, L. Study on the flow and collision characteristics of catalyst particles in FCC reactor. *Powder Technol.* **2024**, *438*, 119642. [CrossRef]
53. Zhu, C.; Liu, N.; Zhang, M.; Li, Z.; Li, Y.; Shi, X.; Lan, X. Multimode residual monitoring of particle concentration in flue gas from Fluid Catalytic Cracking regenerator. *Control Eng. Pract.* **2025**, *156*, 106227. [CrossRef]
54. Liu, N.; Hu, M.; Wang, J.; Ren, Y.; Tian, W. Fault detection and diagnosis using Bayesian network model combining mechanism correlation analysis and process data: Application to unmonitored root cause variables type faults. *Process Saf. Environ. Prot.* **2022**, *164*, 15–29. [CrossRef]
55. Liu, N.; Wang, J.; Sun, S.; Li, C.; Tian, W. Optimized principal component analysis and multi-state Bayesian network integrated method for chemical process monitoring and variable state prediction. *Chem. Eng. J.* **2022**, *430*, 132617. [CrossRef]

processes

MDPI

Article

Vehicle Routing Perfection for Fresh Agricultural Products Distribution Under Carbon Emission Regulation and Customer Satisfaction

Xiaoyong Zhu [1,*], Yu Liang [2], Chao Wu [3] and Yongmao Xiao [4]

[1] School of Economics and Management, Shaoyang University, Shaoyang 422000, China
[2] Shaoyang University Library, Shaoyang University, Shaoyang 422000, China; liangyu0007@163.com
[3] School of Safety and Management Engineering, Hunan Institute of Technology, Hengyang 421002, China; wuchaohnit@126.com
[4] School of Computer and Information, Qiannan Normal University for Nationalities, Duyun 558000, China; xym198302@163.com
* Correspondence: zhuxysyu@126.com

Abstract: Under the background of "carbon peaking and carbon neutrality", carbon reduction is not only a realistic need for the high-quality development of the national economy, but also a key way for the sustainable development of enterprises. Cold chain logistics has the characteristics of high energy consumption and high carbon emissions. Fresh distribution requirements of fresh agricultural products may increase carbon emissions in the cold chain distribution process. Based on customer satisfaction and aiming at minimizing carbon emission and comprehensive distribution cost, this paper establishes an optimization model of cold chain logistics distribution routes for fresh agricultural products. First of all, a two-objective optimization model is proposed considering customer satisfaction maximization and comprehensive cost minimization, including fixed cost, fuel cost, carbon emission cost, cargo damage cost, and time window penalty cost. And when constructing customer satisfaction function, we mainly pay attention to the time factor. Secondly, a hybrid ant colony algorithm, which combines improved ant colony algorithm and local search algorithm 3-opt, is designed to solve the model. Thirdly, a hybrid ant colony algorithm is applied to the simulation example of fresh agricultural products distribution through experimental simulation, and the results are compared with the traditional ant colony algorithm and improved ant colony algorithm. Finally, the results reveal that the average total cost of the hybrid ant colony algorithm is lower than that of the traditional ant colony algorithm and the improved ant colony algorithm, achieving nearly 8.159% and 4.622% cost savings, respectively, and in terms of customer satisfaction, the new algorithm is slightly better than the other algorithms. The results show that the method can effectively optimize the cold chain logistics distribution routes of fresh agricultural products, and reduce carbon emission and distribution cost.

Keywords: fresh agricultural product; path optimization; hybrid ant colony algorithm; carbon emissions; customer satisfaction

1. Introduction

As necessities for residents, fresh agricultural products play a vital role in daily consumption. The demand for fresh agricultural products is steadily growing with the improvement of urbanization in China, the upgrading of consumption standards, and the

https://doi.org/10.3390/pr13030605

rise of e-commerce. According to research data from Sullivan and the China Business Industry Research Institute, the domestic market size in China elevated from CNY 3.67 trillion in 2017 to CNY 5.64 trillion in 2021, with an average annual compound growth rate of 11.34% [1]. It is expected to reach CNY 6.62 trillion by 2024. Compared to industrial and other agricultural products, fresh agricultural products are particularly distinct due to their perishability [2]. As globalization and market economies continue to advance, cold chain logistics has become a critical means to guarantee food safety, fulfill consumer needs, and promote agricultural modernization. With the increasing significance of fresh agricultural products in daily life, their circulation efficiency and quality assurance are garnering growing attention [3].

The rapid development of cold chain logistics of fresh agricultural products has led to stringent requirements for distribution. Unlike traditional logistics, cold chain logistics for agricultural products have higher standards of product quality and timeliness, resulting in increased costs. As commodities, fresh agricultural products are characterized by their depreciation in value over time. On the one hand, the cultivation and production of fresh agricultural products are subject to seasonal cycles. These products only retain commercial value if delivered to consumers within specified time frames. Their value diminishes as the consumption cycle passes [4]. On the other hand, improper handling during storage and transportation accelerates spoilage. Consumer satisfaction decreases when the timeliness of goods exceeds their expectations. Therefore, cold chain logistics is indispensable for the harvesting and delivery of fresh agricultural products. During cold chain warehousing, transportation, and distribution stages, it is essential to fully account for the shelf life of fresh agricultural products and consumer expectations, ensuring that these products can be delivered at their optimal quality to consumers in the demand cycle.

The emergence of e-commerce has brought significant opportunities for the development of cold chain logistics, enhancing the consumption of fresh agricultural products [5–7]. In recent years, with the advancement of artificial intelligence, digital technology, and cold chain technology in the fresh agricultural e-commerce sector, the online consumption of fresh agricultural products has grown swiftly. Consequently, this poses high demands for cold chain logistics. Freshness and convenience are crucial factors influencing consumer purchasing decisions for fresh agricultural products [8]. Freshness mainly relies on effective management of cold chain logistics, including temperature and humidity control. However, cold chain logistics can cause energy loss across various aspects, such as warehousing, processing, packaging, and transportation, adversely impacting the environment. Some scholars have argued that cold chain logistics for fresh agricultural products should be optimized from a low-carbon perspective [9–13]. Cold chain logistics has the characteristics of high energy consumption and high carbon emissions. In global carbon emissions statistics, transportation accounts for 14% of total carbon emissions, with road transport contributing 70% of that figure [14]. Developing strategies to reduce carbon emissions and overall costs has become a hot research topic in cold chain logistics [15]. Therefore, it is pivotal to optimize cold chain delivery, mitigate carbon emissions, and lower the transportation cost.

Against the backdrop of achieving "peak carbon emissions and carbon neutrality", reducing carbon emissions is not only an inevitable requirement for the high-quality development of the national economy but also a critical path for the sustainable development of enterprises. According to the International Energy Agency (IEA), the global transportation sector is responsible for 25% of total carbon emissions. Although China accounts for 10% of this share, it faces significant challenges and untapped potential in reducing emissions. As a blue ocean in the field of cold chain logistics, the cold chain delivery of fresh agricultural products is growing rapidly and demands rigorous time and temperature control. This has made it a focal point for energy conservation and emission reduction research. Enterprises must comply with stringent environmental regulations, reduce carbon emissions, and operate legally and compliantly. Meanwhile, they should establish a positive corporate image and win customer trust by providing green and low-carbon delivery services. In the long run, measures of energy conservation and emission reduction can not only drop operating costs but also enhance market competitiveness and generate additional benefits. Therefore, optimizing the distribution routes of cold chain logistics for fresh agricultural products is crucial for achieving both energy conservation and emission reduction goals while improving customer satisfaction.

With the surge in demand for fresh agricultural products, the high energy consumption, high carbon emissions, and customer time-sensitivity of cold chain logistics distribution have become increasingly prominent issues. However, existing studies mostly focus on single-cost optimization, which struggles to balance economic efficiency and customer experience, and traditional algorithms are prone to local optima under dynamic constraints. How can we construct a multi-objective collaborative optimization model to quantify the interaction effects of carbon emissions and time window constraints? How can we design efficient solution algorithms to address the NP-hard nature of large-scale route combinations? This paper addresses these issues by proposing a bi-objective optimization framework and a hybrid intelligent algorithm, aiming to provide decision support for a green cold chain logistics system that balances low-carbon emission reduction and service quality improvement. Based on the perishable nature of fresh agricultural products and the impact of carbon emissions on the economic and social benefits of enterprises, this paper incorporates the costs incurred from product freshness reduction and vehicle carbon emissions into the total cost accounting. Meanwhile, considering the constraints of vehicle load capacity and service time window, a cold chain delivery optimization model for fresh agricultural products is developed to better comply with practical needs. This research aims to optimize the cold chain delivery of fresh agricultural products. Strategies of reducing delivery time, minimizing product damage, and ensuring product quality are used to indirectly improve customer satisfaction without altering the customer satisfaction function, thereby minimizing carbon emissions and maximizing customer satisfaction. First, an optimization model is established, incorporating customer satisfaction and comprehensive costs (including the fixed cost, costs of fuel and carbon emissions, product damage cost, and penalty costs from freshness reduction and outside service time windows). Moreover, the customer satisfaction function is constructed and mainly focuses on the time factor. Second, a hybrid ant colony algorithm is designed, integrating a correction strategy and the local search 3-opt algorithm to reduce invalid paths and solve the model. The feasibility of the model and the effectiveness of the algorithm are verified through practical cases. The results validate that the proposed algorithm can efficiently reduce distribution costs, mitigate carbon emissions, and enhance customer satisfaction. This paper offers a valuable reference for the optimization of cold chain logistics for fresh agricultural products.

The paper consists of six parts. The rest of the paper is structured as follows: Section 2 presents the literature review. The researched problem description and model construction, and the algorithm design are discussed in Sections 3 and 4, followed by presentation of the case simulation and analysis in Section 5. In the Section 6, some conclusions are drawn.

2. Literature Review

The vehicle routing problem (VRP) was first proposed by Dantzig and Ramser in 1959. Subsequently, based on this foundational concept, enormous research outcomes have been yielded [16]. Solomon et al. highlighted the importance of considering time constraints in vehicle delivery [17]. Jabali et al. developed the VRP model incorporating soft time windows [18]. Shi Jianli and Zhang Jin introduced uncertainty factors into the

VRP model [19]. Ling Haifeng and Gu Junhui explored the multi-yard open vehicle routing problem with soft time windows [20]. Li Shunyong et al. analyzed the impact of traffic congestion and proposed a multi-channel time-varying network low-carbon vehicle routing model [21]. Ning Tao et al. proposed an interference management logistics model, which effectively reduced the impact of interference events on costs [22–25]. In recent years, the optimization of cold chain delivery has garnered widespread attention from scholars. Many studies have focused on the influence of time window constraints in optimizing logistics paths [26,27]. Wang Jun et al. explored the vehicle routing problem with fuzzy customer demands and established a multi-objective fuzzy probability constrain programming model that minimizes vehicle travel distance and distribution service delays [28]. Zhang Xiaonan et al. blurred both the demand and travel time, introducing an opportunity-constrained pre-optimization model for variable compensation that considers time window constraints [29].

The distribution of fresh products has been extensively investigated. Amorim et al. optimized perishable food delivery routes. However, the literature only focused on ordinary vehicles and did not address the issue of product damage during transportation [30]. Solomon et al. applied time window constraints to vehicle routing, providing numerous data sets for the later verification of models for the delivery vehicle routing of fresh agricultural products with time windows [31]. Yu et al. proposed a delivery problem for fresh food, considering quality. They solved the path-designing problem using a biogeographical mobile robot path-planning algorithm [32]. Byung et al. developed the problem of delivery routes for fresh food trucks and refrigerated vehicles, with the objective function of maximizing overall customer satisfaction, though their model did not take environmental factors into account [33]. Meanwhile, due to the perishable nature of products in cold chain logistics, customers have specific requirements regarding product freshness. In order to improve customer satisfaction, it is crucial to incorporate product freshness into the optimization of cold chain logistics routes. M. Rabbani et al. employed the freshness of perishable food as a criterion to determine the optimal distribution path, investigating the multi-distribution center cold chain vehicle routing problem with the objectives of minimizing the transportation cost, vehicle travel time, and product perishability [34]. Shao Juping et al. introduced a coefficient of freshness loss and established a multi-objective optimization model for fresh agricultural product distribution, considering customer satisfaction and total costs [35]. Zhang Yaming et al. developed a multi-distribution center vehicle routing model based on customer service time and product quality requirements [36].

During cold chain logistics, carbon emissions are generated by vehicles and various refrigeration equipment within the vehicles. Traditional VRP models should consider not only economic benefits but also reducing carbon emissions from social and environmental perspectives. A. Tiwari et al. optimized cold chain vehicle paths, minimizing both vehicle travel distance and carbon dioxide emissions [37]. R. S. Kumar et al. examined the multi-objective multi-vehicle routing problem with time windows, aiming to minimize total operating costs and emissions [38]. Ge Xianlong et al. proposed a calculation method for carbon emissions and established an integer programming model based on load and working time constraints of open pollution path problems [39]. Tang Jinhuan et al. built a nonlinear mixed integer programming model considering vehicle travel time and carbon emissions [40]. Ma Qiuzhuo et al. created a distribution path optimization model using vehicle load as a factor affecting fuel consumption and carbon dioxide emissions [41]. Based on energy conservation and emission reduction in cold chain logistics, Wang Zhiyi et al. established a distribution path optimization model with the goal of minimizing carbon emissions [42]. Considering varying energy consumption levels during transportation and unloading caused by the opening of refrigerated doors, Kang Kai et al. constructed an optimization model for fresh agricultural product distribution paths that includes carbon emissions [43]. Ma Qiuzhuo et al. established an optimal VRP model for urban distribution, verifying the impact of capacity changes on comprehensive costs and carbon emissions [44]. Rao Weizhen et al. incorporated the impact of road slope changes on carbon emissions and established a low-carbon distribution model [45]. Chen Zhi et al. accounted for the impact of delivery times and travel routes on carbon emissions, established a low-carbon vehicle delivery model, and verified the effectiveness of the model [46].

Meanwhile, significant progress has been made in solving the VRP. Ye Yong and Zhang Huizhen designed a wolf pack strategy for multiple distribution centers [47]. Yan Rui et al. addressed a two-dimensional vehicle routing problem that includes time windows and multiple vehicle field factors and proposed a quantum particle swarm algorithm to solve it [48]. Huber and Geiger designed a variable neighborhood search algorithm for solving VRP models [49]. Camm et al. used dynamic programming algorithms to tackle the VRP [50]. Tu Haining et al. formulated a vehicle delivery model considering the demand for timely logistics and solved it using an ant colony algorithm [51]. Xia Yangkun et al. introduced a dual-objective vehicle path model that accounts for working hours and soft time windows, employing an adaptive tabu search algorithm for its solution [52]. Franceshette et al. integrated factors such as departure time, vehicle speed, and capacity into their model, addressing it with heuristic algorithms, improved particle swarm optimization techniques, and tabu search algorithms, respectively [53–55]. Fan Houming et al. constructed a multi-objective optimization model incorporating fuzzy requirements and fuzzy time windows, solving it with a combined neighborhood search genetic algorithm and a hybrid genetic algorithm, respectively [56,57]. Rao Weizhen et al. proposed a low-carbon collaborative VRP model, which was solved using a two-stage algorithm consisting of a greedy algorithm and a large neighborhood algorithm [58]. Kang Kai et al. and Bao Huifang et al. developed a model that considered refrigeration costs and transportation-induced carbon emissions and addressed it with a hybrid ant colony algorithm [43,59]. Fang Wenting et al. focused on the comprehensive costs of carbon emissions, established a multi-objective optimization model, and proposed a hybrid ant colony algorithm for its solution [60]. Zhao Zhixue et al. established a cold chain logistics model that accounts for traffic congestion and solved it by improving the ant colony algorithm based on the characteristics of time-varying networks [61].

In summary, extensive research has been conducted on low-carbon cold chain distribution, yet limitations remain. Studies on the optimization of cold chain logistics vehicle routes for fresh agricultural products primarily emphasized economics, with little consideration of green benefits related to energy conservation and emission reduction. Environmental factors are barely incorporated into route optimization models. Although some scholars addressed carbon emissions during delivery, they often overlooked customer satisfaction. Additionally, the calculation of transportation cost typically focused on unit distance, neglecting other factors such as load capacity. Assumptions like a constant vehicle speed during delivery further diverge from real-world logistics.

The ant colony algorithm is widely used to solve the VRP. However, it is prone to stagnating in local optima. In response to these issues, this paper proposes a cold chain delivery optimization model for fresh agricultural products, considering costs, customer satisfaction, and carbon emissions. The model incorporates time windows and calculates the transportation cost based on the fuel consumption rate, enhancing the alignment with real-world conditions. Due to the challenges posed by complex delivery routing environments, the traditional ant colony algorithm suffers from slow convergence and unsatisfying results, making it difficult to escape from local optimal traps. This paper introduces a hybrid ant colony algorithm integrating the correction strategy and the pheromone update mechanism with adaptive volatility intensity. These improvements accelerate convergence, increase robustness, and enhance search efficiency, effectively avoiding local optima. The proposed algorithm demonstrates superior performance in optimizing cold chain delivery for fresh agricultural products.

3. Problem Description and Model Construction

3.1. Problem Description

For simplicity, this paper omitted product collection.

The problem of optimizing cold chain delivery of fresh agricultural products can be described as follows: a fresh agricultural product distribution center in a city serves multiple customers using a fleet of vehicles, all equipped with the same model of refrigeration equipment. These vehicles deliver fresh agricultural products to customers and return to the distribution center after completing specific tasks. The demand and geographic location of each customer are known. Each customer imposes certain restrictions on the delivery time window.

The problem aims to identify the most efficient distribution routes under vehicle load capacity limitations while meeting customer demand. Figure 1 illustrates the basic distribution diagram of this problem [43].

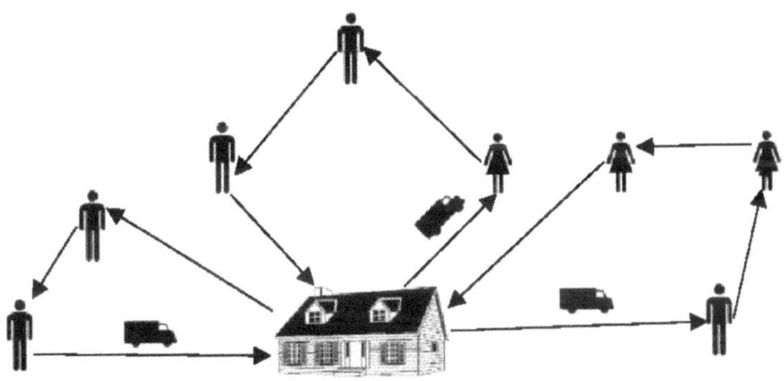

Figure 1. Delivery diagram.

The following assumptions further define the problem. These hypotheses are based on references [62–65] to ensure scientificity and rationality.

Assumption 1. *There is a single logistics center in this area that delivers fresh agricultural products. The center is equipped with a sufficient fleet of refrigerated trucks of the same type, each with a fixed maximum capacity. The transportation cost is constant. The refrigerated compartments are suitable during delivery.*

Assumption 2. *The geographical locations of the logistics center and customers are known, along with demand volumes, required delivery times, and unloading times. The demand of a single customer is within the maximum capacity of the vehicle and cannot be split. This denotes that one customer can only be served by one refrigerated truck and only be served once, with his demand fully met.*

Assumption 3. *Vehicles depart from the distribution center with no maximum mileage restrictions. They will return to the center only after completing their delivery. Trucks exclusively deliver goods without picking up additional items en route. The total demand on any delivery route must not exceed the truck's maximum load capacity. The delivery is conducted in a predetermined order.*

Assumption 4. *Fresh agricultural products have a limited shelf life. They should be delivered within specific time frames required by customers. Therefore, delivery time windows, which are soft* *time windows, are given. The refrigerated trucks must arrive within hard time windows specified by the customers. Failure to meet these windows will incur additional penalty costs.*

Assumption 5. *Although vehicle speeds vary in real-time during delivery, for simplicity, it is assumed that the speed remains constant on the same road segment and may differ across segments. Excluding traffic congestion, vehicle speeds are estimated using a triangular distribution function.*

3.2. Parameter Description

The parameters of the model are defined based on the description and assumptions of the problem. Firstly, a set of delivery networks $G = [N, A]$ are defined, where $N = \{0, 1, 2, \ldots, n, n+1\}$ represents the set of all nodes, **0** and $n + 1$ signify the distribution center, $N = 0$ denotes vehicles departing from the distribution center, and $N = n + 1$ indicates vehicles returning to the distribution center; $N' = \{1, 2, \ldots, n\}$ signifies the set of customers; $A = \{(i, j): i, j \in N, i \neq j\}$ represents the set of paths; and K denotes the total number of delivery vehicles. The definitions of other parameters are shown in Table 1.

Table 1. Definitions of parameters in the model.

Parameter	Definition
x_{ijk}	Binary decision variable. If vehicle k serves customer i and then goes to serve customer j, $x_{ijk}= 1$; otherwise, $x_{ijk}= 0$
y_{ik}	Binary decision variable. If vehicle k provides delivery service to customer i, $y_{ik}= 1$; otherwise, $y_{ik}= 0$
d_{ij}	The distance from customer i to customer j, in km
Q	The maximum load capacity of the delivery vehicles, in t
q_i	The demand from customer i, in t
q_{ijk}	The load carried by vehicle k from customer i to customer j, in t
P_1	The fixed cost of delivery vehicles, in CNY/vehicle
P_2	The price of fuel used by delivery vehicles, in CNY/L
P_3	The real-time carbon trading price at the carbon exchange, in CNY/t
P_4	The price of transported fresh agricultural products, in CNY/t
T_i	The service time of the delivery vehicle at customer i, in min
λ	The coefficient relating vehicle speed to fuel consumption
fu^1_{ijk}	Fuel consumption during vehicle operation, in L
fu^2_{ijk}	Fuel consumption of the refrigeration system, in L
∂_1, ∂_2	Fuel consumption per hour of the refrigeration system during driving and pre-cooling processes, in L/h
ω	Carbon emission coefficient, i.e., the carbon emission per unit of fuel, in kg/L
t_{0k}, t_{ik}	t_{0k}: The time when vehicle k leaves the distribution center; t_{ik}: The time when vehicle k arrives at customer i
fr_i	The freshness of fresh agricultural products at customer i
ε_1	The decay rate of fresh agricultural products at a specific cold chain temperature during delivery
ε_2	The decay rate of fresh agricultural products when the temperature rises during unloading
∂	The coefficient of freshness reduction for agricultural products
(E_i, L_i)	The ideal service time window for customer i (hard time window)
(E_i', L_i')	The acceptable service time window for customer i (soft time window)
μ_1, μ_2	The penalty factor for arriving earlier or later than the hard time window within the soft time window, in CNY/h

3.3. Cost Analysis of Low-Carbon Delivery Optimization and Customer Satisfaction

(1) Fixed cost of vehicles

The fixed cost of delivery vehicles is usually constant, including wear and tear during vehicle operation, depreciation expenses, and driver wages. The cost mainly involves vehicle utilization and is independent of mileage or the number of customers served. The fixed cost function Z_1 is expressed as Formula (1):

$$Z_1 = p_1 \sum_{j=1}^{N} \sum_{k=1}^{K} x_{0jk} \tag{1}$$

where the decision variable X_{0jk} = 1 represents that vehicle k travels from the distribution center to customer j (otherwise, it is 0).

(2) Fuel and carbon emission costs during delivery

Fuel and carbon emission costs incurred during delivery are calculated based on fuel consumption. The fuel cost is determined by multiplying fuel consumption by the unit price of fuel. The cost of carbon emissions is the product of the carbon dioxide emission coefficient of fuel, fuel consumption, and carbon trading price. Fuel consumption during transportation mainly comprises two parts: one for vehicle operation and the other for refrigeration, maintaining the low temperature of the compartment.

Fuel consumption during vehicle operation is complex and influenced by multiple factors. According to similar studies [66–68], the factors encompass product weight, driving speed, mileage, vehicle type, and road surface conditions. For simplicity, this paper overlooks vehicle type and road surface conditions and only considers product weight, driving speed, and mileage. Based on reference [69], product weight has a linear relationship with fuel consumption. Using the load estimation method, fuel consumption per unit distance during normal driving is θ_0 when the load is zero and θ_1 when fully loaded. When the load is q_{ik}, the fuel consumption per unit distance during normal driving is shown in Equation (2).

$$\theta(q_{ik}) = \theta_0 + \frac{\theta_1 - \theta_0}{Q} q_{ijk} \tag{2}$$

The distance refers to the fuel consumption caused by changes in speed, which is proportional to the square of the speed [16]. The proportionality coefficient is denoted by λ (λ = $0.5C_dPS$), where C_d is the air resistance coefficient, P represents air density, and S denotes windward area.

According to reference [18], this paper estimates the speed of vehicles within a reasonable range using a triangular probability distribution. The probability density function of the triangular distribution is expressed by Formula (3), where v_u represents the free flow speed, v_d is the congestion speed, and v_m denotes the possible speed.

$$f(v_{ijk}) = \begin{cases} 0 & , \ v_{ij} \prec v_d \\ \frac{2(v_{ij}-v_d)}{(v_u-v_d)(v_m-v_d)} & , \ v_d \prec v_{ij} \prec v_m \\ \frac{2(v_u-v_{ij})}{(v_u-v_d)(v_m-v_d)} & , \ v_m \prec v_{ij} \prec v_u \\ 0 & , \ v_{ij} \succ v_d \end{cases} \tag{3}$$

Consequently, the fuel consumption of vehicles during transportation is shown in Formula (4).

$$fu_{ijk}^1 = \rho(Q_k)d_{ijk} + \lambda v_{ij} d_{ijk} \tag{4}$$

The refrigeration cost includes the refrigeration energy during transportation and the additional refrigeration energy during unloading. The energy is derived from fuel, making fuel consumption an appropriate measure for the refrigeration cost. Assuming that the fuel consumption per unit time during the operation and pre-cooling process of the refrigeration system is ∂_1 and $\partial_2 (\partial_1 > \partial_2)$, respectively, the fuel consumption for refrigeration is shown in Formula (5).

$$fu_{ijk}^2 = \partial_1 d_{ijk} / v_{ij} + \partial_2 T_j \tag{5}$$

The fuel cost $\mathbf{Z_{21}}$ is shown in Formula (6).

$$Z_{21} = P_2 \sum_{k=1}^{K} \sum_{i=1}^{N} \sum_{j=1}^{N'} \left(fu_{ijk}^1 + fu_{ijk}^2 \right) x_{ijk}, \quad i \neq j \tag{6}$$

The cost of carbon emissions $\mathbf{Z_{22}}$ is expressed by Formula (7).

$$Z_{22} = P_3 \omega \sum_{k=1}^{K} \sum_{i=1}^{N} \sum_{j=1}^{N'} \left(fu_{ijk}^1 + fu_{ijk}^2 \right) x_{ijk}, \quad i \neq j \tag{7}$$

(3) The damage cost of freshness reduction during delivery

General VRP models typically consider the damage to goods caused by collisions during loading and unloading. The freshness of fresh agricultural products deteriorates during delivery, leading to damage costs. This paper assumes that the cost of product damage is attributed to temperature variations during transportation and unloading. The spoilage rate ε of fresh agricultural products is related to the characteristics of the products and the surrounding temperature. A spoilage rate function is adopted to quantitatively analyze the loss of goods [70]. When calculating the cost of product damage, this paper focuses on time-dependent damage, including losses incurred during transportation and unloading.

When the delivery vehicle departs from the distribution center and heads towards customer i, assume that ε_1 represents the spoilage rate of the product at a specific temperature during delivery. During unloading, air convection will increase the temperature inside the carriage, resulting in a higher spoilage rate. Let ε_2 denote the spoilage rate at this time ($\varepsilon_2 > \varepsilon_1$). When the refrigerated truck arrives at customer i, the cost $\mathbf{Z_3}$ of product damage during delivery is shown in Formula (8):

$$Z_3 = P_4 \sum_{k=1}^{k} \sum_{i=1}^{N} y_{ik} q_i (1 - \ell^{-\varepsilon_1 (t_{ik} - t_{0k})}) + Q_{in} (1 - \ell^{-\varepsilon_2 T_i}) \tag{8}$$

(4) The penalty cost of freshness reduction during delivery

Unlike regular ***VRP***, cold chain ***VRP*** mainly involves perishable goods. Consequently, when establishing a cold chain vehicle routing model for agricultural products, the penalty cost caused by freshness decrease during distribution must be considered. According to reference [69], the quality of fresh agricultural products exhibits an exponential change over time. If the initial freshness of the product at the distribution center is assumed to be 100%, the freshness fr of the product upon service at customer ***i*** is expressed by Formula (9):

$$\begin{aligned} \overline{fr_i} &= fr_0^{-\partial[\max(E_i, t_i)]} \\ t_i &= \sum_{k=1}^{K} \sum_{i=1}^{N} x_{ij} (T_i + d_{ij}/v_{ij} + \max(E_i, t_i), \forall j \in N' \end{aligned} \tag{9}$$

Therefore, the penalty cost $\mathbf{Z_4}$ caused by freshness decrease is directly proportional to the price of agricultural products and the customer demand, as shown in Formula (10).

$$Z_4 = p_4 \sum_{i=1}^{N} q_i \max(fr_i - \overline{fr_i}, 0) \tag{10}$$

(5) The penalty cost of service time windows

The service time window $(\boldsymbol{E_i}, \boldsymbol{L_i})$ represents the time range within which the customers can accept. The penalty cost $\mathbf{Z_5}$ is expressed by Formula (11):

$$Z_5 = \mu_1 \sum_{i=1}^{N} \max(E_i - t_i, 0) + \mu_2 \sum_{i=1}^{N} \max(t_i - L_i, 0) \tag{11}$$

(6) Customer satisfaction

If the refrigerated truck arrives within the hard time window **(E_i, L_i)**, the customer satisfaction is 1. **(E_i', L_i')** denotes the soft time window with a customer satisfaction rate of 0. If the refrigerated truck arrives outside the hard time window but within the soft time window, customer satisfaction will be calculated based on the deviation from the hard time window. According to reference [68], customer i's satisfaction can be obtained from Equation (12).

$$CS_i = \begin{cases} 0, & t_{ik} \prec E_i' \\ \frac{t_{ik}-E_i'}{E_i-E_i'}, & E_i' \prec t_{ik} \prec E_i \\ 1 & E_i \prec t_{ik} \prec L_i \\ \frac{L_i'-t_{ik}}{L_i'-L_i}, & L_i \prec t_{ik} \prec L_i' \end{cases} \quad (12)$$

The satisfaction of all customers is denoted by $\sum CS_i$.

3.4. Mathematical Model Formulation

The optimization model for fresh agricultural product delivery considering carbon emissions and customer satisfaction is expressed as follows:

$$\min Z = (Z_1 + Z_2 + Z_3 + Z_4 + Z_5) \quad (13)$$

$$\max CS = \sum_{i=1}^{N} CS_i/N \quad (14)$$

$$st. \quad x_{ijk} \in \{0,1\}, y_{ik} \in \{0,1\}, \forall i,j \in N, i \neq j \quad (15)$$

$$\sum_{i=1}^{N} x_{0ik} \leq 1, \sum_{i=1}^{N} y_{0ik} \leq 1, k \in K \quad (16)$$

$$\sum_{k=1}^{K}\sum_{j=1}^{N'} x_{ijk} = 1, \forall i \in N \quad (17)$$

$$\sum_{i=0}^{N}\sum_{j=1}^{N'} x_{ijk}q_i \leq Q, \forall i \in N \quad (18)$$

$$q_{jk} = \sum_{k=1}^{K}\sum_{i=1}^{N} (q_{ik} - q_i)x_{ijk}, \forall j \in N \quad (19)$$

$$\sum_{i=1}^{N}\sum_{k=1}^{K} x_{i(n+1)k}(q_{ijk} - q_i) = 0 \quad (20)$$

$$\sum_{i=1}^{N} (\max(E_i, t_i) + S_i + d_{i(n+1)}/v_{ij})x_{i(n+1)k} \leq 240 \quad (21)$$

$$t_{jk} = t_{ik} + x_{ijk}t_{ij}, \forall i,j \in N', i \prec j \quad (22)$$

$$E_i \leq t_{ik} \leq L_i, \forall i \in N \quad (23)$$

Equation (13) indicates that the integrated cost of distribution is minimized. Equation (14) indicates that customer satisfaction is maximized. Equation (15) is a constraint on the value of decision variables. Equations (16) and (17) represent the basic conditions for delivery vehicles to depart and customers to receive services, respectively. Each vehicle can only depart once at the same time. Customer i is only served by one vehicle. Equation (18) ensures that the sum of demand from all customers on the delivery path of a vehicle does not exceed the vehicle's maximum carrying capacity Q. Equation (19) denotes that when the vehicle arrives at customer j, it carrying capacity q_{kj} is the total carrying capacity of vehicle k at the time of departure minus the total demand for delivered customers. Equation (20) represents that the total load of a delivery vehicle when departing from the distribution center is equal to the total demand of the customers served by the vehicle. Equation (21) is designed to prevent driver fatigue by limiting the maximum time for a single delivery route to four hours, which requires that the completion time of each vehicle's delivery path should not exceed this limit. Equation (22) is the continuity constraint for delivery, where t_{ik} represents the time when the vehicle arrives at node i, t_{jk} signifies the time when the vehicle arrives at node j, and t_{ij} is the time it takes for the vehicle to travel along path (i, j). Equation (23) is a time window constraint, representing the maximum acceptable service time window for the customer, and t_{ik} is the time when vehicle k arrives at customer i.

4. Algorithm Design

4.1. Review of Traditional Ant Colony Algorithm

The ant colony algorithm was first proposed by Marco Dorigo. It is known for its simple structure, strong adaptability, and robustness, making it suitable for research on delivery path planning.

Supposing that m ants are searching for food in a network with n nodes, they choose their forward path based on the state transition probability formula. Let P_{ij} represent the transition probability of an ant from nodes i to j. The classical ant colony algorithm is shown in Expression (24):

$$P_{ij}(t) = \begin{cases} \dfrac{[\tau_{ij}(t)]^{\alpha}[\eta_{ij}(t)]^{\beta}}{\sum\limits_{l \in allowed_i}\left[[\tau_{il}(t)]^{\alpha}[\eta_{il}(t)]^{\beta}\right]}, & l \in allowed_i \\ 0, & \lambda \notin allowed_i \end{cases} \tag{24}$$

where $allowed_i$ represents the set of nodes that node i is allowed to select in the next step; α is the pheromone heuristic factor; β denotes the heuristic function factor; and $\tau_{ij}(t)$ signifies the pheromone concentration from node i to node j at time t. The constant heuristic function from node i to node j at time t is defined as $\eta_{ij}(t) = \eta_{ij} = 1/d_{ij}$, where d_{ij} represents the Euclidean distance between node i and node j, and $\eta = (\eta_{ij}) \in Rn \times n$ is the heuristic matrix of the ant colony network.

After one cycle of route searching, the pheromone concentration is updated for the next cycle according to Formulas (25)–(27):

$$\tau_{ij}(t+1) = (1-\rho) \times \tau_{ij}(t) + \Delta\tau_{ij}(t,t+1) \tag{25}$$

$$\Delta\tau_{ij}(t,t+1) = \sum_{k=1}^{K} \Delta\tau_{ijk}(t,t+1) \tag{26}$$

$$\Delta\tau_{ijk}(t,t+n) = \begin{cases} \frac{Q'}{L_k}, & \text{ant k passes through path (i,j)} \\ 0, & \text{otherwise} \end{cases} \tag{27}$$

where ρ denotes the volatilization coefficient of pheromones; $\Delta\tau^{k}{}_{ij}$ $(t, t + 1)$ represents the pheromone released by the k-th ant on path (i, j) from time t to $t + 1$; L_k is the length of the path taken by the k-th ant in this cycle; Q' is a constant. Each ant repeats the search and updates parameters according to the above rules until it finds the optimal path from the starting point to the endpoint.

4.2. Ant Colony Algorithm Optimization

The traditional ant colony algorithm has high robustness and optimization ability, with satisfying path search performance. However, it suffers from slow convergence and is prone to getting stuck in local optima. Therefore, this paper optimizes this algorithm from three aspects: improving the heuristic and probability transfer functions, enhancing the pheromone update strategy, and incorporating the local search 3-opt algorithm.

(1) Improvement of the heuristic and probability transition functions

The ant colony algorithm calculates state transition probability based on the heuristic function and pheromone concentration, guiding ants to select paths. In the traditional ant colony algorithm, the heuristic function $\tau_{ij}(t)$ is typically the reciprocal of the distance between nodes i and j. In this paper, the improved algorithm takes a cost-based approach when selecting paths. Therefore, the heuristic function is defined as the reciprocal of the average of the comprehensive delivery cost, as shown in Formula (28).

$$\eta_{ij} = \frac{1}{M_{ij}} \tag{28}$$

where M_{ij} signifies the comprehensive cost of transporting goods from customer i to customer j, including the fixed cost, fuel and carbon emission costs, product damage cost, and penalty costs of freshness reduction and time windows.

This paper adopts the modified Dijkstra algorithm to improve the ant colony transition probability. The algorithm incorporates a set, ***set*1**, into the next set of candidate nodes for ants. This expands the search range, enhances the global search capability, and reduces the number of iterations needed in path planning. In order to prevent premature convergence, a continuously increasing value q_0 is introduced. Before each selection, a random number q_i is generated. The ant selects the next point based on a comparison of q_i and q_0. If q_i is less than or equal to q_0, the path with the highest correction cost (i.e., the lowest actual cost) is chosen from the available paths. Otherwise, the selection is made using the roulette method. The reference [68] improved the state transition probability, as shown in Formula (29).

$$P_{ij}^{k} = \begin{cases} \text{argmax}(\tau_{ij}^{\alpha}\eta_{ij}^{\beta}), & q_i \leq q_0 \\ \frac{\tau_{ij}^{\alpha}\eta_{ij}^{\beta}}{\sum\limits_{i=0}^{N}\sum\limits_{j=0}^{N}\tau_{ij}^{\alpha}\eta_{ij}^{\beta}}, & q_i \succ q_0, j \in openlist_k \\ 0, & otherwise \end{cases} \tag{29}$$

where P_{ij}^{k} denotes the transition probability of the ***k*-th** ant from node i to node j; $\tau_{ij}(t)$ is the concentration of the pheromone from point i to point j; α and β are the heuristic factors for $\tau_{ij}(t)$ and $\eta_{ij}(t)$, respectively; $openlist_k$ represents the improved Dijkstra's ***set*1** set, including the nodes that allow ant k to choose in the next step. When the node with the highest transition probability is found, the ant will migrate to that node and store it in ***set*2**.

(2) Improvement of pheromone updating strategy

In ant colony algorithms, the pheromone volatilization coefficient ρ affects the convergence speed and global search ability. Therefore, in this paper, ρ ($\rho = \cos N_{max}$), where N_{max} denotes the maximum number of iterations) is adaptively adjusted with the number of iterations. In the early stage of the algorithm, ρ is relatively large. The search space is expanded to avoid falling into local optima. After the ants complete their rounds, the ants with the lowest total cost in all rounds and with the lowest total cost in the current round are selected to update the pheromone through paths. In the later stage, ρ decreases to accelerate the convergence speed. The adaptive adjustment rule of the pheromone $\Delta\tau^{k}{}_{ij}$ left by ants on each path is shown in Formula (30):

$$\Delta\tau_{ij}^{k} = \begin{cases} \frac{d_{ij}^{k} Z}{\sum\limits_{k=1}^{K}\sum\limits_{i=1}^{N}\sum\limits_{j=1}^{N} x_{ijk} d_{ij}^{k}}, & \text{the ant passes through path (i, j)} \\ 0, & \text{otherwise} \end{cases} \tag{30}$$

The rule for updating the pheromone on the optimal path is shown in Formula (31):

$$\tau_{ij}^{k+1} = (1-\rho)\tau_{ij}^{k} + \Delta\tau_{ij}^{k} \tag{31}$$

The rule for updating the pheromone on non-optimal paths is shown in Formula (32):

$$\tau_{ij}^{k+1} = (1-\rho)\tau_{ij}^{k} \tag{32}$$

(3) Local search 3-opt algorithm

In response to the problem of ant colony algorithms easily getting stuck in local optima, this paper introduces the 3-opt algorithm for local search optimization. It is assumed that the optimal traversal order L for the current iteration is $1 \to 2 \to 3 \to 4 \to 5 \to 6 \to 7$. Three nonadjacent nodes, 2, 4, and 7, are randomly selected from the path. There are five exchange methods for these nodes, corresponding to five traversal orders. The order with the shortest total route length is compared with L. If the travel distance of the new order is shorter, the optimal strip traversal order is replaced. In summary, the 3-opt algorithm can effectively jump out of local optima and improve the global search capability of the ant colony algorithm through node swapping.

4.3. Hybrid Ant Colony Algorithm

Based on the hybrid ant colony algorithm, the proposed ant colony algorithm for cold chain delivery optimization of fresh agricultural products is outlined as follows:

(1) Initialize parameters, such as the number of customers, service time window for each customer, service time, demand, and distance matrix between customers. Additionally, define the algorithm's iteration count, maximum iteration count, and number of ants (vehicles).
(2) Create an initialization tabu table, record the customers visited by each ant, and place all ants at the distribution center as the starting point.
(3) Ant path construction. Based on the current concentration of pheromones, heuristic information, and the tabu table, ants start searching for paths. Improved probability transition rules are used to calculate the probability of ants choosing the next node.
(4) Update the tabu table. Mark visited customers as inaccessible. Repeat this process until each ant has traversed all customers and returned to the distribution center, forming a complete path and proceeding to step (5). If not, return to step (3).
(5) Path evaluation and 3-opt optimization. Using the updated pheromone method in this paper, update the pheromone and iterate through the paths. Evaluate the optimal path for a single iteration and the corresponding transportation cost C (0) and customer satisfaction S (0). If the maximum number of iterations has been reached, conduct step (6). Otherwise, return to step (3).
(6) Apply the 3-opt local search. Implement the 3-opt local search to optimize the path, updating the cost and customer satisfaction.
(7) Global pheromone update. Based on the path quality of all ants in the current iteration (such as the evaluation based on costs or satisfaction), output the optimal path obtained from the 3-opt search. Update the global information table using the improved pheromone update strategy.

(8) Iteration termination judgment. If the iteration termination criteria are met, proceed to step (9). Otherwise, return to step (6).
(9) End the algorithm and output the optimal result.

Integrating the 3-opt algorithm within the ant colony optimization (ACO) framework can enhance path search efficiency and solution quality through a collaborative hybrid strategy. Specifically, the process begins with ACO generating initial feasible paths, followed by the introduction of 3-opt as a local search operator to refine these paths. The implementation is divided into three stages:

(1) Pheromone-Guided Phase: ACO generates an initial solution set based on pheromone concentration and heuristic rules, focusing on exploring the global solution space.
(2) 3-Opt Neighborhood Optimization Phase: For each path constructed by the ants, three non-consecutive nodes are selected for edge exchanges. By removing three existing edges and reconnecting them, new paths are generated. The objective function values (e.g., comprehensive costs and customer satisfaction) are calculated, and improved non-dominated solutions are retained.
(3) Dynamic Feedback Mechanism: The pheromone increments from the 3-opt optimized paths are fed back into the ACO update rules to reinforce high-quality paths. An adaptive probability threshold controls the triggering frequency of 3-opt to balance computational efficiency and precision.

To further reduce computational complexity, a tabu list can be designed to avoid redundant searches, or sub-paths requiring optimization can be selected based on path similarity. Experimental validation shows that this hybrid strategy effectively breaks local optima through the multi-node exchange mechanism of 3-opt. Particularly in large-scale, high-dimensional cold chain distribution problems, it significantly reduces inefficient detours and lowers transportation costs and carbon emissions while dynamically meeting customer satisfaction under time window constraints.

5. Simulation and Analysis

5.1. Column Data and Parameter Settings

This study compared and analyzed the traditional, improved, and hybrid ant colony algorithms through MATLAB(R2016b) simulation software. In order to reflect the scale effect in this test, the wide spatial distribution of the test data was ensured to evaluate the performance and efficiency of the algorithms across multiple scenarios. Simulation experiments were conducted to analyze the minimum comprehensive cost and similar satisfaction of path planning under the same number of different sub-regions, verifying the performance and convergence speed of the proposed hybrid ant colony algorithm, as well as its effectiveness in identifying the optimal path.

A customer corresponds to a fresh supermarket in an urban area. The distribution center is represented by 0. M refrigerated trucks depart to serve supermarkets numbered 1, 2, 3, . . . , N, and return to the center after delivery. Once a supermarket is assigned a number, it remains fixed. The simulation began by generating an initial population using a random method. The population includes 100 sets of data randomly distributed within a square area with a side length of 100 km. Each set consists of 30 data, where "0" represents the distribution center and "1–29" denotes the store to be serviced. The basic demand information (such as demand volume, time window requirements, required service time, and unloading time) of each supermarket in one set of data is shown in Table 2.

According to parameter tuning experience, the optimal number of ant colonies is typically set to 1.5 times the target number, the pheromone constant Q' is chosen within

[10, 10,000], the pheromone factor α is set between (0, 5], the heuristic function factor β is selected within (0, 5), and the maximum iteration number is chosen between [100, 500].

Table 2. Basic demand information of each supermarket.

Code	Demand Volume/kg	Expected Time Window	Acceptable Time Window	Service Time/Min
0	—	0:00–24:00	0:00–24:00	0
1	2.3	4:10–5:10	3:00–6:00	20
2	2.0	4:20–5:20	3:00–6:00	15
3	1.4	4:30–5:35	3:00–6:00	10
4	1.7	4:10–5:10	3:00–6:00	15
5	2.3	4:00–5:00	3:00–6:00	10
6	1.3	4:00–5:00	3:00–6:00	10
7	2.2	4:40–5:40	3:00–6:00	15
8	1.8	4:00–5:00	3:00–6:00	15
9	1.4	4:20–5:30	3:00–6:00	15
10	0.4	4:00–5:00	3:00–6:00	10
11	1.5	5:45–6:45	4:00–7:00	20
12	2.3	5:55–6:55	4:00–7:00	10
13	0.5	5:05–6:10	4:00–7:00	10
14	2.1	5:05–6:10	4:00–7:00	15
15	1.2	5:10–6:10	4:00–7:00	20
16	2.3	5:00–6:00	4:00–7:00	20
17	2.3	5:35–6:35	4:00–7:00	20
18	0.4	5:35–6:35	4:00–7:00	10
19	2.0	5:45–6:45	4:00–7:00	20
20	2.3	5:50–6:50	4:00–7:00	20
21	1.1	3:30–4:30	3:00–5:00	15
22	2.0	3:25–4:25	3:00–5:00	15
23	1.2	3:50–4:50	3:00–5:00	10
24	0.8	3:45–4:45	3:00–5:00	20
25	1.6	3:30–4:35	3:00–5:00	15
26	0.9	3:30–4:35	3:00–5:00	20
27	1.6	3:20–4:20	3:00–5:00	10
28	1.4	3:50–4:50	3:00–5:00	15
29	2.0	3:45–4:50	3:00–5:00	15

Upon multiple experimental analyses, the ant colony algorithm parameter settings are as follows: the number of ants m is set to 45. The pheromone constant Q' is 50, the pheromone factor α is 2.5, the heuristic function factor β is 4, the regional heuristic factor γ is 1, and the maximum iteration number is set at 300. When the pheromone weight parameter e is 0.2, q_0 is 0.3. The other parameter settings are shown in Table 3.

Table 3. Parameters of the model.

Parameter	Value	Unit	Parameter	Value	Unit
Q	10	t	∂_2	8	L/h
P_1	800	CNY/vehicle	ω	2.7	kg/L
P_2	7.2	CNY/L	ε_1	0.003	—
P_3	97.57	CNY/t	ε_2	0.005	—
P_4	10,000	CNY/t	∂_1	0.04	—
λ	0.86	—	μ_1	50	CNY/h
∂_1	3.5	L/h	μ_2	100	CNY/h

The distribution center employs a specific brand of medium to large refrigerated trucks for cold chain delivery, with a fixed usage cost of CNY 1000 per vehicle and no limitation on the number of delivery vehicles. The total weight of the delivery vehicles is 16,540 kg, with an empty weight of 5580 kg and a maximum load capacity of 10 tons. As the delivery is scheduled from 3:00 a.m. to 7:00 a.m., the temperature is set to be 20 °C outside and 5 °C inside the cabin. The service time windows indicate that the required service periods for each supermarket are relatively early, thus avoiding peak traffic hours. Therefore, the study overlooked the impact of traffic congestion on the operating speed of delivery vehicles. For the speed of refrigerated trucks, a speed matrix following the triangular probability distribution of Formula (3) was generated. The most likely speed was randomly selected within the interval [30, 90]. In the formula, $v_d = 30$, $v_m = 60$, and $v_u = 90$, where the speed unit is km/h.

Parameters of the brand's large refrigerated trucks were obtained from the literature. The estimated fuel consumption during driving is 0.186 L/km with no load and 0.475 L/km when fully loaded. The windward area is approximately 6.8×2.4 m^2, with an air density of 1.29 kg/m^3 and an air resistance coefficient of about 0.3. The average carbon price in the national trading market is CNY 97.57/ton of carbon dioxide [71], and the diesel unit price is CNY 7.2/L.

5.2. Model Calculation Results and Analysis

Simulation experiments were conducted using the traditional, improved, and hybrid ant colony algorithms. Based on the data presented in Tables 2 and 3, the comparative results are shown in Table 4. In the simulation, 100 sets of data were randomly generated. All position points were randomly distributed within a square area, ensuring the wide spatial distribution of the test data. Based on these data, the performance and efficiency of the delivery algorithm were verified across various scenarios. The comprehensive costs and overall satisfaction values on each set of data (a total of five sets) are shown in Table 5. The average values of the algorithms are compared in Table 6.

Table 4. Comparison of solutions from the algorithms.

Algorithm	Optimal Path	Total Cost (CNY)	Customer Satisfaction
Traditional Ant Colony Algorithm	0-20-7-18-14-11-0	2291.66	6
	0-24-23-5-10-17-0	2702.05	6
	0-2-28-15-12-1-0	1815.58	6
	0-21-13-27-29-19-0	2585.18	5
	0-4-22-6-16-8-0	1414.25	6
	0-25-9-26-3-0	1645.58	6
	Total Cost/Total Satisfaction	12,454.30	35
Improved Ant Colony Algorithm	0-24-20-23-18-10-8-11-0	2830.65	5
	0-25-7-5-14-17-0	2036.38	8
	0-2-22-13-15-12-19-0	2665.45	7
	0-21-28-26-27-29-1-0	2464.82	6
	0-4-9-6-16-3-0	2239.74	7
	Total Cost/Total Satisfaction	12,237.04	33
Hybrid Ant Colony Algorithm	0-29-20-24-11-8-0	1826.36	6
	0-23-28-27-26-18-16-7-0	2610.68	6
	0-2-13-15-12-1-0	2635.5	8
	0-6-3-10-21-22-4-19-0	1865.52	7
	0-25-5-9-14-17-0	1955.59	8
	Total Cost/Total Satisfaction	10,893.65	35

The results of the simulation experiments indicate that: (1) the hybrid algorithm proposed in this paper, which combines the improved ant colony algorithm with the 3-opt algorithm, solves the optimal delivery path that requires five vehicles. This is consistent with the improved ant colony algorithm. In comparison, the traditional ant colony algorithm needs six vehicles to fulfill the delivery requirements. (2) From the cost-benefit analysis, the average total cost of the hybrid ant colony algorithm is lower than that of the traditional ant colony algorithm and the improved ant colony algorithm, achieving nearly 8.159% and 4.622% cost savings, respectively. In terms of customer satisfaction, the hybrid algorithm is similar to the traditional ant colony algorithm and is superior to the improved ant colony algorithm. Although the improved ant colony algorithm has a lower average total cost than the traditional algorithm, its customer satisfaction is not optimal. (3) The improvement in customer satisfaction obtained by the hybrid algorithm is modest.

However, the average total cost is significantly diminished. This reduction in operating costs benefits the development of enterprises, decreases the consumption of environmental resources, and mitigates carbon pollution. It practices the concepts of green and low-carbon logistics and sustainable development and responds to the inherent requirement and focus of creating new productivity to promote high-quality development.

Table 5. Test results of the algorithms.

Dataset	Algorithm	Total Cost (CNY)	Customer Satisfaction	Running Time/S
1	Traditional Ant Colony Algorithm	12,554.42	35.2	13.6
	Improved Ant Colony Algorithm	12,346.56	33.2	11.8
	Hybrid Ant Colony Algorithm	10,996.38	35.2	9.6
2	Traditional Ant Colony Algorithm	12,638.93	34.8	12.6
	Improved Ant Colony Algorithm	12,425.83	32.8	11.2
	Hybrid Ant Colony Algorithm	11,253.27	34.8	9.4
3	Traditional Ant Colony Algorithm	12,416.54	33.8	11.9
	Improved Ant Colony Algorithm	12,345.18	31.8	10.3
	Hybrid Ant Colony Algorithm	11,022.83	33.8	8.9
4	Traditional Ant Colony Algorithm	12,857.36	36.5	14.3
	Improved Ant Colony Algorithm	12,426.93	34.5	12.6
	Hybrid Ant Colony Algorithm	10,982.46	36.8	10.5
5	Traditional Ant Colony Algorithm	12,345.33	34.2	12.5
	Improved Ant Colony Algorithm	12,057.82	32.2	10.8
	Hybrid Ant Colony Algorithm	10,698.17	35.1	8.4

Table 6. Comparison of average values obtained from the algorithms.

Algorithm	Average Total Cost (CNY)	Average Total Satisfaction
Traditional Ant Colony Algorithm	12,544.48	34.92
Improved Ant Colony Algorithm	12,306.56	32.92
Hybrid Ant Colony Algorithm	10,974.46	35.12

6. Conclusions

With the promotion of energy conservation and emission reduction in the logistics industry, the VRP under low-carbon conditions has become a hot topic. This paper delves into the optimization of fresh agricultural product delivery routes considering carbon emissions and customer satisfaction. An optimization model is constructed incorporating factors such as the fixed vehicle cost, transportation cost, product damage cost, refrigeration cost, carbon emission cost, and penalty costs. The heuristic function, probability transfer function, and pheromone update strategy of the traditional ant colony algorithm are improved. The 3-opt local optimization strategy is integrated into the algorithm to design a hybrid ant colony algorithm. The construction and simulation research of a low-carbon cold chain logistics system for fresh agricultural products can achieve low-carbon development while solving the long-standing problems in the cold chain logistics of fresh agricultural products, obtaining a synergistic effect where 1 + 1 > 2. This, in turn, can better promote rural economic development, increase economic income in the fresh produce industry, alleviate ecological pressure, and achieve sustainable development. It aligns with the national low-carbon industry development plan and the requirements for agricultural modernization, holding significant theoretical and practical importance. The practicality of this approach is further verified through case studies. Similarly, other scholars, such as Mario Fargnoli et al., have optimized maintenance operations and the supply of critical spare parts using the SLCM method, demonstrating the effectiveness of integrating circular economy principles with supply chain management and balancing economic benefits, customer value, and environmental performance [72]. The primary conclusions are as follows.

Firstly, given the perishable nature of fresh agricultural products in cold chain logistics and the increasing demand for low-carbon and green logistics, the penalty cost of freshness reduction and the cost of carbon emissions are incorporated into fixed costs, fuel costs, and penalty costs of time windows, building an optimization model of vehicle distribution paths in cold chain logistics to minimize the overall cost. This approach enhances the model's practical relevance, ensuring it better aligns with real-world logistics needs.

Secondly, this paper proposes a cold chain delivery optimization model for fresh agricultural products, considering costs, customer satisfaction, and carbon emissions. Under the environmental pressure of contemporary society, enterprises of cold chain logistics must strike a balance between lowering logistics costs (choosing the shortest logistics paths) and fulfilling their social responsibility to protect the environment and ensure customer satisfaction. They should minimize carbon emissions generated during cold chain distribution, reduce environmental costs, establish a good corporate image, win customer trust, enhance social recognition, and achieve the dual goals of energy conservation and emission reduction while improving customer satisfaction.

Thirdly, this paper redesigns the heuristic information related to time window and temperature requirements in cold chain logistics for fresh agricultural products, adjusts the pheromone update and selection strategies in the traditional ant colony algorithm, and integrates the 3-opt algorithm as a local search strategy to improve the quality of solutions. Through these improvements, a hybrid ant colony algorithm that combines the improved ant colony algorithm with the 3-opt algorithm is introduced. This algorithm overcomes the limitations of individual algorithms and demonstrates superior performance in the optimization of cold chain delivery for fresh agricultural products.

Fourth, to ensure the accuracy of the case simulation results and to ensure that the test results can generally reflect the economies of scale of the delivery algorithm, this study randomly generated 100 sets of data. All location points were randomly distributed within a square area, effectively ensuring the widespread spatial distribution of the test data. Based on these data, the performance and efficiency of the proposed algorithm were verified across multiple scenarios. The scale effect is manifested in two aspects: First, the increase in data volume ensures the statistical significance of the test results. Second, the growing complexity of distribution networks fully demonstrates the algorithm's ability to handle large-scale data. This provides a basis for evaluating the algorithm's performance in practical scenarios.

Case studies verify the effectiveness of the optimization model and its solving algorithm. This paper provides scientific theoretical guidance for the optimization of cold chain logistics, which is conducive to addressing practical cold chain distribution problems of agricultural products. However, space for improvement exists. Future research could consider incorporating demand randomness to enhance the model's realism. Additionally, uncertain interference events, such as traffic congestion or vehicle malfunctions, can impact overall costs and customer satisfaction. Therefore, how to reduce the impact of these uncertainties is an important direction for future studies. Finally, further comparative analysis of algorithmic parameters is warranted, along with expanding the study to more complex scenarios, such as those involving multiple distribution centers and vehicle models. These advancements would provide more targeted, operational, and scientifically grounded guidance, promoting the sustainable development of cold chain logistics.

In the era of big data, the cold chain logistics of fresh agricultural products must also consider how to integrate cloud computing, AI, and machine learning into future research. Future research can further integrate cloud computing resources, artificial intelligence (AI), and machine learning (ML) technologies to enhance the dynamic optimization capabilities and generalization performance of the algorithms. First, leveraging the elastic computing resources of cloud computing to build a distributed optimization framework, and processing large-scale cold chain logistics network data in parallel to improve the solution efficiency of the hybrid ant colony optimization (ACO) algorithm. For example, deploying a multi-agent collaborative mechanism based on a cloud platform can process dynamic information such as traffic conditions and temperature and humidity monitoring in real-time, achieving online dynamic adjustment of delivery routes.

Second, introducing deep reinforcement learning (DRL) technology to improve the heuristic rules of traditional ACO algorithms. By training agents to autonomously explore the balance between cost and customer satisfaction under complex constraints, the algorithm's adaptability to multi-objective conflicts can be enhanced. Meanwhile, graph neural networks (GNNs) can be used to predict and model customer spatiotemporal demands, embedding the prediction results into the route optimization model to enhance the forward-looking decision-making ability of cold chain delivery.

Additionally, a federated learning-based carbon emission collaborative computing model can be established. Under the premise of ensuring data privacy, this model can integrate the carbon emission characteristics of cold chain enterprises from multiple regions to optimize the accuracy of global carbon cost estimation.

Future research needs to focus on solving key issues such as the fusion of multi-source heterogeneous data, real-time response latency of algorithms, and dynamic allocation of multi-objective weights. For example, scholars such as Danqi Cai et al. have studied dynamic path planning and real-time decision-making in Intelligent Transportation Systems (ITS), providing efficient data-driven tools that can be further extended to the field of logistics delivery route optimization [73]. By deeply integrating AI and cloud computing, a smart cold chain logistics system with self-learning capabilities can be constructed, promoting the intelligent upgrade of green supply chains.

Author Contributions: Software, X.Z. and Y.L.; conceptualization, X.Z., C.W. and Y.X.; methodology, X.Z., C.W. and Y.X.; formal analysis, Y.L., Y.L. and C.W.; investigation, X.Z. and C.W.; writing—original draft preparation, X.Z., C.W. and Y.L.; writing—review and editing, C.W., Y.L. and Y.X. All authors have read and agreed to the published version of the manuscript.

Funding: This research was funded by the Education Department of Hunan Province (Project number: 23A0541); the Hunan Provincial Natural Science Foundation (Project number: 2024JJ7121); the 14th Five-Year Plan of Educational Science in Hunan Province (Project number: XJK24CGD035); the Hunan Provincial Natural Science Foundation (Project number: 2022JJ50214), and an aid program for science and technology innovative research teams in higher educational institutions of Hunan Province.

Data Availability Statement: The original contributions presented in the study are included in the article; further inquiries can be directed to the corresponding author.

Conflicts of Interest: The authors declare that they have no known competing financial interests or personal relationships that could have appeared to influence the work reported in this paper.

References

1. Forecast and Analysis of the Scale and Structure of China's Fresh Food Retail Market in 2023. Available online: https://www.askci.com/news/chanye/20230117/0910102673917781237843 68.shtml (accessed on 12 February 2025).
2. Wang, A.Y.; Wu, Y.J. Development status, problems and countermeasures of fresh logistics: A case study of Hema Fresh Production. *World Econ. Res.* **2023**, 2, 84–91. [CrossRef]
3. Wang, S.Y.; Fan, X.Q.; Ma, X.L.; Jiang, Y.M. Three-level cold chain inventory optimization model considering commodity freshness and quantitative loss. *J. Syst. Manag.* **2019**, 2, 409–416.
4. Yu, Y.L.; Feng, Y. Supply chain decision-making of fresh agricultural products under different cold chain service models. *Chin. J. Manag. Sci.* **2019**, 9, 135–143.

5. Wang, A.; Gao, J. Impact of cold chain logistics development on consumption of fresh agricultural products: From the perspective of consumption scale and consumer price. *Bus. Econ. Res.* **2023**, *15*, 105–108.
6. Ni, W.H.; Chen, T. Research on integrated service platform of fresh agricultural products cold chain logistics based on block-chain. *Jiangsu Agric. Sci.* **2021**, *23*, 213–218.
7. Wang, L.; Hu, X.Y. Symbiotic co-evolution and evaluation of cold chain logistics and fresh agricultural products e-commerce. *J. Shanghai Mar. Univ.* **2023**, *1*, 50–56.
8. Meng, S.H.; Liu, W.Z.; Sun, Y. Spatial difference and influencing factors of green logistics efficiency of fresh agricultural products. *J. Commer. Econ.* **2024**, *19*, 75–79.
9. Yuan, Y.K.; Chen, J.M.; Dan, B. Study on cold chain logistics efficiency and its convergence under carbon constraint: A case study of fresh agricultural products. *Sci. Technol. Manag. Res.* **2020**, *14*, 260–267.
10. Zhu, Y.; Shi, X.D.; Yang, Y.; Xu, J.; Yao, G.X. Analysis on Influencing factors of low-carbon cold chain logistics Collaborative development of fresh agricultural products in Jiangsu Province: Based on interpretive structure model method. *J. Chin. Agric. Mech.* **2023**, *4*, 222–227.
11. Ning, T.; Gou, T.; Liu, X.D. Research on Cold chain logistics strategy Simulation of fresh agricultural products considering low-carbon constraints. *J. Syst. Simul.* **2022**, *4*, 141–149.
12. Yang, B.; Wang, Z.B. The impact of green technology innovation on green logistics efficiency of fresh agricultural products: A regulatory effect based on industrial agglomeration. *China Bus. Mark.* **2023**, *1*, 62–72.
13. Wan, F.J.; Qiu, Z.C. Efficiency evaluation of cold chain logistics of fresh Agricultural products based on three-stage DEA. *Sci. Technol. Ind.* **2022**, *11*, 322–326.
14. Piecyk, M.I.; Mckinnon, A.C. Forecasting the Carbon Foot-Print of Road Freight Transport in 2020. *Int. J. Prod. Econ.* **2010**, *1*, 31–42. [CrossRef]
15. Shuai, C.M.; Zhang, Y.K. Difference Analysis of Chinese Consumers Willingness to Pay for Low-Carbon Products-Scenario Experimental Data Based on Carbon Labels. *China Soft Sci.* **2013**, *7*, 61–70.
16. Dantzig, G.B.; Ramser, J.H. The truck dispatching problem. *Manag. Sci.* **1959**, *1*, 80–91. [CrossRef]
17. Solomon, M.M.; Desrosiers, J. Survey Paper–Time window constrained routing and scheduling problems. *Transp. Sci.* **1988**, *22*, 1–81. [CrossRef]
18. Jabali, O.; Leus, R.; Van Woensel, T.; De Kok, T. Self-imposed time windows in vehicle routing problems. *OR Spectrum* **2015**, *37*, 331–335. [CrossRef]
19. Shi, J.L.; Zhang, J. Research on VRP Model and algorithm for Batch Distribution with random demand Points. *Control Decis.* **2017**, *2*, 213–222.
20. Ling, H.F.; Gu, J.H. Multi-yard Open Vehicle Scheduling with soft time window. *Comput. Eng. Appl.* **2017**, *14*, 232–239.
21. Li, S.Y.; Dan, B.; Ge, X.L. Low Carbon Vehicle Routing Optimization Model and Algorithm in Multi-Channel Time-Varying Network. *Comput. Integr. Manuf. Syst.* **2019**, *2*, 454–468.
22. Ning, T.; Wang, X.P.; Hu, X.P. Interference Management Decision Model for Delivery Vehicle Scheduling under Driving Delay. *Syst. Eng. Theory Pract.* **2019**, *5*, 1236–1245.
23. Wang, M.; Zhao, L.D. Cold chain investment and pricing decisions in a fresh food supply chain. *Int. Trans. Oper. Res.* **2021**, *2*, 1074–1097. [CrossRef]
24. Ning, T.; Wang, X.P. Study on Disruption Management Strategy of Job-Shop Scheduling Problem Based on Prospect Theory. *J. Clean. Prod.* **2018**, *194*, 174–178.
25. Ning, T.; Gou, T.; Zhang, P. Disruption Management Strategy of Terminal Logistics under Accidental Travel Time Delay Based on Prospect Theory. *J. Intell. Fuzzy Syst.* **2019**, *6*, 8371–8379. [CrossRef]
26. Liu, J.L.; Ma, Z.J. Multi-depot open vehicle routing problem with time windows based on vehicle leasing and sharing. *Syst. Eng. Theory Pract.* **2013**, *3*, 666–675.
27. Zhou, R.; Shen, W.L.; Liu, M.Z.; Zhao, H. Hybrid discrete particle swarm optimization algorithm for integrated loading and unloading vehicle routing problem with time window. *China Mech. Eng.* **2016**, *4*, 494–502.
28. Wang, J.; Li, B. Dynamic management of fuzzy requirements VRPTW based on multi-objective optimization. *Chin. J. Manag.* **2013**, *2*, 238–243+279.
29. Zhang, X.N.; Fan, H.M.; Li, J.F. Chance-constrained model and algorithm for LRP with multiple fuzzy variables under change-reward. *Syst. Eng. Theory Pract.* **2016**, *2*, 442–453.
30. Amorim, P.; Parragh, S.N.; Sperandio, F.; Almada-Lobo, B. A rich vehicle routing problem dealing with perishable food: A case study. *TOP* **2012**, *22*, 489–508. [CrossRef]
31. Agi, M.A.; Faramarzi-Oghani, S.; Hazır, Ö. Game theory-based models in green supply chain management: A review of the literature. *Int. J. Prod. Res.* **2021**, *15*, 4736–4755. [CrossRef]
32. Yu, H.H.; Mu, C.C.; Cheng, L.C. Distribution planning for perishable foods in cold chains with quality concerns: Formulation and solution procedure. *Trends Food Sci. Technol.* **2017**, *61*, 80–93.

33. Byung, D.S.; Young, D.K. A vehicle routing problem of both refrigerated and general-type vehicle for perishable food products delivery. *J. Food Eng.* **2016**, *169*, 61–71.
34. Rabbani, M.; Farshbaf-Geranmayeh, A.; Haghjoo, N. Vehicle routing problem with considering multi-middle depots for perishable food delivery. *Uncert. Supply Chain Manag.* **2016**, *4*, 171–182. [CrossRef]
35. Shao, J.P.; Cao, Q.; Shen, M.Y.; Sun, Y. VRP Model and algorithm with time window in distribution of fresh agricultural products. *Ind. Eng. Manag.* **2015**, *1*, 122–127+134.
36. Zhang, Y.M.; Li, N. Research on VRP Model Optimization of Cold Chain Logistics Based on Elite Single-parent Genetic Algorithm. *Math Pract. Theory* **2016**, *4*, 87–96.
37. Tiwari, A.; Chang, P.C. A block recombination approach to solve green vehicle routing problem. *Int. J. Prod. Econ.* **2015**, *164*, 379–387. [CrossRef]
38. Kumar, R.S.; Kondapaneni, K.; Dixit, V.; Goswami, A.; Thakur, L.; Tiwari, M. Multi-objective modeling of production and pollution routing problem with time window: A self-learning particle swarm optimization approach. *Comput. Ind. Eng.* **2016**, *99*, 29–40. [CrossRef]
39. Ge, X.L.; Miao, G.Q.; Tan, B.C. Research on optimization modeling and algorithm of open pollution path problem. *Ind. Eng. Manag.* **2015**, *4*, 46–53.
40. Tang, J.H.; Ji, S.F.; Shen, G. Vehicle Routing optimization considering Carbon emissions in Time-varying networks. *Syst. Eng.* **2015**, *9*, 37–44.
41. Ma, Q.Z.; Wang, J.; Song, H.Q. Low carbon VRP with multiple vehicles in a small urban area: A case study of the regional collection network of Zhuhai Express Company. *J. Ind. Eng. Eng. Manag.* **2016**, *4*, 153–159.
42. Wang, Z.Y.; Lu, J.Y. Optimization of Vehicle Distribution Path in Cold Chain Logistics considering low carbon. *Sci. Technol. Manag. Res.* **2017**, *17*, 228–232.
43. Kang, K.; Han, J.; Pu, W.; Ma, Y.F. Research on low-carbon Distribution Path Optimization of Cold chain Logistics of Fresh Agricultural Products. *Comput. Eng. Appl.* **2019**, *2*, 259–265.
44. Ma, Q.Z.; Wang, J.; Song, H.Q. Small-Area Low-Carbon Multi-Vehicle Low-Carbon VRP in Urban Areas: A Case Study of Zhuhai Express Co, Ltd.'s Regional Receiving Network. *J. Manag. Eng.* **2016**, *4*, 153–159.
45. Weizhen, R.; Chun, J.; Xinhua, W.; Feng, L. Low Carbon VRP Problem Model and Solution Strategy Considering Road Slope Factor. *Syst. Eng. Theory Pract.* **2014**, *8*, 2092–2105. (In Chinese)
46. Chen, Z.; Jiang, Z.J.; Liu, Y. Optimization of Low Carbon Logistics Paths in Different Links Based on Improved Ant Colony Algorithm. *Eco. Econ.* **2019**, *12*, 53–59+66.
47. Ye, Y.; Zhang, H.Z. Wolf Pack Algorithm for Multi-Distribution Center Vehicle Routing Problem. *App. Res. Comput.* **2017**, *9*, 2590–2593.
48. Yan, R.; Zhu, X.N.; Zhang, Q.; Qi, Y.Y.; Lin, Y.Z. Research on Model and Algorithm of Multi-depot Vehicle Routing Problem with Time Window Considering Two-dimensional Bin Packing Constraint. *Chin. J. Manag. Sci.* **2017**, *7*, 67–77.
49. Huber, S.; Geiger, M.J. Order matters—A variable neighborhood search for the swap-body vehicle routing problem. *Eur. J. Oper. Res.* **2017**, *2*, 419–445. [CrossRef]
50. Camm, J.D.; Magazine, M.J.; Kuppusamy, S.; Marin, K. The demand weighted vehicle routing problem. *Eur. J. Oper. Res.* **2017**, *1*, 151–162. [CrossRef]
51. Tu, H.N.; Xu, X. Research on Material Distribution Path Optimization Based on Improved ant Colony Algorithm. *Mach. Des. Manuf.* **2017**, *8*, 265–268.
52. Xia, Y.K.; Fu, Z. Distribution Vehicle Routing Problem in Supermarket Chains with Soft Time Windows. *Inf. Control* **2018**, *5*, 599–605.
53. Franceschette, A.; Honhon, D.; Woensel, T.V.; Bektaş, T.; Laporte, G. The Time-Depend Pollution-Routing Problem. *Transp. Res. Part B Methodol.* **2013**, *56*, 265–293. [CrossRef]
54. Xiao, Y.; Konak, A. The Heterogeneous Green Vehicle Routing and Scheduling Problem with Time-Varying Traffic Congestion. *Transp. Res. Part E Logist. Transp. Rev.* **2016**, *88*, 146–166. [CrossRef]
55. Wen, L.; Catay, B.; Eglese, R. Finding a Minimum Cost Path Between a Pair of Nodes in a Time-Varying Road Network with a Congestion Charge. *Eur. J. Oper. Res.* **2014**, *236*, 915–923. [CrossRef]
56. Houming, F.; Pengcheng, L.; Jiaxin, W. VRP and Hybrid Variable Neighborhood Search Algorithm for Random Simultaneous Collection of Goods with Demand for Goods Collection. *Syst. Eng. Theory Pract.* **2019**, *10*, 2646–2659.
57. Fan, H.M.; Wu, J.X.; Geng, J.; Li, Y. Vehicle Routing Problem with Fuzzy Demand and Time Widow and Hybrid Genetic Algorithm Solution. *J. Syst. Manag.* **2020**, *1*, 107–118.
58. Weizhen, R.; Zhongfei, D.; Bingcheng, W.; Yan, Y. Research on the Theoretical Boundary of Distance and Energy Saving of Cooperative Vehicle Routing Problem. *J. Syst. Manag.* **2019**, *4*, 697–707.
59. Bao, H.-F.; Fang, J.; Zhang, J.-S.; Wang, C.-S. Low-Carbon Cold Chain Distribution Route Optimization Based on an Improved Ant Colony Algorithm. *J. Syst. Simul.* **2024**, *36*, 183–194.

60. Fang, W.T.; Ai, S.Z.; Wang, Q.; Fan, J.B. Optimization of cold chain logistics distribution route based on hybrid ant colony algorithm. *China Manag. Sci.* **2019**, *11*, 107–115.
61. Zhao, Z.X.; Li, X.M.; Zhou, X.C.; Liu, C.S. GVRP Research on Cold Logistics City Distribution Considering Traffic Congestion. *Comput. Eng. Appl.* **2020**, *1*, 224–231.
62. Fan, Y.; Kleuver, C.D.; Leeuw, S.D.; Behdani, B. Trading off cost, emission, and quality in cold chain design: A simulation approach. *Comput. Ind. Eng.* **2021**, *158*, 107442. [CrossRef]
63. Wang, L.; Dan, B. Research on fresh agricultural product preservation strategy considering quality and quantity loss control. *Chin. J. Manag. Sci.* **2023**, *8*, 100–110.
64. Jing, X.; Yu, Z.; Dai Panqian, Y.G. Optimization research on cold chain logistics process of fresh agricultural foods. *J. Chin. Agric. Mech.* **2024**, *2*, 115–121.
65. Zhang, R.; Pang, M.H.; Liu, B. Route optimization and effect test of combined vehicle and UAV distribution of fresh Agricultural Products. *J. Henan Agric. Univ.* **2024**, *1*, 96–105.
66. Li, W.X.; Zhang, C.M.; Ma, C.X. An optimization model and solution algorithm of multi-objective vehicle path under low carbon conditions. *J. Transp. Inf. Saf.* **2020**, *1*, 118–126+144.
67. Kwon, Y.J.; Choi, Y.J.; Lee, D.H. Heterogeneous fixed fleet vehicle routing considering carbon emission. *Transp. Res. Part D Transp. Environ.* **2013**, *23*, 81–89. [CrossRef]
68. Deng, M.J.; Li, Y.H.; Xu, B.; Wu, L.Q. Distribution Path optimization of pharmaceutical cold chain logistics considering carbon emissions and customer satisfaction. *J. Railway Sci. Eng.* **2024**. [CrossRef]
69. Deng, H.X.; Zhou, J.; Hu, Y. Optimization model of the cold chain logistics distribution path of fresh agricultural products considering carbon emission. *J. Chongqing Uni. Technol (Nat. Sci.)* **2023**, *2*, 289–297.
70. Li, L.; Fan, X.J. Comparative Study on Pricing Strategies of Fresh Agricultural Product Supply Chains Dominated by Retailers. *Chin. J. Manag. Sci.* **2015**, *12*, 113–123.
71. National Carbon Emission Trading Market Clearing and Settlement Information Daily. Available online: http://www.tanjiaoyi.com/article-81675-1.html (accessed on 12 February 2025).
72. Fargnoli, M.; Haber, N.; Tronci, M. Case Study Research to Foster the Optimization of Supply Chain Management through the PSS Approach. *Sustainability* **2022**, *14*, 2235. [CrossRef]
73. Cai, D.; Chen, K.; Lin, Z.; Li, D.; Zhou, T.; Ling, Y.; Leung, M.-F. JointSTNet: Joint Pre-Training for Spatial-Temporal Traffic Forecasting. *IEEE Trans. Consum. Electron.* **2024**, *70*, 1234–1245. [CrossRef]

processes

MDPI

Article

A Comprehensive Optimization Framework for Diesel Filtration in Nuclear Emergency Systems: Integrating Genetic Algorithms, State-Space Networks, and Computational Fluid Dynamics

Lanqi Zhang [1], Yupan He [1], Yong Zhou [2], Guoying Jiang [2], Xiangnan Chu [3], Qi Ma [3], Fengyi Liu [4] and Haotian Ye [4,*]

1 China Nuclear Power Operations Co., Ltd., Shenzhen 518026, China; zhanglanqi@cgnpc.com.cn (L.Z.); heyupan2023@163.com (Y.H.)
2 Daya Bay Nuclear Power Operations and Management Co., Ltd., Shenzhen 518124, China; zhouyong@cgnpc.com.cn (Y.Z.); jiangguoying@cgnpc.com.cn (G.J.)
3 Yangjiang Nuclear Power Co., Ltd., Yangjiang 529941, China; chu_xiangnan@126.com (X.C.); maqicgn@sina.com (Q.M.)
4 School of Chemical Engineering, Dalian University of Technology, Dalian 116024, China; 2543594304@dlut.edu.cn
* Correspondence: htye@dlut.edu.cn

Abstract: Aging diesel fuel in emergency storage tanks at nuclear power plants requires regular filtration to remove impurities. Due to the stringent safety requirements of nuclear power plants, high standards are set for operational timeliness and reliability. This study proposes an efficient optimization framework that integrates MATLAB-based genetic algorithm optimization, state-space networks (SSNs), and computational fluid dynamics (CFD). A FLUENT simulation model is used to simulate the internal flow field of the diesel storage tank, while the SSN comprehensively analyzes flow distribution and filtration path strategies. Global parameter optimization is achieved using a genetic algorithm (GA). The framework improves filtration efficiency by 7.9%, reducing filtration time from 464,000 s to 417,600 s. Water impurity levels decreased by 22.2% (from 0.00045 to 0.00035), and mechanical impurities decreased by 33.3% (from 0.000015 to 0.000010). The findings not only enhance the reliability of emergency diesel generator systems and support nuclear safety operations but also provide a solid foundation for further innovations in emergency fuel filtration technologies.

Keywords: genetic algorithm; state-space network; CFD simulation; optimization; diesel filtration

1. Introduction

Nuclear power plants play a pivotal role in the global energy system, offering significant power generation capacity and low carbon emissions [1]. These attributes make them uniquely advantageous in addressing climate change and reducing dependence on fossil fuels. However, the operation of nuclear power plants faces far stricter safety requirements compared to other energy sources, as any potential risks could have severe impacts on the surrounding environment and public health. Ensuring the safe and reliable operation of nuclear power plants is therefore a critical task, relying on a robust set of emergency management measures. Among these, the emergency diesel storage system serves as a core component of the plant's emergency response framework [2].

 https://doi.org/10.3390/pr13030648

The emergency diesel storage system provides backup power during unexpected power outages or equipment failures, ensuring the normal operation of critical equipment, particularly reactor cooling systems. This function is vital for maintaining the safety and stability of the entire nuclear power plant. However, due to the prolonged storage requirements of emergency diesel, its quality is prone to degradation. Over time, diesel can absorb moisture from the air, leading to increased water content, while tank material oxidation and equipment wear may introduce mechanical impurities such as fine particles, oxides, and other solid contaminants [3]. The accumulation of these impurities and moisture not only reduces diesel combustion efficiency but also causes wear on transport systems, filtration devices, and the injection systems of diesel generators, increasing the risk of equipment failure and reducing system reliability and service life [4]. Regular inspection and filtration of stored emergency diesel to remove water and impurities are therefore critical. This process ensures that the diesel meets usage standards, reduces operational costs and safety risks for nuclear power plants, and extends the service life of emergency equipment, thereby providing a reliable foundation for the long-term stable operation of nuclear power plants.

Traditional filtration methods often lack systematic optimization strategies, leading to prolonged filtration times, reduced overall system efficiency, and increased safety risks. To enhance the efficiency and safety of diesel filtration, it is crucial to adopt optimization strategies based on the principles of process systems engineering. By systematically optimizing diesel storage and filtration strategies, it is possible to significantly improve operational efficiency and system safety, ultimately supporting the long-term stable operation of nuclear power plants [5].

Filtration time, as a critical performance indicator, has seen significant advancements in recent years with the maturation of CFD-based fluid flow optimization and the application of machine learning algorithms for filtration time prediction and optimization. These approaches demonstrate substantial potential in enhancing filtration efficiency. However, achieving rational design of filtration paths in complex systems remains a challenge that requires further in-depth research [6,7].

In process design and optimization, selecting the appropriate superstructure representation method is crucial for achieving efficient and reliable designs [8]. In recent years, with the emergence of various optimization methods, researchers have proposed a variety of superstructure representation approaches to address the challenges of designing complex chemical processes. Each method has its advantages and disadvantages, making them suitable for different optimization tasks and design needs.

As shown in Table 1, The Generalized Modular Representation Framework (GMF) method simplifies the superstructure generation process by identifying favorable structural alternatives without listing all possible configurations. The modular design concept of GMF provides high adaptability and flexibility. However, when dealing with complex systems, GMF still requires significant computational resources and has a complex mathematical model [9,10]. The Phenomena Building Block (PBB) method explores a wide design space by combining various chemical and physical phenomena, allowing for the exploration of multiple equipment combinations. However, a major challenge for PBB is how to select an appropriate optimization objective function. Due to its involvement with innovative equipment combinations, PBB lacks standardized design specifications, and there may be operational and reliability issues in the application of designs [11]. The unit, port, conditioning stream representation (UPCS) method simplifies flow paths and reduces redundancies, offering a clearer superstructure representation that effectively streamlines the design process while avoiding unnecessary bypass flows. However, the design space of UPCS is still somewhat limited, particularly when dealing with highly complex multi-task systems, where it may not cover all necessary design options, potentially overlooking some optimization solutions [5]. The building block-based superstructure (BBS) method, using a two-dimensional grid structure for control volume representation, is particularly suited for new equipment design and the integration of non-traditional processes, offering a high degree of design flexibility. However, the high flexibility of BBS leads to more complex mathematical issues in the solving process, particularly for large-scale design problems, which may require substantial computational resources. While the two-dimensional grid structure of BBS is intuitive, balancing the breadth of the design space, the accuracy of physical properties, and computational feasibility remains a challenge in practical applications [12]. The state-task network (STN) method graphically represents material states and task nodes, providing a clear display of the relationships between tasks and materials, which is ideal for multi-task scheduling and equipment selection problems. STN supports variable task-equipment (VTE) allocation, providing greater flexibility [13]. However, STN can become computationally complex when dealing with large-scale systems, potentially leading to excessive solving times, especially in complex designs involving multiple tasks and equipment selections, where interactions between tasks are hard to capture, impacting optimization efficiency.

Table 1. Comparative analysis of frameworks.

Framework	Strengths	Weaknesses
Generalized Modular Representation Framework (GMF)	- Simplifies superstructure generation by identifying favorable structural alternatives - High adaptability and flexibility due to modular design	- Requires significant computational resources for complex systems - Complex mathematical models
Phenomena Building Blocks (PBBs)	- Explores a wide design space by combining various chemical and physical phenomena - Enables exploration of multiple equipment combinations	- Difficulty in selecting appropriate optimization objective functions - Lacks standardized design specifications, leading to potential operational and reliability issues
Unit, Port, Conditioning Stream (UPCS)	- Simplifies flow paths and reduces redundancies - Provides a clear superstructure representation, streamlining the design process	- Limited design space for highly complex multi-task systems - May overlook some necessary design options and optimization solutions
Building Block-Based Superstructure (BBS)	- Uses a two-dimensional grid for control volume representation, suited for new equipment design and non-traditional process integration - Offers high design flexibility	- Increased mathematical complexity for large-scale design problems - Balancing design space breadth, physical property accuracy, and computational feasibility is challenging

The state-space network (SSN) method is particularly suitable for optimizing nuclear emergency diesel filtration systems due to its efficiency, flexibility, and ease of implementation. SSN divides the system into task nodes and material state nodes, using a directed graph to intuitively represent the relationships between tasks and material flow paths. Compared to traditional methods, SSN has lower computational complexity, significantly reducing solving times, especially in parallel task scheduling and resource optimization [14]. Its graphical structure allows designers to intuitively understand the system and flexibly adjust the design, thereby improving the operability of the design process. By integrating it with genetic algorithms (GAs), SSN further enhances the optimization framework, creating a comprehensive and efficient solution for optimizing emergency diesel filtration systems.

A GA, as a global optimization tool, helps to find the optimal filtration time [15], while SSN provides a structured state-task network to support parallel task execution and resource optimization, thus improving the overall efficiency and stability of the system.

Compared to other frameworks, the state-space network (SSN) method offers significant advantages in terms of computational efficiency, flexibility, and ease of implementation. Unlike GMF and BBS, which often involve high computational complexity, SSN simplifies the optimization process through its structured task-node and state-node design, reducing computational demands while maintaining robust optimization capabilities. Additionally, compared to PBB, SSN provides a more intuitive and operable graphical structure without relying on complex physical modeling, and it surpasses UPCS by offering broader design flexibility, making it particularly suitable for optimizing complex multi-task systems. However, SSN also has potential limitations in highly complex systems with numerous states. As the number of states and transitions increases, computational complexity and resource demands can grow significantly, posing scalability challenges and leading to longer optimization times. Moreover, accurately defining states and mapping transitions in highly dynamic systems can be time-consuming and error-prone. Simplifying states to manage complexity may result in the loss of critical details, potentially affecting the accuracy and reliability of optimization results. These considerations highlight the importance of balancing system complexity and feasibility when applying the SSN method, ensuring its strengths are effectively leveraged while addressing its limitations.

Ansys Fluent is a high-performance CFD software that models mass, momentum, and energy transfer in multiphase flow using the control volume method [16]. In this model, Fick's law or turbulent diffusion models are used to describe component diffusion, convection, and chemical reactions between the gas phase, liquid phase, and solid phase. Through the Euler–Lagrange coupling method, Fluent can accurately capture mass exchange and mass transfer effects between phases, thereby predicting component distribution and reaction kinetics in complex flow fields [17]. This method is widely used in chemical, energy, and environmental engineering, providing precise numerical analysis tools for optimizing designs and reaction processes.

In recent years, significant progress has been made in the application of computational fluid dynamics (CFD) for optimizing diesel filtration systems. These studies have focused on improving the efficiency and performance of filtration processes through advanced modeling techniques and simulation tools. For instance, CFD has been widely employed to analyze flow dynamics, pressure drop, and particulate matter behavior within filter structures, offering valuable insights into the optimization of filtration designs. Additionally, recent research highlights the integration of CFD with optimization algorithms, such as machine learning and genetic algorithms, to enhance filtration efficiency while minimizing energy consumption. These advancements underscore the growing importance of CFD-based methods in addressing the challenges of balancing filtration time, energy efficiency, and separation quality. Building upon this foundation, our study aims to further explore and refine CFD applications in diesel filtration to achieve both improved performance and compliance with stringent safety standards [18–20].

The core contribution of this study is the development of an innovative optimization framework focused on minimizing the filtration time of emergency diesel systems. This framework integrates the SSN and GA, optimizing complex processes through precise scheduling while ensuring global optimization, significantly improving system efficiency and reliability. The framework leverages the flexible modeling capability of SSN to effectively optimize emergency diesel filtration strategies, considering material concentration variations and multiple monitoring targets. At the same time, CFD simulations further enhance the accuracy and stability of the filtration process. The CFD model acts as a key function within the genetic algorithm [21], providing filtration time feedback based on the input of the population. It accurately calculates the distribution of substances within the diesel-storage-tank flow field, thus providing strong numerical support for system optimization. Combined with SSN's parallel task scheduling and resource optimization, a GA further performs global optimization of filtration time, ensuring the optimal solution is found under various constraints. This innovative optimization approach not only reduces filtration time but also enhances the operational efficiency of the nuclear emergency diesel system. Moreover, it significantly improves the system's safety and reliability, ultimately enhancing the overall safety and long-term stability of nuclear power plants.

The structure of this paper is organized as follows: Section 2 introduces the hybrid modeling methodology for diesel filtration optimization, covering the framework of SSN, Genetic Algorithm (GA), and CFD, with detailed descriptions of the SSN superstructure, CFD numerical simulation setup, filtration path design, and GA design; Section 3 presents the result analysis, including the optimization analysis of the genetic algorithm and the state-space superstructure analysis; Section 4 concludes with the key findings of the study.

2. Hybrid Modeling Methodology for Diesel Filtration Optimization

2.1. Framework of Hybrid Modeling Using SSN, GA, and CFD

This study proposes a hybrid modeling strategy for the aging diesel filtration process in diesel storage tanks and establishes a systematic framework, as illustrated in the flowchart shown in Figure 1. The methodology consists of three core modules: the SSN design module, which comprehensively designs and analyzes all possible pathways in the filtration process; the GA prediction module, which iteratively evolves the optimal filtration pathway until reaching the predetermined target state; and the CFD simulation module, which captures the distribution of various impurities within the diesel storage tank during the filtration process. Notably, the computational units established in the CFD model not only adhere to the constraints defined by the SSN design but also form the population for GA optimization, thereby establishing an information exchange bridge among the three modules. This creates a tightly coupled collaborative optimization framework [22].

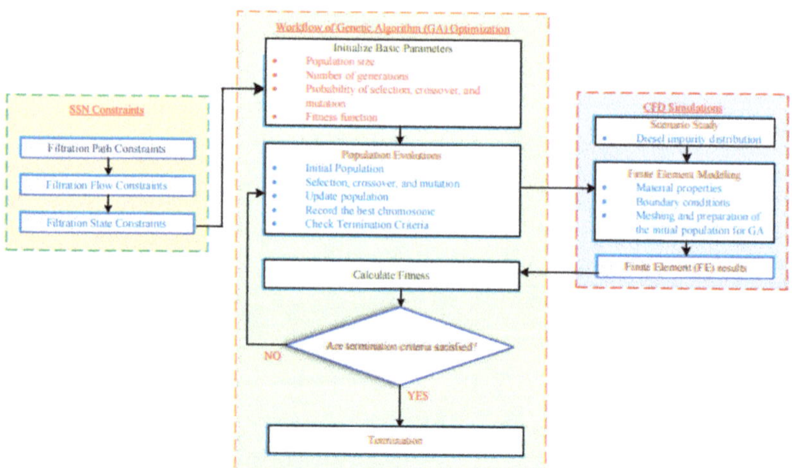

Figure 1. Flowchart of the proposed hybrid modeling strategy.

2.2. SSN Superstructure

When applying the SSN method, it is crucial to balance system complexity and feasibility to ensure its strengths are effectively leveraged while addressing its limitations. The research employs an SSN methodology centered around the following three key components [17]: (1) system solution space construction and definition, (2) mathematical modeling of the problem, and (3) solution algorithm for path optimization. These three parts form the core framework for designing the filtration process path.

To achieve this, the filtration path design in this study is based on the following assumptions:

1. The filtration path is designed as static and does not dynamically adjust with time or changes in tank contamination levels in order to maintain operational stability and simplicity.
2. The key factors affecting diesel degradation are assumed to be an excessive mass fraction of water and mechanical impurities. The filtration process primarily targets these two indicators to ensure the long-term stability and purity of the diesel.
3. Different pollutants require specific filtration devices for treatment, and the filtration processes do not interfere with one another, ensuring the most effective removal of each pollutant.
4. The filtration process is assumed to be ideal, with all pollutants being completely removed by the filtration devices, and the potential impact of flow rate on the filtration process is ignored to simplify the modeling. This assumption is made because the primary objective of this study is to optimize filtration paths and equipment configurations to improve overall system efficiency, rather than to refine the dynamic behavior within the filtration devices. Therefore, treating the flow rate as constant simplifies the analysis, reduces unnecessary complexity, and allows the focus to remain on path optimization as the core issue.
5. Pollutants are assumed to be uniformly distributed throughout the diesel in the tank, with no large-scale sedimentation, ensuring consistency in the filtration process and accuracy in predictions.
6. The feed conditions for the filtration system (such as diesel temperature, pressure, etc.) are assumed to be constant during the design process, simplifying the analysis and modeling of filtration efficiency.
7. The maximum flow rate of the pump used for transport is assumed to be fixed, ensuring that the total inlet flow is always less than the pump's maximum flow rate to prevent overloading and ensure normal pump operation.
8. Each interface of the diesel storage tank is assumed to serve as either an inlet or an outlet, but not both at the same time, ensuring stable operations and clear flow direction.
9. At least one set of inlets and outlets, along with corresponding filtration devices, is guaranteed to exist to ensure the smooth progression of the filtration process.

This study adopts a simplified path design in the SSN model to achieve a stable and efficient filtration solution in the filtration system. The adopted path design assumes relatively simple filtration conditions, allowing for the optimization of paths and equipment configuration. The filtration path defines three ports of the diesel storage tank as free inlets and outlets and identifies three different paths: a water impurity filter, a mechanical impurity filter, and a regular pipeline without any filtration device. As shown in Figure 2, the SSN diagram clearly illustrates the entire flow path and processing sequence.

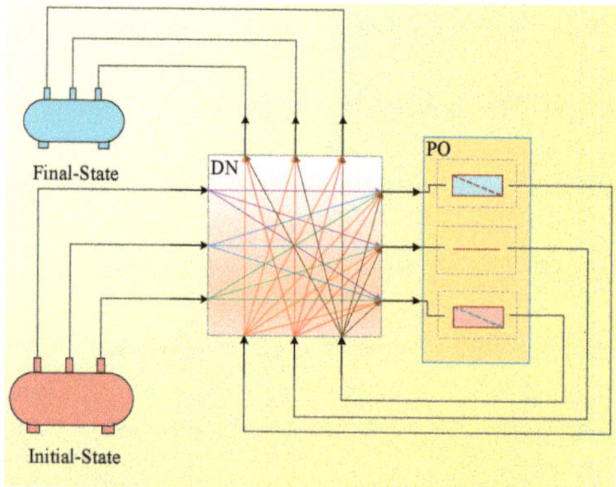

Figure 2. SSN superstructure for diesel filtration pathway.

2.3. *CFD Numerical Simulation Setup*

2.3.1. Physical Model

As shown in Figure 3, this study adopts a scaled physical model of an emergency diesel storage tank from a nuclear power plant, using the actual tank as a prototype. The internal complex components are omitted, and the main structure of the tank is retained. A 1:125 scaled physical model is created using SolidWorks 2021 3D modeling software. The model is centered at the midpoint of the length, width, and height, with three detection points designated as A, B, and C.

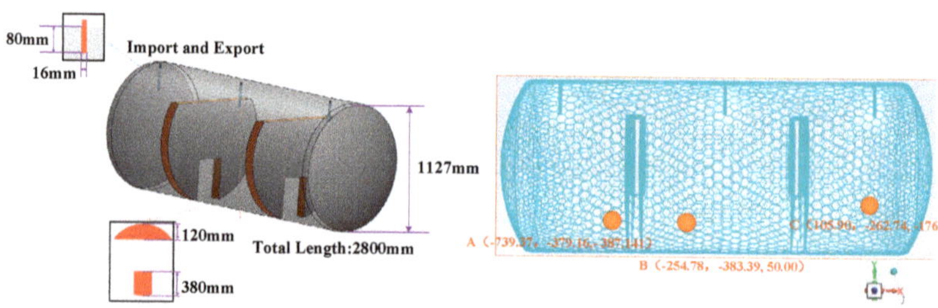

Figure 3. Physical model of diesel storage tank and monitoring point locations.

The model consists of three top-mounted, equal-diameter openings, which can function as either inlets or outlets, a sealed tank body, and two internal baffles. The material of the tank and baffles is entirely made of aluminum alloy. Except for the baffles, the remaining interior space is a blank fluid zone that can accommodate either stored diesel or flowing fluids.

This study focuses on detecting, filtering, and purifying two key impurities in aged diesel: water contaminants and mechanical impurities. The volume fractions of these impurities are compared with standard diesel to ensure that the quality parameters meet the required specifications. As shown in Figure 4, The three-dimensional physical model of the diesel storage tank was meshed using Fluent-Mesh 2022R1 software, with a polyhedral mesh filling the entire structure. To improve computational accuracy, the mesh was refined in the inlet and outlet regions. A total of 124,936 polyhedral meshes were generated, ensuring good mesh quality for the simulation.

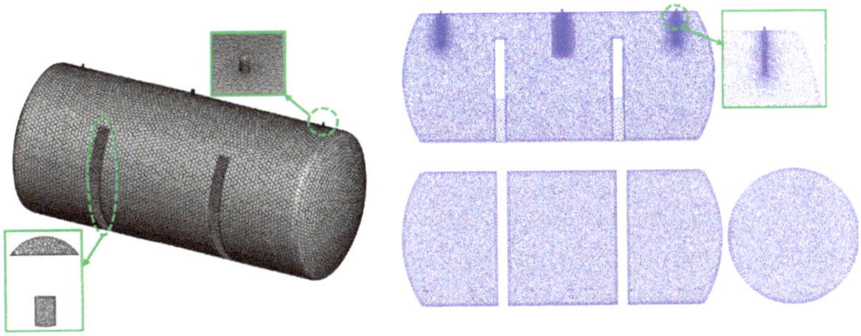

Figure 4. Mesh division of diesel storage tank—external and three cross-sectional diagram views.

2.3.2. Boundary Condition Setup

In this study, a case file was used with a multiphase flow Eulerian model as the primary model, selecting diesel as the main phase, while water impurities and mechanical impurities were treated as the secondary and third phases, respectively. Gypsum powder was used to represent the mechanical impurities. The 'k-epsilon & Realizable' viscous model was applied to simulate the diesel flow process [23]. The k-epsilon model is renowned for its robustness and computational efficiency, making it highly suitable for simulating complex flow phenomena in industrial scenarios. It achieves an excellent balance between accuracy and computational cost, making it particularly effective in capturing the large-scale turbulence and fluid behavior relevant to this study. Moreover, the k-epsilon model has been widely validated in similar applications, such as multiphase flow and impurity distribution in pipelines, providing reliable results to support the optimization of filtration systems [24].

The three inlets were defined as velocity inlets, with velocity and multiphase flow parameters for diesel and impurities sourced from external inputs. If the velocity is greater than zero, it is considered a normal inlet, while a velocity less than zero indicates a diesel outlet. Initialization and local initialization were performed to ensure that, at the initial state, the water impurities and mechanical impurities were evenly distributed within the diesel, with equal volume fractions of impurities in different parts of the tank.

Additionally, this case involves a user-defined function (UDF) that monitors the volume fractions of water impurities and mechanical impurities at the three detection points to check if they meet the stopping criteria (less than the volume fraction threshold). When the concentration at all detection points falls below the threshold, a global variable 'interruptflag = 1' is set to stop the computation and output the final flow time. Otherwise, the calculation continues.

2.4. Filtration Path Design

In this study, the combination of SSNs and GAs is utilized to design an optimized filtration path by achieving the best allocation of state spaces. The following settings are applied to each individual in the population:

(a) Parameter Grouping: Each individual in the population consists of three sets of parameters, each containing three parameters, making a total of nine parameters. These are denoted as A, B, and C, with the subparameters A1, B1, and C1, as well as A2, B2, and C2.

(b) Parameters for Inlet Speeds: The parameters A, B, and C correspond to the velocity at the three inlets, measured in meters per second (m/s). The corresponding equations are as follows:

$$\begin{cases} A+B+C=0 \\ |A|+|B|+|C| \leq 2PM \end{cases} \tag{1}$$

When any of the parameters is positive, the corresponding port is defined as an inlet, with the inlet velocity equal to the given value; when the parameter is negative, the port is an outlet, with the outlet velocity equal to the absolute value of the negative parameter. If the velocity is zero, the port is treated as a wall with no fluid flow. Additionally, PM represents the maximum pumping capacity of the diesel delivery pump. Since the pipe diameter is fixed, PM refers to the inlet velocity (in m/s). This setup ensures flexibility in both flow direction and velocity, allowing for dynamic adjustments to optimize the filtration path and system efficiency.

(c) A1, B1, and C1, as well as A2, B2, and C2, represent the connection parameters for water impurities and mechanical impurities. The constraint for these parameters is given by the following:

$$\begin{cases} sign(A_1) = sign(A_2) = sign(A) \\ sign(B_1) = sign(B_2) = sign(B) \\ sign(C_1) = sign(C_2) = sign(C) \end{cases} \tag{2}$$

where "sign" denotes the sign function, indicating that each port can only have one state as either an inlet or an outlet, ensuring the correct execution of the filtration process. For each inlet/outlet state, the filtration state must also be considered. Each of the A1, B1, and C1 subparameters, as well as A2, B2, and C2, is subject to filtration state constraint equations, which are denoted as A11, B11, and C11, and A21, B21, and C21. Taking A11 as an example, its constraint equation is as follows:

$$A_{11} = \begin{cases} -1, A_1 < 0 \\ 0, A_1 = 0 \\ 1 \text{ else } 2,\ A_1 > 0 \end{cases} \tag{3}$$

(d) A11 refers to the water impurity filtration state at Port 1, which can take four states (−1, 0, 1, or 2) depending on the sign of A1:

- A state of −1 indicates that the port is a diesel outlet, and the system will record relevant parameters.
- A state of 0 indicates that the port is a wall with no flow in or out.
- A state of 1 indicates that the port is a diesel inlet, where diesel, after being completely filtered for water, is reintroduced, meaning the transport path includes a filter.
- A state of 2 indicates that the port is a diesel inlet, where diesel is directly recycled back into the storage tank without filtration, and its volume fraction is the same as the −1 state, meaning the transport path does not include a filter.
- B11 and C11 refer to the water impurity filtration states for Ports 2 and 3, with the same logic as A11.
- A21, B21, and C21 are the constraint equations for the mechanical impurity filtration states, with the same logic as A11, B11, and C11.

We provide a numerical example to clarify the parameter constraints and boundary conditions applied in this study. Assuming a set of parameters [0.007, 0.001, −0.008, 1, 1, −1, 1, 2, −1] corresponding to A, B, C, A1, B1, C1, A2, B2, and C2, the state can be described as follows: Port 1 and Port 2 act as inlets, with inlet velocities of 0.007 and

0.001, respectively; Port 3 serves as an outlet, with an outlet velocity of 0.008. For water impurities, all impurities are filtered through Port 1. For mechanical impurities, those entering through Port 1 are filtered, while those entering through Port 2 bypass filtration and circulate directly.

2.5. Genetic Algorithm Design

The GA in this study incorporates selection, crossover, mutation, generation gap, and elitism strategies to optimize the design and configuration of the emergency diesel filtration system [25–27].

The selection operation adopts the Roulette Wheel Selection method. The fitness ratio of each individual is calculated first, as shown in the following formula:

$$p_i = \frac{f_i}{\sum\limits_{j=1}^{N} f_i}, i = 1, 2, \ldots, N \tag{4}$$

where p_i is the selection probability of each individual, f_i is the fitness of the i-th individual, and N is the number of individuals in the population. The cumulative probability is C_i which is then constructed as follows:

$$c_i = \sum_{j=1}^{i} p_i, c_0 = 0, c_N = 1 \tag{5}$$

A random number $rk{\sim}U$ (0, 1) is generated, and the i-th individual is selected if the following condition is true:

$$c_{i-1} < r_k < c_i \tag{6}$$

As shown in Figure 5, The crossover operation adopts the Double-Point Crossover strategy. The idea is to randomly choose two crossover points and exchange the corresponding segments of the parent genes to generate two new offspring individuals. The specific steps are as follows:

1. Select Crossover Points: Randomly choose two different crossover points $q1$ and $q2$, ensuring $q1 < q2$.
2. Generate Offspring: The parent genes are divided into three parts:
   - The part outside the crossover points $q1$ and $q2$ is retained;
   - The part between $q1$ and $q2$ is exchanged;
   - The two parts are then combined to generate the new offspring.

The mutation operation introduces random perturbations in the genes while combining dynamic mutation probability adjustment. The mutation function disturbs certain genes of an individual to maintain the diversity of the population, preventing the algorithm from falling into local optima. Additionally, the mutation probability is gradually decreased as the number of iterations increases to improve stability and convergence speed. To avoid premature convergence in the search process, a generation gap strategy is employed. This strategy accelerates the updating speed of individuals in the population to enhance the search process and prevent the population from becoming stuck in local optima. To ensure that the best individuals are passed on to the next generation, the elitism strategy is applied. In each generation, the most optimal individual is directly transferred to the next generation without undergoing selection, crossover, or mutation, ensuring that the optimal solution is not lost [28,29].

Figure 5. Mesh illustration of crossover and mutation.

2.6. Case Study Setup

As shown in the CFD modeling, this case study is based on the emergency diesel storage tank of a nuclear power company, focusing on the design of the diesel filtration process. The diesel pump used is the 65YHC3-25 model (The manufacturer is Botou Hengyun Oil Pump Manufacturing Factory, located in Botou City, China), with the parameters provided in Table 2. The pump has a rated flow capacity of 25 m^3/h, and the actual diameter of the tank inlet is 800 mm. After comprehensive consideration, the maximum flow capacity of the diesel pump is set as PM = 0.014 m/s. The key parameters for standard diesel are provided in Table 3.

Table 2. Diesel pump parameters.

Model	Flow Rate Q (m^3/h)	Working Pressure P (MPa)	Speed n (r/min)	Efficiency η (%)	Allowable Suction Height (m)	Motor Power (kW)
65YHC3-25	25	0.6	950	65	5	5.5

Table 3. Quality specifications.

Item	Quality Specification
Color/Grade	3.5
Sulfur Content/(mg/kg)	≤350
Acid Value/(mg KOH/g)	≤0.18
Kinematic Viscosity (20 °C)/(mm^2/s)	3.0–8.0

We obtained a degraded diesel sample with water and mechanical impurity volume fractions of 0.00045 and 0.000015, respectively. The threshold volume fractions for water and mechanical impurities at the three monitoring points are set to 0.00035 and 0.000010, respectively. This means that when the volume fractions of water and mechanical impurities at all three monitoring points are below 0.00035 and 0.000010, the diesel in the tank is considered to meet the required quality standards. The initial impurity distribution in the diesel tank is shown in Figure 6 below. To ensure proper analysis, the specific diesel tank configuration and impurity levels used in this study are based on past engineering projects and represent the preliminary internal requirements of the industry. These parameters reflect typical operational conditions, ensuring the relevance and practicality of the proposed framework.

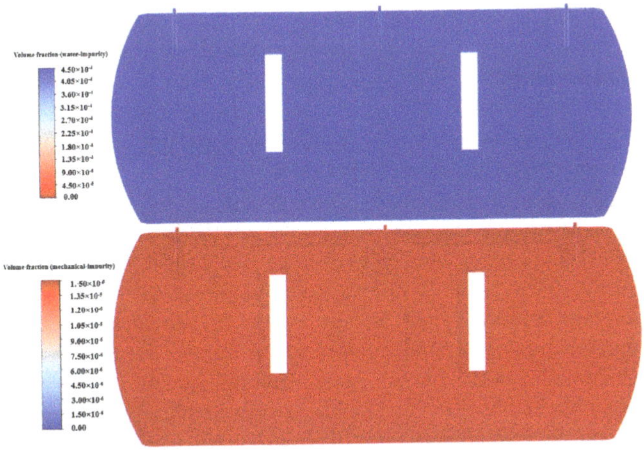

Figure 6. The initial states of water impurities (**top** image) and mechanical impurities (**bottom** image) in the diesel tank.

3. Result Analysis

3.1. Genetic Algorithm Optimization Analysis

The optimization of filtration process parameters was conducted on a PC equipped with an Intel Core i7-13700 processor (16 cores, maximum frequency 5.20 GHz) and 16.0 GB of memory. The population size was set to 20, with a time step of 3600 s. After nine generations of genetic algorithm optimization, the following conclusions were drawn.

As shown in Figure 7, the optimal objective value plot illustrates the optimization process across generations. In the initial optimization phase (from generation 1 to generation 3), the initial optimal objective value was 4.64×10^5, indicating that the population contained some quality solutions at the start. During this stage, the objective value remained stable, suggesting that the algorithm was conducting a broad global search without significantly improving the optimal solution.

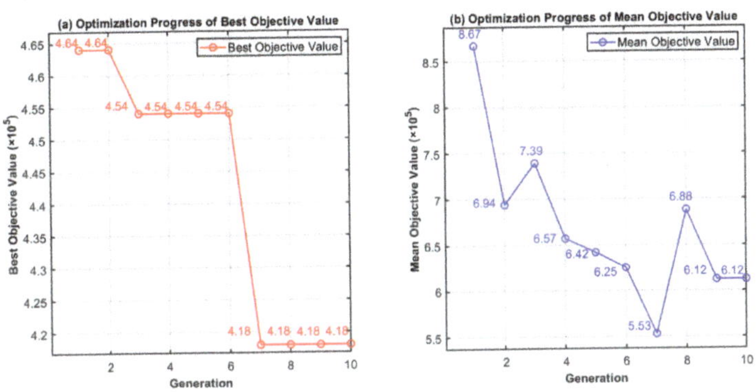

Figure 7. Optimization process of objective values across generations ((**a**) optimal objective values; (**b**) average objective values).

In the mid-optimization phase (from generation 4 to generation 7), the optimal objective value reached 4.54×10^5. At this point, the algorithm entered the local development phase and began to refine the solution quality. By generation 7, the optimal objective value rapidly decreased to 4.18×10^5, marking a significant improvement in the solution.

The reduction from 4.54×10^5 to 4.18×10^5 represents an optimization improvement of approximately 7.9%, demonstrating that the algorithm effectively enhanced the optimal solution. During the convergence phase (from generation 8 to generation 9), the optimal objective value remained stable at 4.18×10^5, indicating that the optimization potential had been fully exploited. The algorithm exhibited excellent convergence and robustness, confirming its effectiveness in solving complex optimization problems.

The convergence analysis of the average objective value shows the trend of mean values across generations, reflecting the dynamic improvement of the overall population's performance. In the initial optimization phase (from generation 1 to generation 3), the mean objective value in generation 1 was 8.67×10^5, which then quickly decreased to 7.39×10^5 in generation 3. This stage indicates the algorithm's strong global exploration capability, quickly identifying high-quality regions in the solution space. The mean reduction from generation 1 to generation 3 was 14.8%, demonstrating significant quality improvement through global search. In the mid-optimization phase (from generation 4 to generation 7), the mean value steadily decreased from 6.42×10^5 to 5.53×10^5, with a reduction of 13.9%. This phase indicates that the algorithm refined the population solutions through local development while enhancing solution diversity. In the oscillation and stabilization phase (from generation 8 to generation 9), the mean value increased to 6.88×10^5 in generation 8, possibly due to an increase in solution diversity during local development. By generation 9, the mean value decreased to 6.12×10^5 and stabilized, indicating that the algorithm gradually converged and that the optimization potential had been fully explored.

In Figure 8, the boxplot presents the distribution of objective values across different generations, highlighting the overall quality change trends and the dispersion characteristics of the population's solutions during the optimization process.

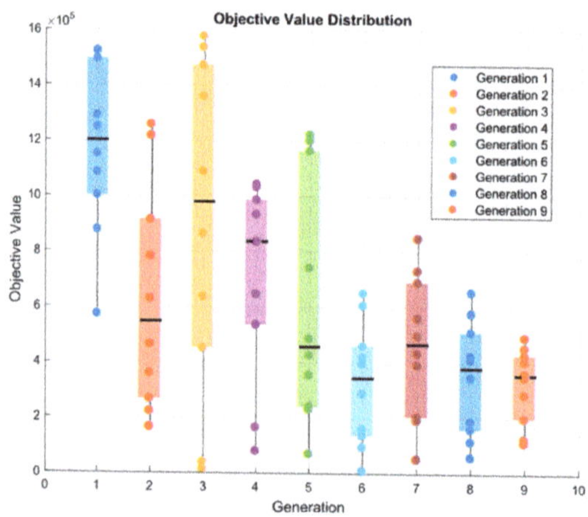

Figure 8. Objective value distribution trend.

In the initial generation (generation 1), the distribution of objective values is relatively wide, with a median of approximately 0.8×10^6 and a large interquartile range, indicating significant variability in the quality of the initial population, while still containing some high-quality solutions. As the optimization progresses, particularly in the mid-phase (generations 5 to 9), the distribution range of objective values narrows significantly, with the height of the boxplot decreasing and the median stabilizing around 0.6×10^6. This indicates that during the local development phase, the algorithm effectively improves the overall quality of solutions while reducing fluctuations. In the early optimization phase (generations 1 to 3), several high-objective-value outliers appear, reflecting the algorithm's extensive global search for potential high-quality solutions. In the mid-phase (generations 4 to 7), lower-objective-value outliers emerge, suggesting that the algorithm strengthens the development of potential high-quality solutions. By the later stages (generations 8 to 9), the number of outliers significantly decreases, and their distribution becomes more concentrated, indicating that the optimization potential has been fully exploited and the population's solution quality has stabilized. Overall, the algorithm demonstrates a balanced and effective optimization process, transitioning from global exploration in the early stage to efficient local development in the mid-stage, and robust convergence in the later stage. In the initial phase, the wide distribution of objective values, ranging from 0.6×10^6 to 2.2×10^6, highlights strong global search capabilities. As optimization progresses, solutions gradually concentrate on lower-objective-value regions while maintaining diversity to avoid local optima. In the later stages, solutions become highly concentrated around 0.6×10^6, with significantly increased distribution density, showcasing the algorithm's powerful convergence capability and stability. Furthermore, the transition in scatter point colors from deep blue in the early generations to orange-red in the later generations visually illustrates the evolution of the population's solutions, from broad distribution to concentrated regions of high-quality solutions, while maintaining a certain sparsity in the lower-objective-value region to ensure the robustness and diversity of the optimization results. In conclusion, the algorithm achieves a well-balanced optimization process, progressing from broad exploration to efficient local development and finally to robust convergence, significantly improving the quality of solutions and overall optimization performance.

3.2. State-Space Superstructure Analysis

The optimal individual obtained is [0.007, 0.007, −0.014, 1, 1, −1, 1, 1, −1], with an optimal objective value of 417,600. Based on this optimal solution, the following conclusions can be drawn: For inlet 1 and inlet 2, the filtration status for both water impurities and mechanical impurities is 1, indicating that the diesel at these inlets undergoes filtration and is returned. For inlet 3, the filtration status for both water impurities and mechanical impurities is −1, indicating that it serves as an outlet and does not undergo filtration. Consequently, the optimal filtration structure can be derived, forming an SSN, as shown in Figure 9.

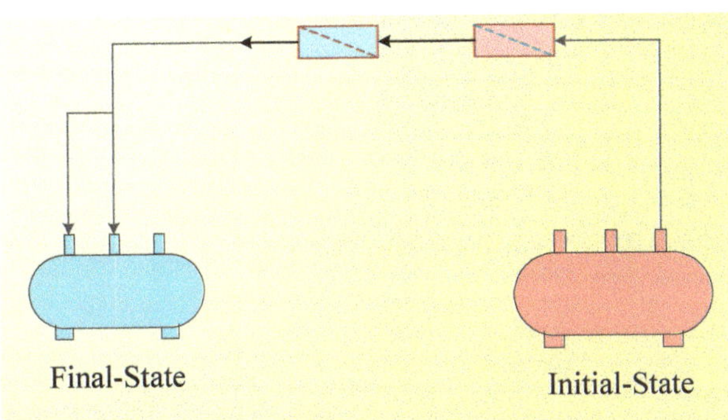

Figure 9. Optimal filtration structure of SSN network.

Under the optimal allocation of flow rate and filtration path, the CFD modulation process is illustrated in the figure. The filtration process at 24 h, 72 h, and the final filtration state are analyzed, as shown in Figures 10–12.

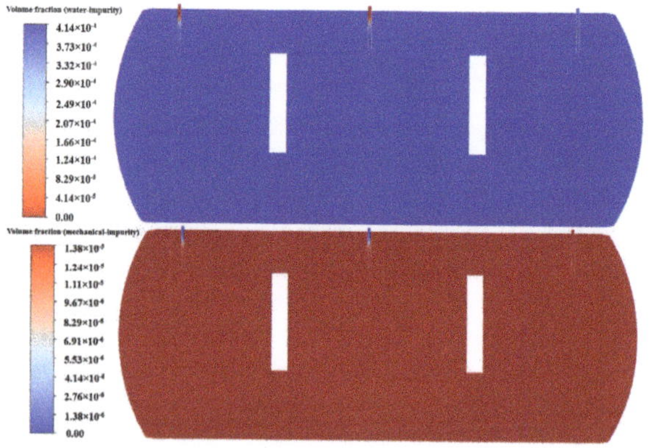

Figure 10. Filtration state at 24 h (**top**: volume fraction of water impurities; **bottom**: volume fraction of mechanical impurities).

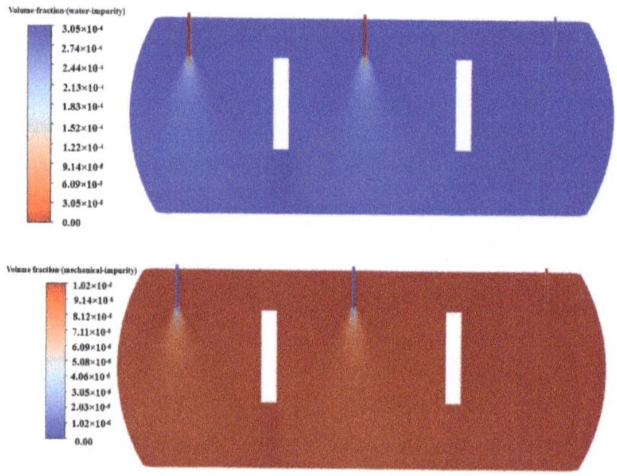

Figure 11. Filtration state at 72 h (**top**: volume fraction of water impurities; **bottom**: volume fraction of mechanical impurities).

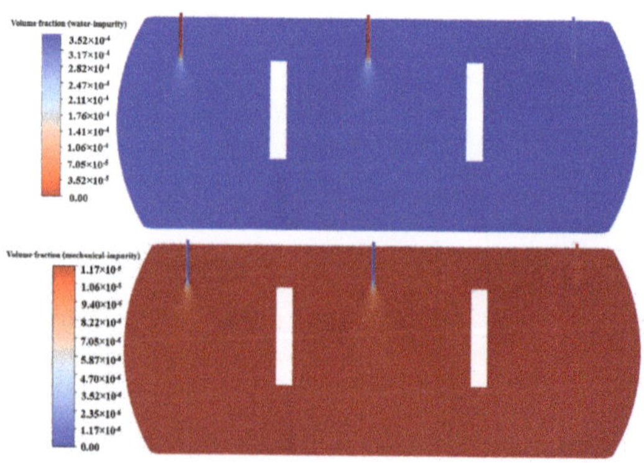

Figure 12. Final filtration state (**top**: volume fraction of water impurities; **bottom**: volume fraction of mechanical impurities).

4. Conclusions

This study presents an innovative optimization framework that integrates state-space networks (SSNs), genetic algorithms (GAs), and computational fluid dynamics (CFD) simulations for optimizing multi-inlet filtration paths in emergency diesel storage tanks of nuclear power plants. The proposed framework enhances both global exploration and local exploitation by dynamically adjusting crossover and mutation probabilities within the GA. The constructed SSN superstructure effectively captures all possible filtration paths, and its performance was validated through CFD simulations. The results demonstrate that, through the optimal configuration, filtration time and impurity volume fractions were significantly reduced, meeting the stringent quality requirements for emergency diesel in nuclear power plants.

The optimal solution obtained through the GA was [0.007, 0.007, −0.014, 1, 1, −1, 1, 1, −1], with an objective value of 417,600 s. CFD simulations under optimal conditions showed a progressive decrease in water and mechanical impurity volume fractions, ultimately achieving values of 0.00035 and 0.000010, respectively, at all monitoring points, thus meeting the quality thresholds. During the GA optimization, the objective value decreased from an initial 4.64×10^5 seconds to 4.18×10^5 seconds, representing a 7.9% improvement. The proposed optimization framework achieves a 7.9% reduction in filtration time, which is particularly significant during the maintenance process of nuclear power emergency diesel storage tanks. This maintenance involves taking the system offline to restore the quality of aged diesel to normal operating levels, with filtration being a critical step that directly impacts the maintenance duration. By reducing filtration time, the framework effectively minimizes system downtime, allowing the system to return to operational readiness more quickly. It also enhances the reliability and resilience of the emergency diesel system while meeting the stringent operational standards of the nuclear power industry. Additionally, the average objective value continuously decreased throughout the optimization, while population diversity was effectively maintained, preventing premature convergence. Future research could explore real-time filtration optimization under dynamic environments or extend the framework to multi-component complex mixtures, addressing broader industrial challenges. The proposed methodology, while focused on nuclear emergency diesel filtration systems, holds potential for broader applications in various industrial contexts. For instance, it could be extended to optimize filtration systems in industrial chemical processing, where impurity control is crucial for maintaining product quality and system efficiency. Additionally, the methodology could be adapted for use in renewable energy storage systems, ensuring the long-term stability and reliability of critical components. Another promising direction is its integration into real-time adaptive control systems for dynamic operational environments, allowing for continuous optimization based on changing system conditions. These potential applications demonstrate the versatility of the methodology and provide a strong foundation for future research and development.

Author Contributions: Conceptualization, L.Z., Y.H. and Q.M.; methodology, G.J., X.C., F.L. and H.Y.; validation, L.Z. and Y.Z.; investigation, G.J. and X.C.; resources, L.Z. and Y.Z.; data curation, Q.M. and Y.H.; writing—original draft preparation, G.J.; writing—review and editing, L.Z. and Y.Z.; visualization, Y.H.; supervision, Y.H.; project administration, Q.M. All authors have read and agreed to the published version of the manuscript.

Funding: This research was supported by the National Natural Science Foundation of China, 52204208.

Data Availability Statement: The original contributions presented in the study are included in the article; further inquiries can be directed to the corresponding author.

Conflicts of Interest: Authors Lanqi Zhang, Yupan He were employed by China Nuclear Power Operations Co., Ltd. Authors Yong Zhou, Guoying Jiang were employed by Daya Bay Nuclear Power Operations and Management Co., Ltd. Authors Xiangnan Chu, Qi Ma were employed by Yangjiang Nuclear Power Co., Ltd. The remaining authors declare that the research was conducted in the absence of any commercial or financial relationships that could be construed as a potential conflict of interest.

References

1. Mathew, M.D. Nuclear Energy: A Pathway towards Mitigation of Global Warming. *Prog. Nucl. Energy* **2022**, *143*, 104080. [CrossRef]
2. Marqusee, J.; Jenket, D. Reliability of Emergency and Standby Diesel Generators: Impact on Energy Resiliency Solutions. *Appl. Energy* **2020**, *268*, 114918. [CrossRef]
3. Groysman, A. *Corrosion in Systems for Storage and Transportation of Petroleum Products and Biofuels: Identification, Monitoring and Solutions*; Springer: Dordrecht, The Netherlands, 2014; ISBN 9789400778849.
4. Baskov, V.; Ignatov, A.; Polotnyanschikov, V. Assessing the Influence of Operating Factors on the Properties of Engine Oil and the Environmental Safety of Internal Combustion Engine. In Proceedings of the Transportation Research Procedia; Elsevier B.V.: Amsterdam, The Netherlands, 2020; Volume 50, pp. 37–43.
5. Mencarelli, L.; Chen, Q.; Pagot, A.; Grossmann, I.E. A Review on Superstructure Optimization Approaches in Process System Engineering. *Comput. Chem. Eng.* **2020**, *136*, 106808. [CrossRef]
6. Davoodi, S.; Al-Rubaii, M.; Wood, D.A.; Al-Shargabi, M.; Mehrad, M.; Rukavishnikov, V.S. Hybrid Machine-Learning Model for Accurate Prediction of Filtration Volume in Water-Based Drilling Fluids. *Appl. Sci.* **2024**, *14*, 35. [CrossRef]
7. Sawant, Y.; Admuthe, L.; Prof, A. Validating Filtration Efficiency of Nonwoven Air Filters through CFD Analysis. *Grenze Int. J. Eng. Technol.* **2024**, *10*, 2992–2997.
8. Yang, D.; Zhou, W.; Liu, L.; Zhang, L.; Du, J. Hydrogen Network Synthesis Integrated with Multi-Stage and Multi-Technology Purification System. *Processes* **2024**, *12*, 2415. [CrossRef]
9. Tian, Y.; Pistikopoulos, E.N. Generalized Modular Representation Framework for the Synthesis of Extractive Separation Systems. In *Computer Aided Chemical Engineering*; Elsevier B.V.: Amsterdam, The Netherlands, 2019; Volume 47, pp. 475–480.
10. Papalexandri, K.P.; Pistikopoulos, E.N. Generalized Modular Representation Framework for Process Synthesis. *AIChE J.* **1996**, *42*, 1010–1032. [CrossRef]
11. Kuhlmann, H.; Skiborowski, M. Synthesis of Intensified Processes from a Superstructure of Phenomena Building Blocks. In *Computer Aided Chemical Engineering*; Elsevier B.V.: Amsterdam, The Netherlands, 2016; Volume 38, pp. 697–702, ISBN 9780444634283.
12. Li, J.; Demirel, S.E.; Hasan, M.M.F. Building Block-Based Synthesis and Intensification of Work-Heat Exchanger Networks (WHENS). *Processes* **2019**, *7*, 23. [CrossRef]
13. Ye, H.; Zou, X.; Zhu, W.; Yang, Y.; Dong, H.; Bi, M. Synthesis and Optimization of Reaction-Separation-Recycle Systems with Complex Distillation Sequences. *Ind. Eng. Chem. Res.* **2020**, *59*, 10509–10530. [CrossRef]
14. Majozi, T.; Zhu, X.X. A Novel Continuous-Time MILP Formulation for Multipurpose Batch Plants. 1. Short-Term Scheduling. *Ind. Eng. Chem. Res.* **2001**, *40*, 5935–5949. [CrossRef]
15. Yi, Y.K.; Anis, M.; Jang, K.; Kim, Y.J. Application of Machine Learning (ML) and Genetic Algorithm (GA) to Optimize Window Wing Wall Design for Natural Ventilation. *J. Build. Eng.* **2023**, *68*, 106218. [CrossRef]
16. Li, S.J.; Zhu, L.T.; Zhang, X.B.; Luo, Z.H. Recent Advances in CFD Simulations of Multiphase Flow Processes with Phase Change. *Ind. Eng. Chem. Res.* **2023**, *62*, 10729–10786. [CrossRef]
17. González-Lavín, G.; García-Merino, B.; Fernández-Maza, C.; Bringas, E.; Gómez-Coma, L.; Fallanza, M.; Ortiz, I. Tailored Euler-Lagrange Modelling of Microfluidic Solid/Liquid Reactive Separations. *Chem. Eng. J.* **2024**, *495*, 153393. [CrossRef]
18. Bao, G.; He, C.; Zi, T.; Li, J.; Liu, X. Analysis of Gas Flow and PM Movement Characteristics inside Diesel Particulate Filter Channel. *Process Saf. Environ. Prot.* **2024**, *186*, 1134–1148. [CrossRef]
19. Di Sarli, V.; Di Benedetto, A. Using CFD Simulation as a Tool to Identify Optimal Operating Conditions for Regeneration of a Catalytic Diesel Particulate Filter. *Appl. Sci.* **2019**, *9*, 3453. [CrossRef]
20. Ramanan, C.J.; Bora, B.J.; Alom, N.; Buradi, A.; Shukla, S. Investigation of Dead-End Filtration of Crude Biodiesel Using Reverse Osmosis Membrane Through CFD Simulation. *Int. J. Energy A Clean Environ.* **2022**, *23*, 49–66. [CrossRef]
21. Chen, Q.; Ji, H.; Zhu, Y.; Yang, X. Proposal for Optimization of Spool Valve Flow Force Based on the MATLAB-AMESim-FLUENT Joint Simulation Method. *IEEE Access* **2018**, *6*, 33148–33158. [CrossRef]
22. Shi, S.; Jiang, B.; Ludwig, S.; Xu, L.; Wang, H.; Huang, Y.; Yan, F. Optimization for Pipeline Corrosion Sensor Placement in Oil-Water Two-Phase Flow Using CFD Simulations and Genetic Algorithm. *Sensors* **2023**, *23*, 7379. [CrossRef] [PubMed]
23. Zhang, Y.; Li, G.; Bai, S.; Sun, K.; Wang, G.; Jia, Y.; Fang, Z. Optimization of a Diesel Injector Nozzle. *Fluid Dyn. Mater. Process.* **2023**, *19*, 2933–2951. [CrossRef]
24. Kim, T.; Sohn, I.; Kim, M. Cost-Effective Framework for Design Optimization Using Metamodel and Evolutionary Algorithm: With a Case Study of Bio-Filtration System. *J. Mech. Sci. Technol.* **2024**, *38*, 3941–3952. [CrossRef]
25. Cui, L.; Zhang, J.; He, D.; Pu, L.; Peng, B.; Ping, X. A Method for Optimizing Production Layer Regrouping Based on a Genetic Algorithm. *Processes* **2024**, *12*, 1881. [CrossRef]
26. Zhang, R.; Zhang, G.; Li, G.; Li, J.; Zhang, N. Optimization Algorithm and Genetic Coding Method for an Oilfield Development Plan Considering Production Constraints. *Processes* **2023**, *11*, 3386. [CrossRef]

7. Xu, Z.; Ning, X.; Li, R.; Wan, X.; Zhao, C. Configuration Optimization of a Shell-and-Tube Heat Exchanger with Segmental Baffles Based on Combination of NSGAII and MOPSO Embedded Grouping Cooperative Coevolution Strategy. *Processes* **2023**, *11*, 3094. [CrossRef]
8. Katoch, S.; Chauhan, S.S.; Kumar, V. A Review on Genetic Algorithm: Past, Present, and Future. *Multimed. Tools Appl.* **2021**, *80*, 8091–8126. [CrossRef]
9. Liu, T.; Ye, X.; Cheng, L.; Hu, Y.; Guo, D.; Huang, B.; Li, Y.; Su, J. Intelligent Pressure Monitoring Method of BP Neural Network Optimized by Genetic Algorithm: A Case Study of X Well Area in Yinggehai Basin. *Processes* **2024**, *12*, 2439. [CrossRef]

processes

MDPI

Article

Deep Neural Network Model Based on Process Mechanism Applied to Predictive Control of Distillation Processes

Zirun Wang [1,2], Hao Wang [1,2] and Zengzhi Du [1,2,*]

1 Center for Process Simulation & Optimization, Beijing University of Chemical Technology, Beijing 100029, China; 2022200161@buct.edu.cn (Z.W.); 2023200157@buct.edu.cn (H.W.)
2 College of Chemical Engineering, Beijing University of Chemical Technology, Beijing 100029, China
* Correspondence: duzz@mail.buct.edu.cn

Abstract: In modern process industries, precise process modeling plays a vital role in intelligent manufacturing. Nevertheless, both mechanistic and data-driven modeling methods have their own limitations. To address the shortcomings of these two modeling methods, we propose a neural network model based on process mechanism knowledge, aiming to enhance the prediction accuracy and interpretability of the model. The basic structure of this neural network consists of gated recurrent units and an attention mechanism. According to the different properties of the variables to be predicted, we propose an improved neural network with a distributed structure and residual connections, which enhances the interpretability of the neural network model. We use the proposed model to conduct dynamic modeling of a benzene–toluene distillation column. The mean squared error of the trained model is 0.0015, and the error is reduced by 77.2% compared with the pure RNN-based model. To verify the prediction ability of the proposed predictive model beyond the known dataset, we apply it to the predictive control of the distillation column. In two tests, it achieves results far superior to those of the PID control.

Keywords: deep learning; process mechanism; data-driven modeling; model predictive control

1. Introduction

With the advent of Industry 4.0, the process industry has an ever-growing and urgent demand for digital twins and smart manufacturing. Aspects such as product quality monitoring, process control, fault diagnosis, and process data soft sensing are directly related to the profitability of plants, all of which require analysis and modeling of the processes. Pallavicini et al. [1] have constructed a digital twin model of a biogas purification absorption unit through an approach that involves experimental data-aided mechanistic modeling. Utilizing this model, they have obtained a substantial amount of reliable data for the evaluation and analysis of the unit's performance. This method has ensured the quantity and accuracy of the data while simultaneously reducing costs and enhancing the speed of analysis. Min et al. [2] have proposed a digital twin framework for the optimization of petrochemical production control based on the industrial Internet of Things and machine learning. They have implemented and evaluated this digital twin framework and methods in a catalytic cracking unit of a petrochemical plant. The results have demonstrated the effectiveness and value of the framework and methods in optimizing production control, leading to a reduction in production costs and an increase in production efficiency.

The classical solution involves constructing simulation models based on process mechanisms using equations grounded in physical knowledge. Mechanistic models can

 https://doi.org/10.3390/pr13030811

be distinguished based on whether they incorporate a time parameter, dividing them into steady-state models and dynamic models. Steady-state models are commonly utilized in the design and optimization of industrial processes, characterizing a stable point of the system. To predict future process data, however, all extensive quantities in the model equations must be associated with the time variable. Dynamic models constructed through such equations are more aligned with the actual industrial processes and can be effectively applied in scenarios such as product quality monitoring and process control. Tanaka et al. [3] have performed steady-state local linearization on a multistage countercurrent extraction process. They characterized the relationship between metal ions and independent operational variables using total differential equations at each stage, and they verified the reliability of this model. The influence of parameters on the process can be visually represented through charts obtained from this model, providing a theoretical foundation for dynamic modeling. Zhang et al. [4] have employed the most classical mechanistic modeling approach to conduct steady-state and dynamic modeling of an ammonia synthesis reactor and the entire process. Based on this, they have investigated control strategies for the entire process. The optimal control strategy obtained can guide plants in reducing costs and increasing efficiency, exemplifying the value and significance of mechanistic modeling.

However, due to the uncertainties inherent in chemical processes, such as the decrease in the catalytic activity of catalysts, pipeline blockages, equipment corrosion, and changes in production procedures, mechanistic models are challenging to apply over an extended period without adjustments to all units. Furthermore, the construction of mechanistic models requires complex model equations, and the modeling process itself demands considerable labor. For large-scale industrial processes, the resource expenditure is substantial, particularly when changes in the production process occur, leading to model mispatch. In such instances, remodeling becomes necessary, which is an onerous task. This results in the limitations of the application of mechanistic models in actual industrial settings.

In recent years, data-driven model construction methods have been increasingly becoming the mainstream approach in research and application. On one hand, thanks to comprehensive industrial big data systems, the collection and storage of industrial process data have become quite straightforward; on the other hand, due to the rapid development of computer hardware computing power and the widespread application of artificial intelligence neural network algorithms, the rapid and efficient analysis of large-scale data has become feasible. Gao et al. [5] proposed a hybrid modeling strategy for the gas-phase catalytic coupling synthesis of dimethyl oxalate from carbon monoxide, which integrates process mechanism and data-driven methods, and its effectiveness has been verified. Savage et al. [6] tackled the inefficiency issues of traditional surrogate modeling and superstructure optimization techniques in dealing with complex process flows such as utility and refrigeration systems, by proposing an inherited data-driven modeling and optimization framework, which has been rigorously validated for its accuracy as well as its advantages in solution efficiency and flexibility. Chen et al. [7], in response to the complex chemical kinetic processes represented by the enzymatic biodiesel production system, have developed a modeling framework that combines prior physical knowledge with data-driven approaches, and have verified its accuracy and reliability. Alauddin et al. [8] have proposed a Process Dynamics-Guided Neural Network (PDNN) model to enhance the model's generalization capability, which maintains superior performance in sparse and low-quality data environments.

In addition, numerous scholars have utilized neural networks to construct surrogate models for dynamic processes and have applied them within model predictive controllers, achieving favorable outcomes. Sharma et al. [9] investigated the tert-amyl methyl ether reactive distillation column and compared three control strategies: PID, Model Predictive

Control (MPC), and Neural Network Model Predictive Control (NMPC). The study found that the NMPC exhibited the best performance. Li et al. [10], addressing the control issues of an intensified continuous reactor, proposed and implemented two nonlinear model predictive control algorithms based on neural networks. Through simulation, it was demonstrated that the algorithm with local linearization performed better. Jian et al. [11] introduced a multi-time-scale process modeling method based on Multi-Time-Scale Recurrent Neural Networks (MTRNNs) and validated the method's effectiveness in multi-time-scale modeling through a Continuous Stirred Tank Reactor (CSTR). Rebello et al. [12] have formulated a set of guidelines for the identification of machine learning models in diverse applications within process systems engineering, which are commonly utilized for simulation or predictive purposes. They demonstrated that relying solely on test performance without considering the specific application context can be misleading. Xu et al. [13] have introduced a nonlinear identification method based on a Self-Adaptive Mechanism Regulated Long Short-Term Memory (SA-MRLSTM) network, which exhibits robustness and commendable generalization capabilities.

However, compared to classical mechanistic models, data-driven modeling methods suffer from a lack of interpretability, which makes it difficult for industrial personnel to place full trust in data-based model construction approaches. As a result, data-driven models face challenges in widespread application in actual industrial facilities. Many researchers have made attempts to fix this issue. Timur et al. [14] have explored various methods of integrating machine learning with process mechanisms and tested different models in a case study on multiphase flow estimation in petroleum production systems. Sun et al. [15] have proposed a comprehensive state-space description system and the corresponding hybrid first-principles/machine learning modeling framework, providing a theoretical foundation for intelligent manufacturing solutions in the process industry. Li et al. [16] have developed a Light Attention–Convolution–Gate Recurrent Unit (LACG) architecture to render data-driven dynamic chemical process models interpretable and have validated its efficacy. The model parameters of the LACG afford the observation of more detailed insights regarding the interactions between variables.

At present, research that combines model-driven and data-driven approaches for dynamic modeling, which holds higher application value, is quite rare. Generally, there are three methods to achieve the integration of mechanistic models and data-driven approaches [17]: The first method involves incorporating process mechanism knowledge when constructing the dataset, imposing physical constraints on input data, using parameters obtained from mechanistic models as inputs during model construction, and adding data generated by the mechanistic model to the dataset as training and test sets. The second method entails adding process mechanism constraints within the hidden layers of deep neural networks, such as modifying the hidden layer structure, incorporating additional intermediate output variables based on the mechanistic model, and altering the construction of the loss function. The third method involves imposing process mechanism constraints at the output stage, making judgments and selections of output parameters that align with chemical engineering knowledge through the experience and expertise of process engineers or the application experience of actual equipment. In our research, we focus on the first method, employing the results of mechanistic models to expand the dynamic range of the model, enhancing the generalization capability of data-driven models, and ensuring interpretability in the structure of input–output relationships.

Therefore, we have conducted research in this area and attempted to apply it to the predictive control of distillation systems. As one of the most typical unit operations in chemical engineering, distillation systems involve complex processes of mass transfer, momentum transfer, and heat transfer. The operational efficiency of distillation systems is intimately associated with the profitability of the production process. Specifically, the utility consumption of condensers and heaters affects the secondary costs of production, while the product quality, which determines the product's market value, is a direct reflection of whether the product meets standards. This is the reason why distillation systems have long been a focus of attention. In this paper, we investigate a benzene–toluene distillation column, using a process mechanism to analyze the distillation system. Based on this analysis, we utilize a neural network with a targeted architecture to construct a predictive model for the distillation column, which is capable of forecasting the operational state of the distillation column over a certain period in the future. By constructing a surrogate model for the dynamic process of the distillation system, we can effectively predict the operational state of the system, thereby reducing energy consumption and increasing profitability.

The remainder of this paper is organized as follows: Section 2 is the Methodology section, where we first introduce the fundamental structures of neural networks such as recurrent neural networks and attention mechanisms. Subsequently, we discuss the improvement strategies for neural networks based on process mechanisms, and a brief introduction of the accompanying model predictive controller is also provided in this section. Section 3 is the Application and Discussion section, where we demonstrate the effectiveness of the proposed model in constructing predictive models for distillation processes and apply the predictive model to predictive control tasks. The conclusions of this study are summarized in Section 4.

2. Methodology

2.1. Recurrent Neural Network (RNN)

Chemical process data belong to time series data, which can be characterized as continuous data in the time domain and exhibit a certain degree of autocorrelation. Consequently, the data can reflect the state of the process system as it changes over time, and there is a relationship between the data of the subsequent time step and that of the preceding time step. In chemical processes, Recurrent Neural Networks (RNNs) are naturally suited for handling such systems and have demonstrated favorable performance. For instance, the Gated Recurrent Unit (GRU) model is capable of fully exploiting the dynamic and nonlinear relationships between input and output variables and has shown good performance in applications. This model primarily consists of a reset gate and an update gate. The update gate is used to compute the update rate z_t, calculated as follows:

$$z_t = \sigma\left(W_z x_t + U_z h_t^{-1}\right) \quad (1)$$

where x_t is the input of the current time step, h_{t-1} is the hidden state of the last time unit, W_z is the input weights, and U_z is the hidden state weights. The update gate helps the model to decide how much information from the previous time step and the current time step needs to be passed on.

Similarly, the reset gate can be represented as the following formula:

$$r_t = \sigma(W_r x_t + U_r h_{t-1}) \quad (2)$$

The reset gate determines how much past information needs to be forgotten by the memory units.

The hidden state of the current time step can be computed by the following:

$$h_t = z_t \odot h_{t-1} + (1 - z_t) \odot \tilde{h}_t \quad (3)$$

Candidate variable $\tilde{h}_t$ can be written as follows:

$$\tilde{h}_t = \tanh(Wx_t + r_t \odot Uh_{t-1}) \tag{4}$$

where W is the input weights, U is the hidden state weights, and $\odot$ is the Hadamard product.

2.2. Attention Mechanism

The attention mechanism was initially proposed by researchers at Google [18] to enhance the recognition efficiency of seq2seq models when performing NLP tasks. The attention mechanism assigns weights to each input variable. The method for calculating the attention weights is as follows:

$$\text{Attention}(Q, K, V) = \text{softmax}\left(\frac{QK^T}{\sqrt{d_k}}\right)V \tag{5}$$

where Q stands for query, which is the query vector; K represents key, which is the key vector; and V denotes value, which is the vector of input parameter values. The vector produced by the aforementioned equation has been processed by the attention mechanism. The allocation ratio of attention weights can be visualized using a heatmap, as illustrated in Figure 1:

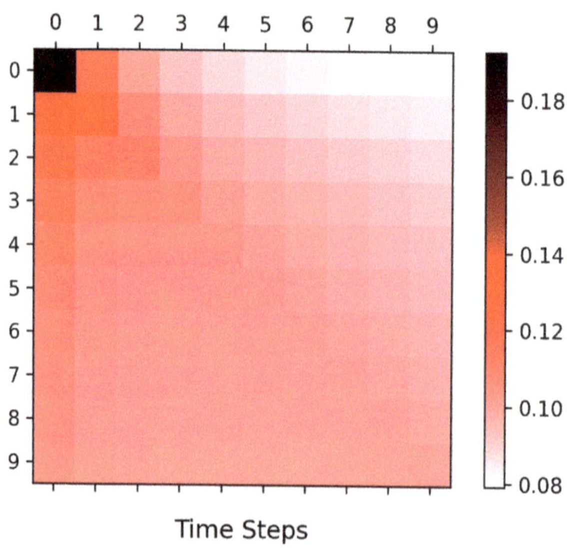

Figure 1. Heatmap of attention weights.

In our work, the method employed is a special case of the attention mechanism known as self-attention. The characteristic of self-attention lies in the fact that the matrices used to compute the vectors Q, K, and V are the same matrix. This allows us to process the various input parameters to obtain attention in the feature dimension. The calculation method is as follows:

$$Q = W_q \times x_t + b_g \tag{6a}$$

$$K = W_k \times x_t + b_k \tag{6b}$$

$$V = W_v \times x_t + b_v \tag{6c}$$

$$\alpha = \text{softmax}\left(\frac{QK^T}{\sqrt{d_k}}\right) \tag{6d}$$

$$\widetilde{x_t} = \alpha V \tag{6e}$$

2.3. Neural Network Architecture

In this study, a neural network sub-model is constructed using Gated Recurrent Units (GRUs) and Attention Mechanism. The structure of this sub-model is depicted in Figure 2. Time series data must first undergo data preprocessing, which will be introduced as follows, and then these preprocessed data are fed into the self-attention layer for the extraction of latent relationships between variables. After the allocation of attention weights, the processed variables are passed through the GRU layer for feature extraction along the temporal dimension. Following the processing by the GRU layer, the time series data are compressed into a single context vector, which is believed to encapsulate all the information of the current time step. The context vector obtained from the previous step is then fed into a Feedforward Neural Network (FNN) composed of three layers of Multilayer Perceptions (MLPs) for computation, and the final results are obtained through the output layer. The activation function between layers is the Exponential Linear Unit (ELU) function. The ELU function is a variant of the Rectified Linear Unit (RELU) function, which is smoother at zero, facilitating gradient computation during backpropagation.

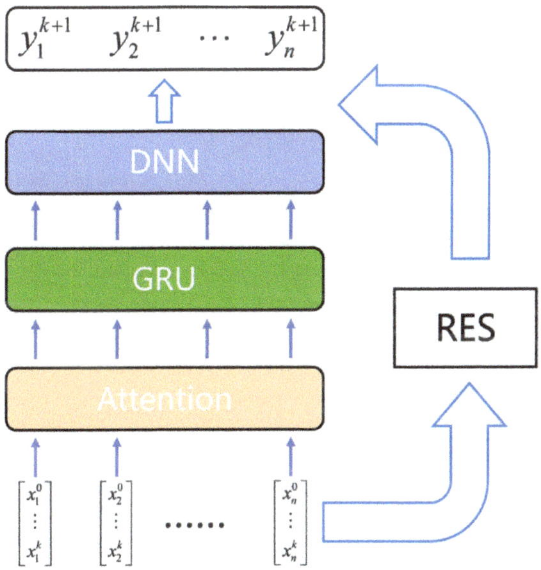

Figure 2. Structure of sub-model.

Repeated experiments have shown that a more complex network structure, a larger network scale, and an increased number of network layers fail to improve the predictive performance of neural networks. Conversely, these factors can lead to more difficult convergence of the neural network, resulting in an inability to improve model accuracy and a decline in predictive efficacy. Therefore, we do not optimize the structure of the sub-models. In the subsequent section, the focus is on the utilization of these sub-models to construct a complete predictive model.

2.4. Dynamic Process Mechanism Model

In chemical engineering processes, there are numerous unit operations, each with its distinct internal structure. To describe these different unit operations, various model equations are required. However, regardless of the type of unit operation model, their dynamic models can be generalized as a system of differential equations and algebraic equations, which can be represented by the following equation set:

$$\frac{\mathrm{d}y_i}{\mathrm{d}\tau} = f(x, y, \tau, c) \tag{7a}$$

$$\Phi(x, y, c) = 0 \tag{7b}$$

$$y|_{\tau=\tau_0} = y_0 \tag{7c}$$

$$c = c(\tau) \tag{7d}$$

where x and y denote all the variables within the process; τ represents time; c is the disturbance variable, expressed as a time-variant function.

However, there are numerous unknown variables in the model equations. A significant number of these variables cannot be directly measured by sensors, such as enthalpy, fugacity, and other thermodynamic parameters, and equipment parameters like heat transfer coefficients, separation efficiency, and pipe friction factors. Consequently, when modeling an actual plant, it is challenging to reduce model errors through parameter calibration, leading to certain dilemmas in the application of mechanistic models. Data-driven modeling methods can utilize specific measurement data from the process for analysis, thereby compensating for the deficiencies of mechanistic models in this regard.

For the purpose of applying dynamic models in predictive control, it is necessary to first deliberate upon the classification of variables. From the perspective of interactions between operators and the system, as well as between the environment and the system, variables can be categorized into four distinct types: state variables, control variables, environmental variables, and equipment parameters.

State variables represent the physical quantities used to characterize the current motion state of a system during the occurrence of transfer phenomena within the system. These variables are influenced by changes in external conditions and by human manipulation, and they are essential for calculating the system's state at the subsequent moment. Examples include the pressure inside a distillation column, the temperature within a reactor, and the flow rate on a particular side of a heat exchanger.

Control variables represent the variables directly influenced by human operational actions, and the variation of these variables can impact the dynamic state of the system. Typical control variables include the opening of a throttle valve, the rotational speed of a compressor, and the power supplied to an electrolytic cell.

The distinction between environmental variables and control variables lies in the fact that environmental variables are not subject to human control and represent external influencing factors. Environmental variables can also impact the dynamics of the system. For instance, environmental variables may include the pressure at the boundary of the system, the content of impurities in the feedstock, and the ambient temperature around the industrial site.

The meaning of equipment parameters is self-evident; these variables are characterized by their independence from human operation and external environmental changes. They remain constant throughout the system's dynamics. However, these variables are essential inputs for the development of mechanistic models. Examples include the heat transfer area and coefficient of a heat exchanger, the flow capacity of a valve, and the volume of a storage tank.

Upon substituting the aforementioned variables into the model equations and reorganizing them, the general set of equations for dynamic models used in predictive control can be represented as follows:

$$\frac{\mathrm{d}x_i}{\mathrm{d}\tau} = f(x, u, \tau, c) \tag{8a}$$

$$\Phi(x, u, c) = 0 \tag{8b}$$

$$x|_{\tau=\tau_0} = x_0 \tag{8c}$$

$$c = c(\tau) \tag{8d}$$

Owing to the requirement for numerical computations, the aforementioned continuous representation is subjected to discretization. Concurrently, in order to enable iterative calculations, it is necessary to convert the representation into an iterative form:

$$x_{k+1} = f(x, u, c)\Delta t + x_k \tag{9a}$$

$$\Phi(x, u, c) = 0 \tag{9b}$$

$$x|_{\tau=\tau_0} = x_0 \tag{9c}$$

$$c = c(\tau) \tag{9d}$$

Ultimately, a formulation has been developed that can be represented as an input-to-output relationship and is amenable to direct iterative computation. By using the receding horizon method, it is possible to calculate the values of variables at future time instances, achieving the effect of over-horizon prediction. Applying this formulation to the input–output structure of machine learning models enables the establishment of a discretized data-driven process dynamic model.

However, it can be observed that within the model equations, there is a requirement for solving algebraic equation sets as well as integrating differential equation sets to determine the variables. In conventional methods, it is typical to first solve the algebraic equations followed by the integration process, resulting in some variables being derived from algebraic equations, while others are obtained through integration. The process of solving mechanistic models and applying them to predictive tasks can be regarded as a forward problem, whereas the use of historical data to identify parameters within the model equations constitutes an inverse problem. Similarly, if we can train a neural network model that is equivalent to the inverse problem using numerical methods, then the forward propagation process of the neural network model is equivalent to solving the mechanistic model. Under this premise, the process of training data-driven models can be viewed as a process of fitting algebraic equations and ordinary differential equations (ODEs).

A combined form of multiple sub-models is proposed, embedding prior mechanistic model knowledge into the neural network model. By incorporating sub-models within the neural network, the equations used to solve for each variable are approximated, thereby replacing the solution process of the mechanistic model and enhancing the interpretability of the neural network model. These concepts will be further elaborated in the subsequent section.

2.5. Strategies for the Enhancement of Multivariate Prediction Networks

In practice, single recurrent neural networks face challenges in multivariate prediction. Thus, a distributed network architecture optimized for such tasks is employed. In the present work, the entire model is replaced with multiple sub-models, each tasked with predicting a single variable. At the output layer of the neural network, all variables to be predicted are aggregated into a single array to serve as the output of the prediction model. The efficacy of this distributed model will be compared and demonstrated in the application section, with the model architecture illustrated in Figure 3.

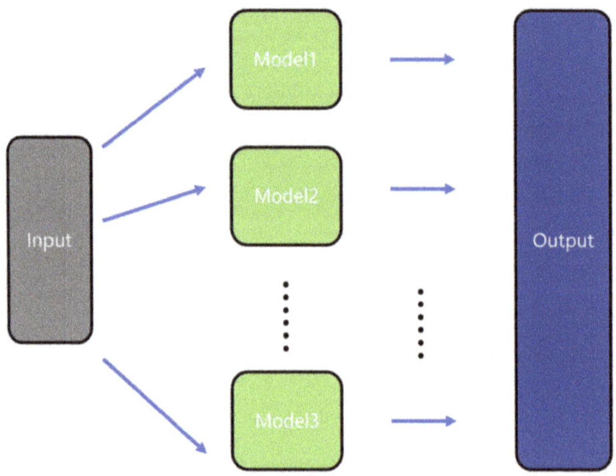

Figure 3. Structure of distributed network.

Simultaneously, it is noted that for multivariate prediction in large-scale systems, the correlation between the variables to be predicted and the input variables is a consideration worth examining. In industrial processes, two-unit operations that are spatially distant may exhibit a time lag within the sampled time steps, such that the relationship between the variables of these two operations may not be captured. For instance, in a distillation sequence consisting of three distillation columns, an increase of 5 °C in the feed temperature may take several hours to propagate to the vapor phase outlet temperature of the last column in the sequence. However, the model does not sample forward for such an extended number of time steps during forward sampling. If machine learning models are directly applied to learn from the dataset without any preprocessing, such fundamentally uncorrelated variables might be mis-learned. To prevent this occurrence, a model mask is introduced at the beginning of the neural network model to enhance the precision of the prediction model and accelerate the convergence of the model.

The mask can be represented as a matrix of size input dimension × output dimension, denoting the connectivity relationship from inputs to outputs, where a value of 1 indicates connectivity and a value of 0 indicates disconnection. When combined with the proposed distributed network, the form of the model can be simply represented by the following equation:

$$mask = [x_1, x_2, \cdots, x_n],\ x_i = 0\ or\ 1,\ i \in (1, 2, \cdots, n) \quad (10)$$

By integrating the distributed network architecture with the model mask, it is possible to employ different sub-models for distinct variables and utilize different input variables for each sub-model, as illustrated in the figure. The advantage of this approach lies in the ability to decouple the prediction tasks for individual variables, and even to optimize the parameters of each sub-model separately, until a model with satisfactory precision is trained.

In the process of solving model equations, there are complex computational procedures. The model equations include tasks such as solving algebraic equations and integrating variables within differential equations. The conventional approach is to first handle the algebraic equations, obtaining results for some variables, and then conduct integration operations to determine the values of the other variables. From a modeling perspective, solving mechanism models for prediction tasks is considered a direct problem, while using historical data to determine the parameters in the model equations is considered an inverse problem.

If a neural network model equivalent to the inverse problem is trained using numerical methods, then the forward propagation process of this neural network model is equivalent to the solving process of the mechanism model. Based on this principle, the training process of data-driven models is actually a process of fitting algebraic equations and ordinary differential equations (ODEs).

To better harness the model's performance, we adopt a combination of multiple sub-models, incorporating prior knowledge of mechanism models into the neural network model. By using the sub-models of the neural network to approximate the equations for solving various variables, we replace the complex solving process of the mechanism model and also enhance the interpretability of the neural network model.

Therefore, for the problem of solving ODEs, we have added residual connections to the network. The fundamental structure of the residual network can be described in the following form:

$$H(x) = F(x) + x \quad (11)$$

where $H(x)$ represents the "optimal function" output by the residual network; $F(x)$ represents the residual function within the residual network.

Residual networks were first proposed by researchers at Microsoft and applied to deep convolutional neural networks for image recognition [19]. Residual connections are capable of alleviating the difficulty of training and reducing the occurrence of overfitting.

Departing from conventional research, some scholars have explored the relationship between residual networks and ODEs/PDEs, providing a mathematical interpretation of residual networks based on differential equations, and proposing the feasibility of using residual networks for dynamic system modeling [20,21]. Indeed, by referencing Equations (9a) and (11), the residual network is also clearly similar in form to ODE, and this connection is easily discernible. By leveraging this relationship, units with residual connections can solve ODEs more efficiently. In other words, the model can more effectively predict variables that would originally need to be solved through integration in the model equations, which are various extensive quantities, such as heat, liquid holdup, etc.

In our research, sub-models with residual connections are used to predict variables that are solved using ODEs in the mechanism model, while the remaining variables are predicted using sub-models without residual connections. This approach is selectable for different variables.

Figure 4 illustrates the architecture of a neural network model for predictive control. The grey dashed lines in Figure 4 represent the input layer connections that are masked by the model mask. Sub-models have the option to include a residual connection module, and each sub-model can freely select the input variables it requires. Finally, the predictions from the sub-models are aggregated into a single output vector.

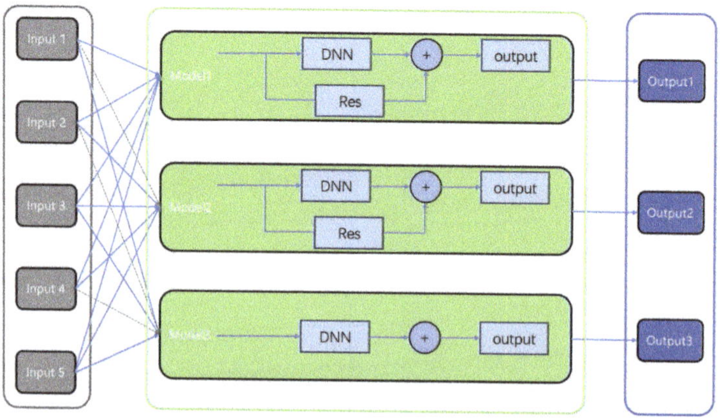

Figure 4. Distributed network model within residual connections.

2.6. Data Acquisition, Dataset Compilation, and Data Preprocessing

Process systems are inherently continuous, both at the microscopic and macroscopic levels. However, when focusing on modeling, particularly when computational solutions are required, the model equations must be discretized in the temporal dimension to facilitate numerical solution methods. The discretization of time necessitates consideration of the corresponding time intervals that arise post-discretization. In the context of mechanistic modeling, this time interval refers to the integration time, whereas for process monitoring and subsequent process data analysis, it refers to the sampling interval during measurement. Different modeling approaches require different time intervals, and in practice, the sampling intervals may vary even for different variables. Therefore, it is essential to align the time scales to facilitate the modeling of input-to-output relationships. Moreover, the length of the time interval must be considered; an excessively long interval may result in data-driven models failing to capture sufficient information, leading to an inadequate approximation of the true process. Conversely, an overly short interval may cause unexpected overfitting due to the measurement noise inherent in the data. In our work, the integration time for the mechanistic model is set at 0.5 s, and the sampling interval for data-driven approaches is uniformly set at 5 s, a determination made through experimentation and constrained by computational speed.

Based on the established sampling rate, a time series dataset comprising 100 thousand data points was acquired and subjected to preprocessing. The min–max normalization method was employed to scale the data within the range of 0 to 1. The min–max normalization method can be described by the following expression:

$$x^* = \frac{x - x_{min}}{x_{max} - x_{min}} \quad (12)$$

Since it is necessary to compute the mean squared error for variables with different data dimensions, normalizing all variables to have the same data dimensionality renders the loss function and accuracy evaluation metrics more meaningful.

Subsequently, the dataset was partitioned into training, validation, and test sets according to a ratio of 7:1.5:1.5. The training set was utilized for the training of the neural network, while neither the validation nor the test set participated in the training process. The validation set served two purposes: first, to facilitate early stopping of the neural network training to prevent overfitting, and second, to optimize the hyperparameters of the neural network and the trainer. The test set was employed post neural network training to comprehensively evaluate the model's fitting efficacy, including the assessment of the predictive performance for each predicted variable, as well as the distribution patterns of errors across the test set.

2.7. Model Predictive Control

To validate the efficacy of the proposed modeling methodology, we will apply the trained data-driven model to a predictive control task. A model predictive controller is utilized to manipulate the temperature and pressure within a distillation column, thereby controlling the concentration of the product.

The model predictive controller employed is the LMPC controller proposed by Wu et al. [22]. The objective of the MPC is to minimize the differences between the target trajectory and the predicted trajectory. The controller proposed by Wu et al. can be described as follows:

$$min \int_{k}^{k+\Delta T} L(\hat{x}(\zeta), \hat{u}(\zeta)) d\zeta \tag{13a}$$

$$s.t.\ \dot{\hat{x}} = F(\hat{x}(k), \hat{u}(k)) \tag{13b}$$

$$\hat{x}(\zeta) - \dot{x}(\zeta) < \varepsilon \tag{13c}$$

$$\hat{u}(k) \in U \tag{13d}$$

$$g(\hat{x}(k), \hat{u}(k)) \leq 0 \tag{13e}$$

where $\hat{x}$ denotes the model predictive trajectory; $\hat{u}$ represents the control action trajectory; $F(\hat{x}(k), \hat{u}(k))$ is the predictive model; Equation (13c) represents the error between the predicted trajectory and the actual trajectory. Wu et al. [23] have demonstrated that when the error of the predictive model is sufficiently small, the closed-loop stability of the controller can be guaranteed. For this reason, model accuracy is of great importance in the MPC. Equation (13d) represents the constraints on the control variables, and Equation (13e) denotes the equality and inequality constraints of the model.

In practical operations, the discrete form of the aforementioned is used, and the predictive model is described using discrete-time state-space equations. For the objective function, the explicit Euler method is employed for computation, which is formulated as follows:

$$min \sum_{i=k}^{np} \left((\hat{y}_i - y_i^t)^T P (\hat{y}_i - y_i^t) + \Delta u_i^T Q \Delta u_i \right) \tag{14a}$$

$$x_{k+1} = F(x_k, \hat{u}_k, c_k) \tag{14b}$$

$$\hat{y}_k = G(\hat{x}_k) \tag{14c}$$

$$U_{min} \leq \hat{u} \leq U_{max} \tag{14d}$$

$$g(\hat{x}_k, \hat{u}_k) \leq 0 \tag{14e}$$

where Equation (14a) represents the quadratic objective function; where n is the prediction horizon; k is the starting time of the prediction; $\hat{y}$ is the predicted trajectory of the controlled variable; y_i^t is the target trajectory; Δu_i is the change in the manipulated variable; P and Q are weight matrices. In the computation, at each sampling time, the controller establishes and solves the optimization problem based on historical data and current input data.

In addressing the optimization problem, it is necessary to repeatedly construct the prediction sequence. The method of constructing the prediction sequence is termed the moving horizon, and the specific procedure is delineated in Equation (15).

$$x_{k+1} = F_{ss}(x_k, c_k, u_k) \tag{15a}$$

$$X_k = [x_1, x_2, \cdots, x_n] \tag{15b}$$

The method employs input variables at time step k to predict the state variables at time step k + 1 and utilizes the manipulated and environmental variables at time step k + 1 to predict the state variables at time step k + 2. This process is repeated in a cyclical manner to construct the prediction sequence X_k. Following the construction of the prediction sequence, the objective function in Equation (14) can be optimized. The resulting optimization problem is a nonlinear optimization problem, which is typically solved iteratively using numerical methods. In this study, the Adam algorithm is employed for its resolution.

The Adam algorithm, known as Adaptive Moment Estimation, is a solution algorithm for nonlinear optimization problems, proposed by Kingma et al. [24] Derived from AdaGrad and RMSProp, Adam is an efficient stochastic optimization algorithm based on gradients. It was initially designed to address a variety of non-convex optimization problems in the field of machine learning, and in this study, it is applied to the solution of general nonlinear optimization problems. By solving the optimization problem constructed in Equation (14), the optimal control action trajectory and its corresponding predictive trajectory can be obtained. Once the solution is completed, only the initial several steps of the control action trajectory are implemented in the controlled system to achieve the optimal dynamic performance.

3. Results and Discussion

3.1. Introduction to Distillation Column

In the section pertaining to the mechanism model, the distillation system is modeled using equations of material balance, phase equilibrium, mole fraction summations, and heat balance (MESH).

Equation of material balance:

$$\frac{dM_i}{dt} = V_{F_{in}} y_{F_{in},i} + L_{F_{in}} x_{F_{in},i} + L_{in} x_{in,i} + V_{in} y_{in,i} - V_{out} y_{V_i} - L_{out} x_{L_i} \tag{16a}$$

$$M_m = V + L \tag{16b}$$

Equation of phase equilibrium:

$$y_{V_i} = K_i x_{L_i} \tag{17a}$$

$$K_i = f_{\text{euqilibrium}}(T, P) \tag{17b}$$

$$H_V = f_{\text{enthalpy}}(T, P, y_V) \tag{17c}$$

$$K_i = f_{\text{enthalpy}}(T, P, x_L) \tag{17d}$$

Equation of mole fraction summations:

$$\sum_{i=1}^{NC} x_i = 1 \tag{18a}$$

$$\sum_{i=1}^{NC} y_i = 1 \tag{18b}$$

Equation of heat balance:

$$H = VH_V + LH_L \tag{19}$$

The system of interest comprises benzene and toluene, with the Peng–Robinson method utilized for phase equilibrium calculations. The distillation column operates at a pressure of 440 kPa, with a pressure drop of 13 kPa across the column. The column is configured with a total of 20 theoretical stages, and the feed is introduced at the 10th theoretical stage. The reflux ratio is 0.278. Steady-state flow rates are presented in Table 1.

Table 1. Distillation process stream data.

Stream Code	Temperature/°C	Pressure/kPa	Molar Flow Rate/kmol·h^{-1}	Molar Concentration of Toluene/%
1	100.00	500.00	500.0	50.00
4	100.00	445.52	500.0	50.00
2	137.52	440.00	250.8	1.22
5	135.40	420.00	250.8	1.22
3	172.83	453.00	249.2	99.13
6	170.61	433.00	249.2	99.13

As the neural network-based dynamic model in this study is designed for prediction control tasks, input and output variables are selected in accordance with the specific information of the control task.

Initially, it is essential to determine the overall control strategy based on the characteristics of the distillation column. Following the research by Konda et al. [25], after calculating the Control Degrees of Freedom (CDOF), it can be observed that for a benzene–toluene distillation column, which operates as a fully condensing reflux column, the control scheme depicted in Figure 5 is required. This scheme involves the control of four target variables, which are the liquid levels in the reflux drum and the column bottom, the column pressure, and the temperature of the sensitive tray. In Figure 5, the blue part represents the distillation section of the distillation column, the red part represents the stripping section of the distillation column, the green dotted line represents the control loop, and the label on the stream corresponds to the stream code in Table 1.

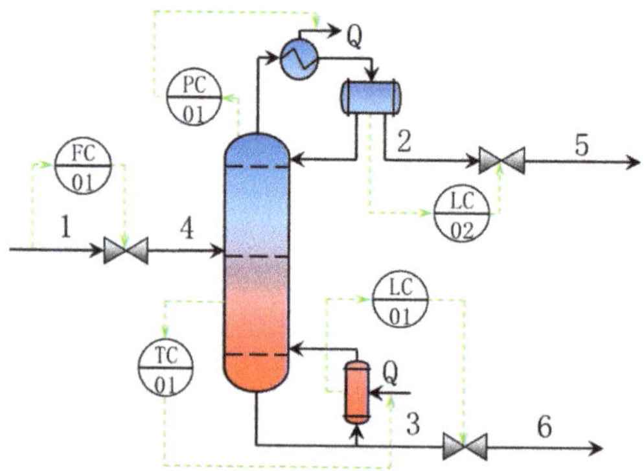

Figure 5. Dynamic simulation flowsheet of distillation process.

In our study, the control strategy employs a hybrid MPC-PID control scheme. For the liquid levels in the flash drum and the column bottom, PID controllers are utilized to manipulate the withdrawal rates. For the column pressure and the temperature of the sensitive tray, an MPC controller is directly employed to manage the cooling and heating utilities at the column top and bottom, thereby controlling the column top pressure and the sensitive tray temperature. The comparison of this scheme with a PID-only control scheme clearly demonstrates the advantages of using predictive models, which will be elaborated upon in the discussion section.

In accordance with the requirements of the control task, there are five state variables required, which are the column top pressure, column bottom pressure, column top temperature, column bottom temperature, and the temperature of the sensitive tray. There are two environmental variables, namely, the feed temperature and feed flow rate. The control variables consist of the cooling capacity at the column top and the heating capacity at the column bottom.

Now that the variables involved in the model have been determined, the state variables, environmental variables, and operation variables at time k are taken as the input parameters of the model, and the state variables at time k + 1 are taken as the output parameters of the model. In this way, a machine learning model with an input dimension of 9 and an output dimension of 5 can be obtained.

After identifying the input and output parameters of the model, according to the method mentioned in Section 2.6 above, all the data of relevant variables within 500,000 thousand seconds are selected to create a dataset, obtaining data sufficient to represent the dynamic range of the system, and these data are used for data-driven modeling.

Additionally, in real chemical industrial processes, the majority of the time is spent in stable operation, which means that the historical operational data from factories does not provide a sufficient range of parameter variations. This limitation precludes the use of existing data for modeling purposes. However, it is possible to conduct mechanistic modeling of process units based on the equipment parameters and operational conditions of the factory, thereby generating highly available dynamic data. Utilizing such data can compensate for the required range of parameter variations.

The hidden layer dimensions of the neural network are set to 32 for all layers. The network is trained using the Adam optimizer with an initial learning rate of 0.001. The learning rate is adjusted using an exponential decay method with a decay factor of 0.98, and the maximum number of training epochs is set to 1000. We have explored the adjustment of the hidden layer dimensions of the neural network, conducting a comparative study with dimensions of 16, 24, 32, and 64. It was found that the performance of the neural network does not improve beyond a hidden layer dimension of 32, leading us to apply a dimension of 32 for all hidden layers. Both training and testing were conducted on a laptop equipped with Intel Core i5-12400 CPU and NVIDIA RTX 3070 GPU.

3.2. The Training of Neural Networks and the Analysis of Outcomes

To ascertain the efficacy of the process-mechanism-based improvement strategies proposed in the previous section, four model architectures were designed in this part. Using the same training and validation sets, neural networks with four different structures were trained. The improvement strategies to be validated are two: Distribution net and Residual connections. The model structures for architectures 1 through 4 are as follows:

Structure 1: Distribution net + Residual connections + RNN

Structure 2: Distribution net + RNN

Structure 3: Residual connections + RNN

Structure 4: RNN only (baseline)

Maintaining all other model parameters and training parameters as constant, the neural networks with four distinct architectures were trained separately. The root mean squared error (RMSE) was employed to compute the average error of the models, while R^2 was used to represent the model accuracy. A lower RMSE value indicates a smaller model error, and an R^2 value closer to 1 signifies higher model accuracy. Table 2 documents the RMSE and R^2 values for the neural networks with four different structures on the test set.

Table 2. The RMSE and R^2 of four model structures.

Model Structure	RMSE	R^2
Structure 1	0.0014972378	0.9999
Structure 2	0.0045016849	0.5436
Structure 3	0.0017783857	0.9987
Structure 4	0.0065688221	0.0784

The results presented in Table 2 suggest that all four architectures achieve a satisfactory level of accuracy, thus validating the superiority of the recurrent neural network model. Upon comparing the accuracy gaps among the four structures, it is observed that Structure 1, which employs both a Distribution net and Residual connections, attains the highest level of precision. The outcomes for Structure 2 and Structure 3 are, to varying degrees, inferior to those of Structure 1. Meanwhile, the performance of Structure 4 is relatively the least favorable. From this set of results, it can be inferred that both the Distribution net and Residual connections contribute to enhancing the predictive efficacy of the neural network.

The following analysis focuses on the neural network derived from Structure 1 as a case study. Figure 6 presents a comparison between the predicted and actual values of the five variables on the test set. It can be observed that for each variable, the predicted values closely align with the actual values, exhibiting a level of model accuracy comparable to that of mechanistic models. On one hand, this demonstrates the potential of recurrent neural networks for processing sequential data. On the other hand, it also attests to the effectiveness of the two improvement strategies based on process mechanisms, thereby validating the feasibility of the data-driven modeling approach that integrates process mechanisms. The data-driven model established using the proposed method can serve as a surrogate, effectively replacing mechanistic models and playing a significant role in various scenarios.

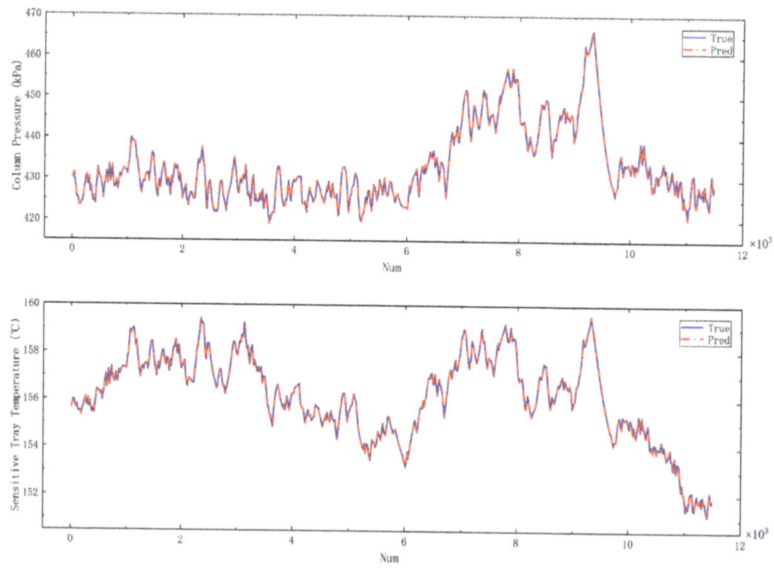

Figure 6. Comparison of predicted and true values on the test set.

3.3. Gaussian Noise Data Test

In order to evaluate the performance of the model under suboptimal dataset conditions, normally distributed noise was incorporated into the dataset, with the process noise adhering to distribution $\upsilon \sim N\ (0,\sigma^2)$. Noise with $\sigma = 0.002$ and $\sigma = 0.005$ was respectively applied to the dataset after normalization, followed by retraining and testing. The results of the training are presented in the Table 3, and Figure 7 illustrates the impact of noise on the prediction accuracy.

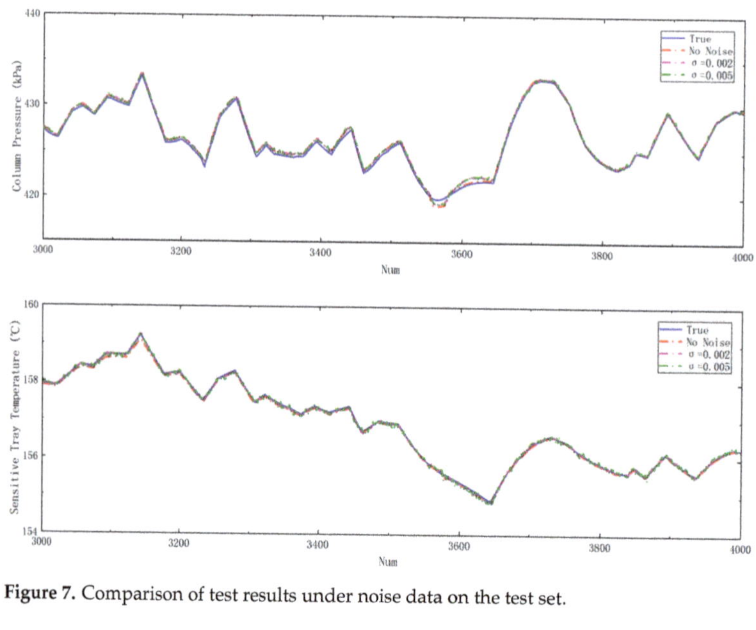

Figure 7. Comparison of test results under noise data on the test set.

Table 3. The RMSE and R^2 under various noise conditions.

Model Structure	RMSE	R^2
No Noise	0.0014972378	0.9999
$\sigma = 0.002$	0.0033924667	0.9987
$\sigma = 0.005$	0.0064540802	0.9969

3.4. Model Predictive Control Using Predictive Model

Once a neural network model with adequate accuracy is obtained, it is integrated into the LMPC framework for the predictive control. The controller performance is evaluated through closed-loop disturbance tests and setpoint tracking tests, respectively, on both MPC and PID control systems. These two tests are capable of characterizing the control system's ability to quickly stabilize against external disturbances in the closed-loop circuit, as well as its capability to respond rapidly to setpoint changes due to real-time optimization and production scheduling. The Integral Square Error (ISE) is employed to reflect the efficacy of the MPC, where the ISE is calculated by integrating the squared error over time for a single controlled variable. The calculation method for the ISE is as follows:

$$ISE = \int e^2 dt \tag{20}$$

The closed-loop disturbance test involves introducing a perturbation of a 5 °C decrease in feed temperature. The closed-loop response curves are depicted in Figure 8. The ISE is calculated for the two controlled variables under the use of MPC and PID controllers, respectively, and the values are presented in Table 4.

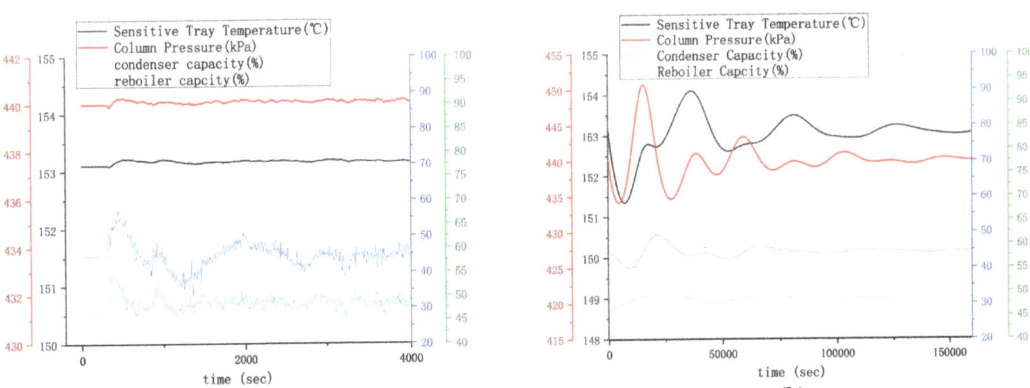

(a) (b)

Figure 8. Dynamic response curves of disturbance testing of distillation process: (**a**) MPC response curve; (**b**) PID response curve.

Table 4. Disturbance test ISE of distillation process.

Controlled Variables	MPC	PID
Sensitive Tray Temperature	4	3398
Column Pressure	63	72,201

The PID control scheme, configured as column top pressure–column top cooling capacity and sensitive tray temperature–column bottom heating capacity, inherently causes significant lag and fails to stabilize rapidly. Modifying the control scheme to column top pressure–reflux flow rate may yield improvement. Due to the extended stabilization time, the calculation of the ISE for the PID controller is limited to the first 4000 s. Upon examining the response curves of the MPC controller, the disturbance in feed temperature has negligible impact on the column pressure and temperature, indicating a robust resistance to disturbances. Under the combined regulation of cooling and heating capacities, the heat is prematurely conducted into the column through the gas–liquid mass transfer process within the column, counteracting the effect of the temperature decrease.

In the setpoint tracking performance test, the setpoint of the column top pressure was increased by 5 kPa. The closed-loop response curves are depicted in Figure 9. The ISE was calculated for the two controlled variables under the use of MPC and PID controllers, respectively, and the values are presented in Table 5.

Table 5. Control point tracking test ISE of distillation process.

Controlled Variables	MPC	PID
Sensitive Tray Temperature	50	576
Column Pressure	8910	92,060

By analyzing the behavior of the MPC, it can be observed that upon a setpoint change, the cooling capacity at the column top rapidly decreases and the heating capacity swiftly increases to achieve a rapid rise in the overall column pressure, followed by a gradual adjustment of the cooling and heating capacities to a new steady state. In the setpoint tracking test, the PID controller takes nearly 100 times as long as the MPC controller to reach stability (correspondingly, the ISE integration time takes up to 10,000 s for this case), whereas the MPC controller achieves stability rapidly. The mass and heat transfer within the distillation column necessitate time, and the effect becomes more significant as the number of tray stages increases, leading to a significant time lag. Additionally, the presence of volume in the column bottom and reflux drum introduces the volume lag. For processes with such a substantial lag, the PID controller is unable to quickly stabilize the distillation column, whereas the MPC, by invoking the predictive model and utilizing multiple variables measured by a distributed control system, is capable of predicting the overall column conditions and responding swiftly and accurately to external changes.

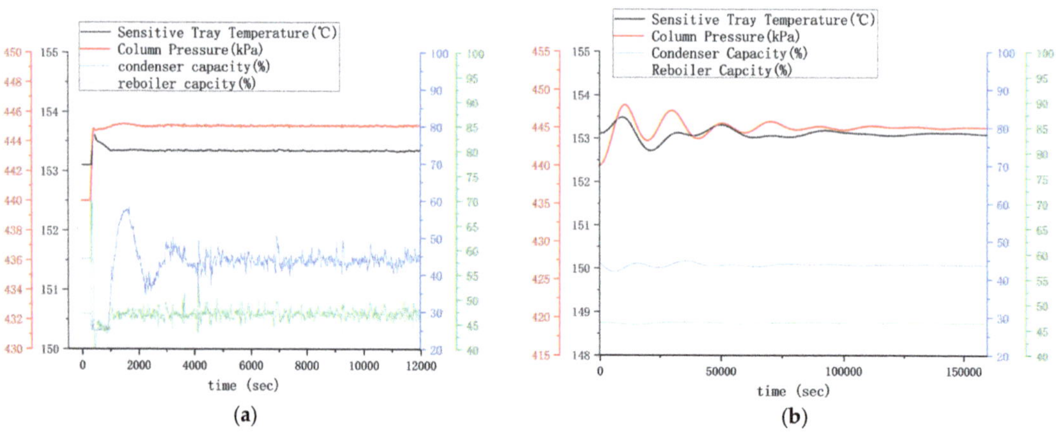

Figure 9. Dynamic response curves of control point tracking testing of distillation process: (**a**) MPC response curve; (**b**) PID response curve.

By examining the log records generated during the control process, the prediction errors of the prediction model within the control process were compiled. Upon comparison, it was found that the prediction errors, as measured by the MAE (Mean Absolute Error), were all below 0.001. Due to the dynamic variation patterns of input variables during the prediction process being smoother than those in the test set, there were no instances of abrupt changes in other system parameters caused by sudden mutations in operational parameters. Consequently, the performance during the prediction process was superior to that of the test set.

In the MPC testing, it was observed that under the condition of utilizing a sufficiently accurate model, the MPC controller demonstrated robust performance against disturbances, as well as the capability to respond rapidly to commands from the scheduling layer. On one hand, this verifies the high precision and strong generalization ability of our process-mechanism-based deep neural network model. On the other hand, it also confirms that this model is fully applicable for tasks involving model predictive control.

The results of the testing were compared with similar studies, selecting the Simultaneous Multistep Transformer Architecture proposed by Park et al. [26], the physics-informed hybrid neural network model based on the Long Short-Term Memory (LSTM) neural network proposed by Zarzycki et al. [27], and the Neural Ordinary Differential Equation proposed by Luo et al. [28].

The Simultaneous Multistep Transformer Architecture proposed by Park et al. [26]. serves as a multistep prediction model. In their study, the transformer-based model was used for comparison with an LSTM-based model. Since the RMSE for each case was not listed in their study, we selected Case III: Temperature Control Lab Device for comparison, which involves a multivariate prediction task similar in form and difficulty to our established scenario. Firstly, in terms of accuracy, both the LSTM and transformer models in Park et al.'s study had MAEs above 1; in our study, the errors calculated during the control process were used to characterize the model's multistep prediction performance, and the relative errors were converted to obtain the MAE, with all MAEs not exceeding 0.05. In terms of computational efficiency, although our proposed model is also based on RNN, it benefits from a neural network framework constructed on automatic differentiation and gradient backpropagation, which forms the MPC architecture. The solution speed for each MPC problem is within 2 s, which is significantly faster than the LSTM-based model used for comparison by Park et al., and comparable to the transformer model used for multistep prediction.

The physics-informed hybrid neural network model based on the Long Short-Term Memory (LSTM) neural network proposed by Zarzycki et al. [27]. is a data-driven modeling approach that incorporates process physical information, which shares some similarities with our approach. In the study by Zarzycki et al., a model was developed for a polymerization reactor to predict the Number-Average Molecular Weight (NAMW). As the model's performance on the test set was not directly provided, and only the control process error was given without units, a direct comparison with our model is infeasible. In terms of computational speed, the model proposed by Zarzycki et al. exhibits rapid computation, completing a single iteration within approximately 140 milliseconds. This may be attributed to the fact that their study employed single-variable prediction, whereas our research involved multi-variable prediction, indicating a difference in model scale, as well as the efficiency of the employed MPC algorithms. Regarding the interpretability and extrapolation capabilities of the model, the Physics-Informed Neural Network (PINN) elegantly addresses this issue. In contrast to our study, which focuses on guiding the construction of data-driven models with process knowledge, the research by Zarzycki et al. presents a true hybrid model.

The Neural Ordinary Differential Equation proposed by Luo et al. [28]. is a novel deep neural network that integrates ordinary differential equations. In terms of model accuracy, the study by Luo et al. reported a Mean Squared Error (MSE) of 3.1×10^{-4} for datasets without noise. Since our research utilized Root Mean Squared Error (RMSE) for measurement, the converted MSE is 2.24×10^{-6}, indicating significantly higher precision. For datasets with noise, due to the inability to quantify the relative scale of noise for comparison, the following are provided for reference only. The model proposed by Luo et al. yielded an MSE of 0.0068 at a 2% noise level and an MSE of 0.03 at a 6% noise level. In contrast, our model achieved an MSE of 1.15×10^{-5} at a 0.2% noise level and an MSE of 4.16×10^{-5} at a 0.5% noise level.

Relative to the model proposed by Luo et al. [28], our model exhibits slight deficiencies in computational speed and resistance to noise, which identifies areas for potential improvement in our research.

4. Conclusions

This study proposes a deep neural network model based on a process mechanism and applies it to the predictive control of distillation processes. Initially, a neural network sub-model based on the GRU + attention mechanism is constructed, with the input–output relationship for machine learning determined according to the mechanism equations of chemical unit operation models. Furthermore, based on the principles of solving ordinary differential equations (ODEs) in mechanism models, improvement strategies such as a Distribution net and Residual connections are proposed. Comparative experiments are designed with four neural network architectures, and the results indicate that the model's accuracy is significantly enhanced after employing the aforementioned improvement strategies. Compared to the RNN-only model, the mean squared error is reduced by 77.2%.

The well-trained model is applied to the predictive control of distillation columns. When compared to the PID control, the integral squared error (ISE) for key variables is significantly decreased in both closed-loop disturbance tests and setpoint tracking tests. For instance, the ISE of our model is reduced by 98.8% compared to the PID control in sensitive tray temperature control, and by 90.3% in the column pressure control. This fully demonstrates the high efficiency and practicality of the model in the control of distillation processes.

In summary, the model proposed in this study is capable of fitting the complex nonlinear behavior of chemical systems with high precision and exhibits strong generalization capabilities. It can serve as a general framework for predictive applications in various unit operations and plant systems. Future research will focus on the optimization of the model structure, expanding its application in more complex industrial processes, and exploring its potential in real-time optimization and fault diagnosis to enhance the overall benefits and safety of industrial production.

Author Contributions: Conceptualization, Z.W.; Methodology, Z.W.; Software, Z.W.; Validation, Z.W. and H.W.; Formal analysis, Z.W.; Resources, Z.W. and H.W.; Data curation, Z.W. and H.W.; Writing—original draft preparation, Z.W.; Writing—review and editing, Z.W., H.W., and Z.D.; Visualization, Z.W. and H.W.; Project administration, Z.W. and Z.D.; Funding acquisition, Z.D. All authors have read and agreed to the published version of the manuscript.

Funding: This research received no external funding.

Data Availability Statement: Data are contained within the article.

Acknowledgments: The authors thank the editors and reviewers for their helpful comments regarding the manuscript and other individuals who contributed but are not listed as authors of this study.

Conflicts of Interest: The authors declare no conflicts of interest.

Abbreviations

The following abbreviations are used in this manuscript:

PID	Proportional-Integral-Derivative Control
MPC	Model Predictive Control
NMPC	Nonlinear Model Predictive Control
CSTR	Continuous Stirred Tank Reactor
RNN	Recurrent Neural Network
GRU	Gated Recurrent Unit
FNN	Feedforward Neural Network
MLP	Multilayer Perception
ELU	Exponential Linear Unit
RELU	Rectified Linear Unit
NLP	Natural Language Processing
ODE	Ordinary Differential Equation
PDE	Partial Differential Equation
LMPC	Lyapunov-based Model Predictive Control
ISE	Integral Square Error

References

1. Pallavicini, J.; Fedeli, M.; Scolieri, G.D.; Tagliaferri, F.; Parolin, J.; Sironi, S.; Manenti, F. Digital Twin-Based Optimization and Demo-Scale Validation of Absorption Columns Using Sodium Hydroxide/Water Mixtures for the Purification of Biogas Streams Subject to Impurity Fluctuations. *Renew. Energy* **2023**, *219*, 119466. [CrossRef]
2. Min, Q.; Lu, Y.; Liu, Z.; Su, C.; Wang, B. Machine Learning Based Digital Twin Framework for Production Optimization in Petrochemical Industry. *Int. J. Inf. Manag.* **2019**, *49*, 502–519. [CrossRef]
3. Tanaka, M.; Koyama, K.; Shibata, J. Steady-State Local Linearization of the Countercurrent Multistage Extraction Process for Metal Ions. *Ind. Eng. Chem. Res.* **1997**, *36*, 4353–4357. [CrossRef]
4. Zhang, C.; Vasudevan, S.; Rangaiah, G.P. Plantwide Control System Design and Performance Evaluation for Ammonia Synthesis Process. *Ind. Eng. Chem. Res.* **2010**, *49*, 12538–12547. [CrossRef]
5. Gao, S.; Bo, C.; Jiang, C.; Zhang, Q.; Yang, G.; Chu, J. Hybrid Modeling for Carbon Monoxide Gas-Phase Catalytic Coupling to Synthesize Dimethyl Oxalate Process. *Chin. J. Chem. Eng.* **2024**, *70*, 234–250. [CrossRef]
6. Savage, T.; Almeida-Trasvina, H.F.; del Río-Chanona, E.A.; Smith, R.; Zhang, D. An Adaptive Data-Driven Modelling and Optimization Framework for Complex Chemical Process Design. In *Computer Aided Chemical Engineering*; Pierucci, S., Manenti, F., Bozzano, G.L., Manca, D., Eds.; 30 European Symposium on Computer Aided Process Engineering; Elsevier: Amsterdam, The Netherlands, 2020; Volume 48, pp. 73–78. [CrossRef]
7. Chen, Y.; Yu, H.; Liu, C.; Xie, J.; Han, J.; Dai, H. Synergistic Fusion of Physical Modeling and Data-Driven Approaches for Parameter Inference to Enzymatic Biodiesel Production System. *Appl. Energy* **2024**, *373*, 123874. [CrossRef]
8. Alauddin, M.; Khan, F.; Imtiaz, S.; Ahmed, S.; Amyotte, P. Integrating process dynamics in data-driven models of chemical processing systems. *Process Saf. Environ. Prot.* **2023**, *174*, 158–168. [CrossRef]
9. Sharma, N.; Singh, K. Model Predictive Control and Neural Network Predictive Control of TAME Reactive Distillation Column. *Chem. Eng. Process. Process Intensif.* **2012**, *59*, 9–21. [CrossRef]
10. Li, S.; Li, Y.-Y. Neural Network Based Nonlinear Model Predictive Control for an Intensified Continuous Reactor. *Chem. Eng. Process. Process Intensif.* **2015**, *96*, 14–27. [CrossRef]
11. Jian, N.L.; Zabiri, H.; Ramasamy, M. Data-Based Modeling of a Nonexplicit Two-Time Scale Process via Multiple Time-Scale Recurrent Neural Networks. *Ind. Eng. Chem. Res.* **2022**, *61*, 9356–9365. [CrossRef]
12. Rebello, C.M.; Marrocos, P.H.; Costa, E.A.; Santana, V.V.; Rodrigues, A.E.; Ribeiro, A.M.; Nogueira, I.B.R. Machine Learning-Based Dynamic Modeling for Process Engineering Applications: A Guideline for Simulation and Prediction from Perceptron to Deep Learning. *Processes* **2022**, *10*, 250. [CrossRef]
13. Xu, B.; Wang, Y.; Meng, Z.; Chen, Y.; Yin, S. An Intelligent Identification Method Based on Self-Adaptive Mechanism Regulated Neural Network for Chemical Process. *J. Taiwan Inst. Chem. Eng.* **2024**, *155*, 105318. [CrossRef]
14. Bikmukhametov, T.; Jäschke, J. Combining Machine Learning and Process Engineering Physics towards Enhanced Accuracy and Explainability of Data-Driven Models. *Comput. Chem. Eng.* **2020**, *138*, 106834. [CrossRef]
15. Sun, B.; Yang, C.; Wang, Y.; Gui, W.; Craig, I.; Olivier, L. A Comprehensive Hybrid First Principles/Machine Learning Modeling Framework for Complex Industrial Processes. *J. Process Control* **2020**, *86*, 30–43. [CrossRef]

16. Li, Y.; Li, N.; Ren, J.; Shen, W. An Interpretable Light Attention–Convolution–Gate Recurrent Unit Architecture for the Highly Accurate Modeling of Actual Chemical Dynamic Processes. *Engineering* **2024**, *39*, 104–116. [CrossRef]
17. Bradley, W.; Kim, J.; Kilwein, Z.; Blakely, L.; Eydenberg, M.; Jalvin, J.; Laird, C.; Boukouvala, F. Perspectives on the Integration between First-Principles and Data-Driven Modeling. *Comput. Chem. Eng.* **2022**, *166*, 107898. [CrossRef]
18. Vaswani, A.; Shazeer, N.; Parmar, N.; Uszkoreit, J.; Jones, L.; Gomez, A.N.; Kaiser, L.; Polosukhin, I. Attention Is All You Need *arXiv* **2023**, arXiv:1706.03762. [CrossRef]
19. He, K.; Zhang, X.; Ren, S.; Sun, J. Deep Residual Learning for Image Recognition. In Proceedings of the 2016 IEEE Conference on Computer Vision and Pattern Recognition (CVPR), Las Vegas, NV, USA, 27–30 June 2016; pp. 770–778. [CrossRef]
20. E, W. A Proposal on Machine Learning via Dynamical Systems. *Commun. Math. Stat.* **2017**, *5*, 1–11. [CrossRef]
21. Ruthotto, L.; Haber, E. Deep Neural Networks Motivated by Partial Differential Equations. *arXiv* **2018**, arXiv:1804.04272. [CrossRef]
22. Wu, Z.; Tran, A.; Rincon, D.; Christofides, P.D. Machine Learning-based Predictive Control of Nonlinear Processes. Part I: Theory. *AIChE J.* **2019**, *65*, e16729. [CrossRef]
23. Wu, Z.; Rincon, D.; Christofides, P.D. Process Structure-Based Recurrent Neural Network Modeling for Model Predictive Control of Nonlinear Processes. *J. Process Control* **2020**, *89*, 74–84. [CrossRef]
24. Kingma, D.P.; Ba, J. Adam: A Method for Stochastic Optimization. *arXiv* **2014**, arXiv:1412.6980.
25. Murthy Konda, N.V.S.N.; Rangaiah, G.P.; Krishnaswamy, P.R. A Simple and Effective Procedure for Control Degrees of Freedom. *Chem. Eng. Sci.* **2006**, *61*, 1184–1194. [CrossRef]
26. Park, J.; Babaei, M.R.; Munoz, S.A.; Venkat, A.N.; Hedengren, J.D. Simultaneous multistep transformer architecture for model predictive control. *Comput. Chem. Eng.* **2023**, *178*, 108396. [CrossRef]
27. Zarzycki, K.; Ławryńczuk, M. Long short-term memory neural networks for modeling dynamical processes and predictive control: A hybrid physics-informed approach. *Sensors* **2023**, *23*, 8898. [CrossRef]
28. Luo, J.; Abdullah, F.; Christofides, P.D. Model predictive control of nonlinear processes using neural ordinary differential equation models. *Comput. Chem. Eng.* **2023**, *178*, 108367. [CrossRef]

processes

MDPI

Article

Multi-Objective Evaluation Strategy Based on Data Envelopment Analysis for Working Fluid Selection in the Organic Rankine Cycle

Luoyu Zhang, Lili Wang, Xiaoyan Sun, Li Xia, Shaohui Tao *, Shuguang Xiang and Siyi Jin

Institute of Process Systems Engineering, Qingdao University of Science and Technology, Qingdao 266101, China; 4022010117@mails.qust.edu.cn (L.Z.); liliw@qust.edu.cn (L.W.); sxy@qust.edu.cn (X.S.); xiali@qust.edu.cn (L.X.); xsg@qust.edu.cn (S.X.); siyijin66@126.com (S.J.)

* Correspondence: tau@qust.edu.cn

Abstract: Currently, in Chinese industry substantial amounts of low-grade waste heat are underutilized. Effectively harnessing these low-temperature waste heat sources is instrumental in promoting energy conservation and emission reduction objectives. The organic Rankine cycle (ORC) serves as an effective method for utilizing low-grade waste heat. The selection of a suitable working fluid is a pivotal aspect of the design of an ORC system. There are many kinds of working fluid and they have complex molecular structures, which increases the difficulty of screening working fluids. A novel approach is proposed based on data envelopment analysis (DEA) for multi-objective evaluation of working fluids. This method takes into account the thermodynamic performance of the working fluid in the ORC (thermal efficiency, net power output, exergy efficiency), economic aspects (investment cost, exergy loss cost), and environmental considerations (exergy environmental factors, CO_2 emission reduction). DEA offers a distinct advantage by objectively balancing these conflicting objectives through data-driven optimization, eliminating the need for subjective weight assignment and enabling simultaneous evaluation of thermodynamic, economic, and environmental metrics in working fluid selection. A total of 62 different working fluids were evaluated in the integrated technology. Heptane working fluid screened out by DEA was compared with working fluid R245fa, a fluid commonly used in existing literatures. The exergy loss of the Heptane working fluid is reduced by 5.02%, the thermal efficiency is increased by 0.24%, and the net output work is increased by 2.04%. The proposed evaluation method introduces a novel perspective for the efficient screening of working fluids in the ORC system for low-temperature waste heat power generation.

Keywords: organic Rankine cycle; data envelopment analysis; working fluid selection; multi-objective evaluation

1. Introduction

Energy is the cornerstone of contemporary civilization, a pivotal driver for economic development, and an essential material foundation indispensable for the survival and advancement of all entities. With societal development and progress, almost all sectors require increased energy, and the relentless exploitation and utilization of finite energy sources, such as coal and oil, have been ongoing. This unbridled exploitation has led to the depletion of reserves. The global energy issue has emerged as a prominent challenge faced collectively by nations across the globe. According to data from the BP 2022 Statistical

Yearbook of World Energy, global primary energy consumption experienced a notable resurgence in 2021, escalating by almost 6%. China's primary energy consumption showed an annual growth of 7.1%, accounting for approximately 70% of the nation's total energy consumption. Energy scarcity and environmental pollution become critical bottlenecks constraining the development of ecological civilization and the low-carbon progression of the national economy. As a developing nation, China has proactively implemented measures to meet its energy conservation and carbon reduction targets. Analyzing energy consumption trends from 2010 to 2021 reveals a relatively modest increase in coal consumption (refer to Figure 1) [1]. Nevertheless, China's energy consumption rate remains elevated, with electricity consumption increasing annually. The advancement of high-efficiency, energy-saving, and emission-reducing technologies holds profound strategic significance in propelling national economic development, mitigating environmental pollution, and expediting the transformation of the energy landscape.

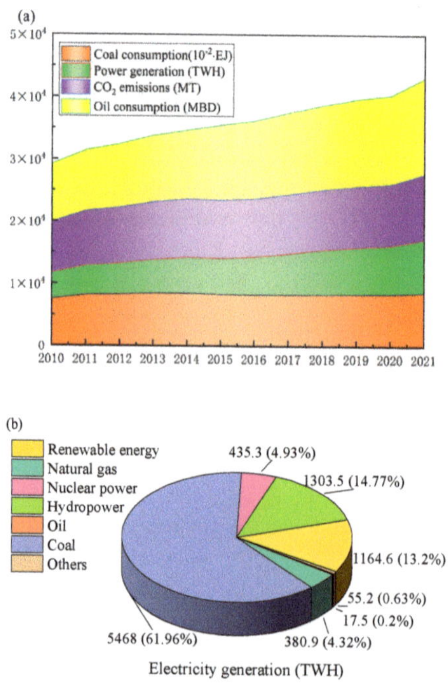

Figure 1. (**a**) Energy consumption in China from 2010 to 2021; (**b**) Power generation of different types of energy in 2021.

Within industrial production, there is inefficient use of low-temperature waste heat, which contributes to energy scarcity. To address this, the proficient recovery and utilization of low-temperature waste heat is a strategic approach for its recovery. In recent years, low-temperature waste heat power generation technology has received considerable attention [1,2]. Notably, the ORC is a Rankine cycle employing low-boiling-point organic matter as the working fluid, offering advantages such as a broad range of applicable heat source temperatures, simple equipment, and safety and environmental benefits [1,3]. The properties of the organic working fluid have a significant impact on system performance [4], which highlights the crucial importance of choosing the right organic working fluid for the effective application of ORC technology [5,6]. Xie Pan et al. [7] conducted an assessment of working fluid compatibility at different waste heat temperatures, utilizing thermal efficiency as an evaluation criterion. Guo Hao et al. [8], employing an exergic analysis methodology, evaluated diverse working fluids based on exergic efficiency and the exergic loss of individual components under specific heat source temperatures. Notably, common research practices indicate a tendency among scholars to employ one or several thermodynamic characteristics, consolidating them into a singular objective function through linear weighting for the optimization of working fluid performance, which can be subject to inherent subjectivity.

DEA is a quantitative method used to assess the relative efficacy of comparable homogeneous units. The establishment of a linear programming model, based solely on input and output indices, enables the derivation of quantitative evaluation results through solution procedures [9]. The DEA methodology presents distinct advantages, particularly in avoiding subjective biases, simplifying the evaluation process, and enabling quantitative assessments. The methodology offers distinct advantages over traditional evaluation approaches [10]:

Eliminates Subjectivity: Unlike weighted aggregation methods, DEA does not require predefined weights for evaluation criteria, thus avoiding the subjectivity inherent in subjective weight assignment [11].

Handles Multi-Dimensionality: DEA effectively manages complex scenarios with multiple inputs and outputs, making it particularly suitable for evaluating systems with interdependent performance indicators.

Data-Driven Objectivity: By directly analyzing raw data without assuming functional relationships, DEA provides a transparent and empirical basis for efficiency assessment [12].

Benchmarking Capability: DEA identifies efficient DMUs as reference points, enabling underperforming units to benchmark and improve their operations.

These characteristics make DEA an ideal tool for comprehensive performance evaluation in systems where multiple conflicting objectives need to be balanced, such as the organic Rankine cycle (ORC) working fluid selection problem addressed in this study.

This paper proposes a new multi-objective evaluation method for the selection of organic Rankine cycle (ORC) working fluids, and applies this method in the ethylene glycol plant [13,14]. First, we conduct ORC process simulations and optimize all candidate working fluids to determine the operating conditions that yield optimal thermal efficiency for each fluid. Under these conditions, we calculate thermodynamic, economic, and environmental indices for each candidate working fluid.

Finally, we perform a multi-objective evaluation to select a working fluid with optimal overall performance. This analytical method establishes a new criterion for screening organic working fluids in ORC applications. The main innovations and contributions include the following:

(1) A new methodology based on data envelopment analysis (DEA) for multi-objective assessment of workflows.
(2) The evaluation of 62 different working fluids and quantitative assessment of the advantages and disadvantages of these fluids in ORC systems.
(3) The evaluation methodology presented in this paper can be extended to the design of hybrid working fluids, thereby saving significant costs related to simulation and experimental studies. It provides a new perspective for the effective screening of working fluids in ORC systems for low-temperature waste heat power generation.

2. Optimization of the Orc System

Low-temperature waste heat power generation technology has the capability to transform the thermal energy within waste gases and liquids generated during industrial production into electrical energy. The ORC system includes an expander, a working fluid pump, and a heat exchanger, as delineated in Figure 2. The operational principle involves pressurizing a low-temperature and low-pressure working fluid through the pump and sending the medium into the heat exchanger. Within the heat exchanger, the working fluid absorbs heat from the heat source, transitioning into a high-temperature and high-pressure gas. Subsequently, the working fluid enters the expander, whereby a portion of the working fluid's internal energy is converted into mechanical energy.

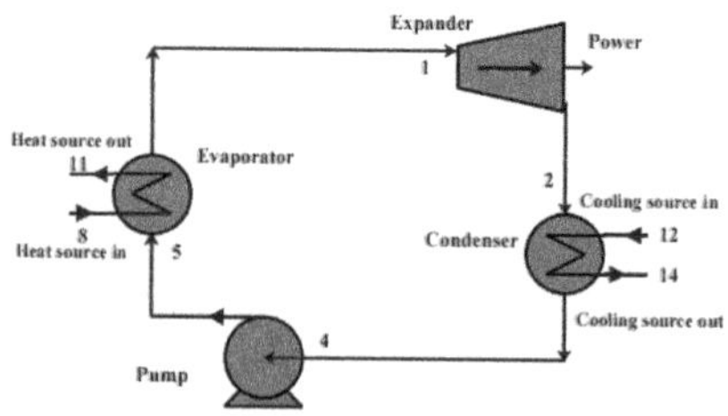

Figure 2. A schematic diagram of the basic organic Rankine cycle system.

In the context of the ORC system for low-temperature waste heat power generation, the cyclic thermal efficiency serves as a pivotal indicator, elucidating the efficacy of converting low-temperature waste heat into electrical energy. Consequently, leveraging the specified low-temperature heat source, the cyclic thermal efficiency is designated as the objective function, with the pump and expander pressures serving as control variables. The Aspen Plus process simulation software is employed for the simulation and optimization of the ORC process. This approach facilitates the derivation of optimal operating conditions for the working fluid, ensuring maximal thermal efficiency.

This ORC system operating status optimization diagram starts by analyzing the characteristics of the ORC system model. Then, it progresses through two parallel processes: one is selecting optimization objectives and establishing the objective function, the other is analyzing variable parameters that influence optimization objectives and determining manipulation variables. After setting constraints, an optimization algorithm is chosen. Subsequently, a convergence check is conducted. If the operation does not converge, the algorithm is adjusted or constraints are reconsidered. Once convergence is achieved, the optimized operating conditions of the ORC system are obtained. The optimization procedure is illustrated in Figure 3.

Optimization is predicated on the following assumptions and conditions:

(1) The ORC system maintains a steady-state flow.
(2) Friction and pressure losses associated with the flow of organic working fluids in the system are disregarded.
(3) Isentropic efficiencies of the expander and pump are set at 0.8.
(4) No pressure drops are considered for the evaporator and condenser, with the evaporating temperature fixed at 120 °C.

The specific optimization model is outlined as follows:

Objective function:

$$max\eta = (W_{Exp} - W_{Pum})/Q_{Eva} \tag{1}$$

Constraint condition:

(1) To ensure that the working fluid steam can be condensed by cooling water, the outlet temperature of the expander is set higher than the cooling water temperature:

$$T_{Exp,out} \geq 40\ ^\circ\mathrm{C} \tag{2}$$

(2) According to the cyclic process of the ORC system, the evaporation pressure is higher than the condensation pressure, meaning the outlet pressure of the expander is lower than the inlet pressure of the expander:

$$P_{Exp,out} < P_{Exp,in} \tag{3}$$

(3) The maximum pressure in the system is set below the corresponding subcritical pressure of the working fluid:

$$P_{Pum} \leq 0.9P_c \tag{4}$$

(4) To safeguard the expander from damage and extend its lifespan, the outlet flow rate is set to be devoid of liquid phase, and the inlet pressure is lower than the saturated vapor pressure of the working fluid at the evaporation temperature:

$$\alpha_{Exp,out} = 1,\ P_{Eva,in} \leq P^{s}_{T_{Eva}} \tag{5}$$

(5) To ensure that the expander's outlet steam is neither superheated nor fails to condense successfully, the outlet pressure is set higher than the saturated vapor pressure of the working fluid at 40 °C:

$$P_{Exp,out} \geq P^{s}_{T=40^\circ\mathrm{C}} \tag{6}$$

Following optimization, the operational conditions corresponding to the optimum thermal efficiency of the ORC system are determined for the working fluid, subsequently enabling the calculation of various performance parameters.

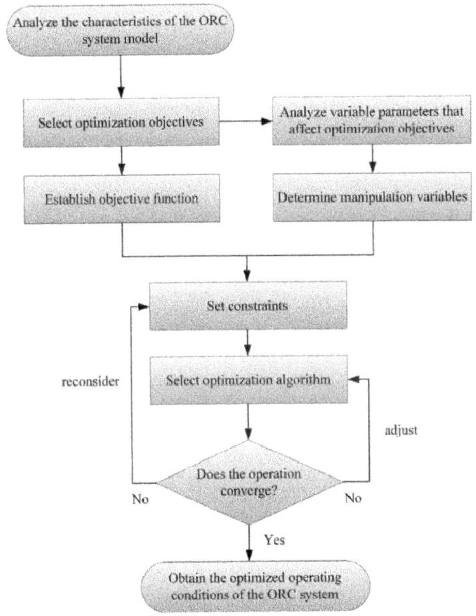

Figure 3. ORC system operating condition optimization diagram.

3. Calculation of Evaluation Indicators

3.1. Thermodynamic Model

The system's thermal efficiency, exergic efficiency, and net output work are outcomes derived from the principles of the first and second laws of thermodynamics. These metrics serve to elucidate the efficacy of waste heat recovery within the system. Specifically, the system's thermal efficiency is articulated as follows:

$$\eta_{sys} = \frac{W_{net}}{Q_{Eva}} \tag{7}$$

Q_{Eva} (kW) signifies the heat absorbed by the working fluid at intermediate pressure within the evaporator, while W_{net} (kW) denotes the system's net output work, computed as follows:

$$Q_{Eva} = m_{wf}(h_1 - h_5) \tag{8}$$

$$W_{net} = W_{Exp} - W_{Pum} \tag{9}$$

where m_{wf} represents the mass flow rate of the organic working fluid (kg·s^{-1}) and hi signifies the enthalpy value at each point measured (J·kg^{-1}). W_{Exp} (kW) quantifies the work accomplished through the entropy expansion of the working fluid within the expander, while W_{Pum} (kW) denotes the work expended due to the entropy compression of the working fluid within the pump.

The exergic efficiency of the system is as follows:

$$\eta_{exg} = \frac{W_{net}}{E_{sys}} = \frac{W_{net}}{W_{net} + I_{sys}} \tag{10}$$

$$I_{sys} = I_{Eva} + I_{Exp} + I_{Con} + I_{Pum} \tag{11}$$

E_{sys} is the exergic loss of the system and I_{sys} is the total exergic loss of all components in the system. The exergic loss of all components can be calculated as follows:

Exergic loss of evaporator I_{Eva} (kW):

$$I_{Eva} = E_{hot,in} + E_{cold,in} - E_{hot,out} - E_{cold,out} \tag{12}$$

Exergic loss of expander I_{Exp} (kW):

$$I_{Exp} = E_{Exp,in} - E_{Exp,out} - W_{Exp} \tag{13}$$

Exergic loss of condenser I_{Con} (kW):

$$I_{Con} = E_{Con,out} + E_{XQ} - E_{Con,in} \tag{14}$$

$$E_{XQ} = Q_{Con}\left(1 - \frac{T_0}{T_m}\right) \tag{15}$$

$$T_m = \frac{(T_{Con,out} - T_{Con,in})}{\ln \frac{T_{Con,out}}{T_{Con,in}}} \tag{16}$$

Exergic loss of pump I_{Pum} (kW):

$$I_{Pum} = E_{Pum,in} - E_{Pum,out} + W_{Pum} \tag{17}$$

where the subscripts delineate distinct parameters: "hot" denotes the heat flow from the heat exchanger, "*cold*" designates the cold flow, "out" signifies the output of the unit component, and "*in*" corresponds to the input. "E_{XQ}" symbolizes the heat from a constant-temperature heat source Q, kW. "E_i" represents the exergy of component i, kW. "T_0" denotes the ambient temperature, K, while "Tm" signifies the temperature of the heat source, K.

3.2. Economic Model

Investment costs predominantly encompass both equipment costs and annual operating costs. The equipment cost, specifically pertaining to the expander, evaporator, condenser, and pump, can be determined through the application of the subsequent formula [15]:

$$Y = \left(\frac{Q}{Q_B}\right)^M Z_B f_P f_T f_m \tag{18}$$

Y signifies the cost of each component (USD). For pumps and expanders, Q represents the work (kW), while for evaporators and condensers Q denotes the heat transfer area (m^2). Q_B represents the basic size of the unit component, M corresponds to the cost index of the unit component, and Z_B represents the benchmark cost of the unit component. The associated cost calculation coefficient for each unit component [16,17] is outlined in Table 1.

Table 1. Cost calculation for each unit component in the system.

Unit Component	Basic Size	Cost Index	Benchmark Costs	f_P	f_T	f_m	Material
expander	250	0.46	98,400	1	1.6	1	carbon steel
pump	4	0.55	9840	1.5	1	1	carbon steel
evaporator	80	0.68	32,800	1.5	1.6	1	carbon steel
condenser	80	0.68	32,800	1	1.6	1	carbon steel

Annual operating expenditures comprise both annual electricity consumption and annual water consumption. The annual cost attributed to electricity consumption is denoted as V (USD) and determined by the subsequent formula:

$$V = C_{ele} W_{Pum} t \tag{19}$$

C_{ele} represents the unit price of electricity, set at 0.11 USD/kWh, and t denotes the annual operating time of the system device, calculated as 8000 h [18].

Annual water charges are recorded as U (USD) and are calculated as follows:

$$U = C_{cw} Q_{Con} t \tag{20}$$

C_{cw} is the unit price of water charge, set at 0.35 USD/GJ, Q_{Con} is the condenser heat load (GJ/h), and t is the system operating time in a year, calculated as 8000 h.

The ORC system will have exergic loss during operation, and the economic waste caused by exergic loss is called the exergic loss cost, which is calculated as follows:

$$C_{D_{sys}} = C_{D_{Exp}} + C_{D_{Eva}} + C_{D_{Con}} + C_{D_{Pum}} \tag{21}$$

$C_{D_{sys}}$ represents the overall exergic loss cost of the system, while $C_{D_{Exp}}$, $C_{D_{Eva}}$, $C_{D_{Con}}$, and $C_{D_{Pum}}$ represent the exergic loss cost of the expander, evaporator, condenser, and pump, respectively. The formula is as follows:

$$C_{D_k} = 3600\frac{I_k C_{F_k}}{10^6} \tag{22}$$

C_{D_k} indicates the exergic cost of a unit part k, I_k indicates the exergic cost of a unit part K, and C_{F_k} indicates the average cost of fuel used per exergic unit.

For the convenience of explanation, based on the fact that the investment cost dominates compared with the operation and maintenance cost [18], this study has employed a simplified model that ignores the operation and maintenance cost. If it is necessary to further establish a more accurate and rigorous model, the operation and maintenance cost should be taken into account.

3.3. *Environmental Model*

The organic Rankine cycle (ORC) system stands as a means to repurpose industrial waste heat, geothermal energy, and other energy sources, concurrently diminishing CO_2 emissions in accordance with energy conservation and emission reduction imperatives. Consequently, the reduction in CO_2 emissions assumes significance as a pivotal metric for gauging the operational efficacy of the system. Exergic environmental factors, including CO_2 emission reduction, are chosen to appraise the merits and drawbacks of the organic working medium within the ORC system. The calculation model for these indices is articulated as follows:

System CO_2 emission reduction [19]:

$$m_{CO_2} = W_{net}\beta_{CO_2} \tag{23}$$

where β_{CO_2} is the CO_2 emission coefficient: that is, the amount of CO_2 emitted during the production of unit electric energy, usually 0.872 kg/(kWh) [20].

The exergic environmental factors of the system are as follows:

$$f_{ei} = \frac{I_{sys}}{E_{sys,in}} = \frac{I_{sys}}{E_{Eva,in} + E_{Exp,in} + E_{Con,in} + E_{Pum,in}} \tag{24}$$

The exergy environmental factor serves as an indicator of the potential impact on the environment. A smaller exergy value implies a reduced environmental impact.

4. Multi-Objective Evaluation Strategy

The DEA method is an outcome of interdisciplinary research involving operations research, management science, and mathematical economics. Primarily applied for assessing the relative efficiency of homogeneous decision-making units (DMUs) with multiple inputs and outputs, DEA stands out as a non-parametric method.

As a non-parametric methodology that does not require predefined weights [21], DEA automatically optimizes the combined weight of the input and output through linear programming. It is completely data-driven and avoids subjective assignment biases. Its advantages lie in not requiring preset weights. It is suitable for multi-input and multi-output scenarios and objectively reflects the relative efficiency of decision-making units.

Its principal advantage over other methods lies in its exemption from assuming the production function and estimating pertinent parameters in advance. Instead, it yields relatively objective and practical quantitative evaluation results based directly on the original input and output data. The fundamental models of DEA are categorized into the Charnes–Cooper–Rhodes (CCR) model, based on constant returns to scale [22], and the Bankers–Charnes–Cooper (BCC) model, based on variable returns to scale [23]. The efficiency calculated by the CCR model typically represents overall efficiency, while that derived from the BCC model signifies pure technical efficiency. In this study, we mainly focused on overall efficiency and chose the CCR model to comprehensively evaluate the working fluid of the ORC system.

Consider a scenario where there exists n decision units (DMUs), each possessing m inputs and s outputs. Denoted as input X_j and output Y_j, respectively, we express these as $X_j = (x_{1j}, x_{2j}, \ldots, x_{mj})^{\mathrm{T}}$, $Y_j = (y_{1j}, y_{2j}, \ldots, y_{sj})^{\mathrm{T}}$, and $j = 1, 2, \ldots, n$. For a designated $\mathrm{DMU_k}$, with k = 1, 2, ..., n, the dual programming of the CCR model, essential for assessing its validity, is formulated as follows [13]:

$$\begin{cases} Min\left[\theta - \varepsilon\left(\sum\limits_{i=1}^{m} S_i^+ + \sum\limits_{r=1}^{s} S_r^-\right)\right] \\ s.t. \sum\limits_{j=1}^{n} \lambda_j x_{ij} + S^- = \theta x_{ik}, S^- \geq 0, \\ \sum\limits_{j=1}^{n} \lambda_j y_{rj} - S^+ = y_{rk}, S^+ \geq 0, \lambda_j \geq 0. \end{cases} \tag{25}$$

where ε denotes a non-Archimedean infinitesimally small quantity (typically an arbitrarily small positive number, often set at 10^{-6}); θ represents the effective value of DMUk, signifying the efficient utilization degree of output relative to input; λ is the combination ratio of the jth decision unit within an effectively reconstructed DMU combination relative to DMUk; and S^+ denotes the slack variable, while S^- represents the residual variable, indicating potential input redundancy and output deficiency of the DMU. A DMU is deemed optimal when θ equals 1, and both S^+ and S^- are zero. In cases where θ equals 1 but S^+ or S^- is non-zero, the DMU is considered a weak DEA efficient. In such instances, within a system comprising DMUs, the input index can reduce S^- while maintaining the original output or increase the output S^+ without altering the original input. When θ is less than 1, the DMU is categorized as non-DEA efficient, with smaller values indicating a greater distance from the optimal frontier.

Utilizing the DEA method, a multi-objective evaluation of the organic working fluid within the ORC system is conducted, considering performance, economy, and environmental aspects. The evaluation framework is delineated in Figure 4. The objective of this assessment is to identify the optimal working fluid from a myriad of options for the system. As depicted in the figure, the evaluation strategy encompasses three assessment systems: thermodynamic and economic performance, thermodynamic and environmental performance, and thermodynamic, economic, and environmental performance. Through a cross-combination approach, working fluid evaluation outcomes catering to diverse user requirements are ascertained.

The working fluid is a crucial working fluid for realizing waste heat power conversion. It can transfer heat from a lower temperature to a higher temperature. The existing literature reports that there are more than 140 kinds of working fluids suitable for ORC systems. Based on the concept of low-carbon environmental protection, in order to better achieve the goal of low-carbon and near-zero emissions, according to the content of the Kigali Amendment to the Montreal Protocol, the application of hydrofluorocarbon CFCs, HCFCs, PFCs, and other working fluids has been gradually canceled, which can meet the ORC system. The working fluid is gradually reduced to more than 60 kinds. The 62 working fluids selected in this paper take into account environmental factors (global warming potential ozone depletion potential (ODP), global warming potential (GWP)), basic physical properties of working fluids, and other factors, and study the comprehensive performance of these working fluids in ORC systems, as detailed in Table 2.

This DEA-based multi-objective evaluation strategy for ORC working fluids starts by selecting a working fluid set, obtaining its basic physical properties, and building both the ORC system and its optimization model, iterating the optimization process until convergence to acquire optimal operating parameters. It then incorporates economic performance indicators (investment cost, effective energy loss cost), thermodynamic performance indicators (thermal efficiency, exergy efficiency, net output power), and environmental performance indicators (effective energy environmental factors, CO_2 reduction emissions). Finally, through three DEA evaluations (DEA evaluation 1, DEA evaluation 2, DEA evaluation 3), the strategy integrates these multi-objective indicators and outputs comprehensive evaluation results. The procedure is illustrated in Figure 4.

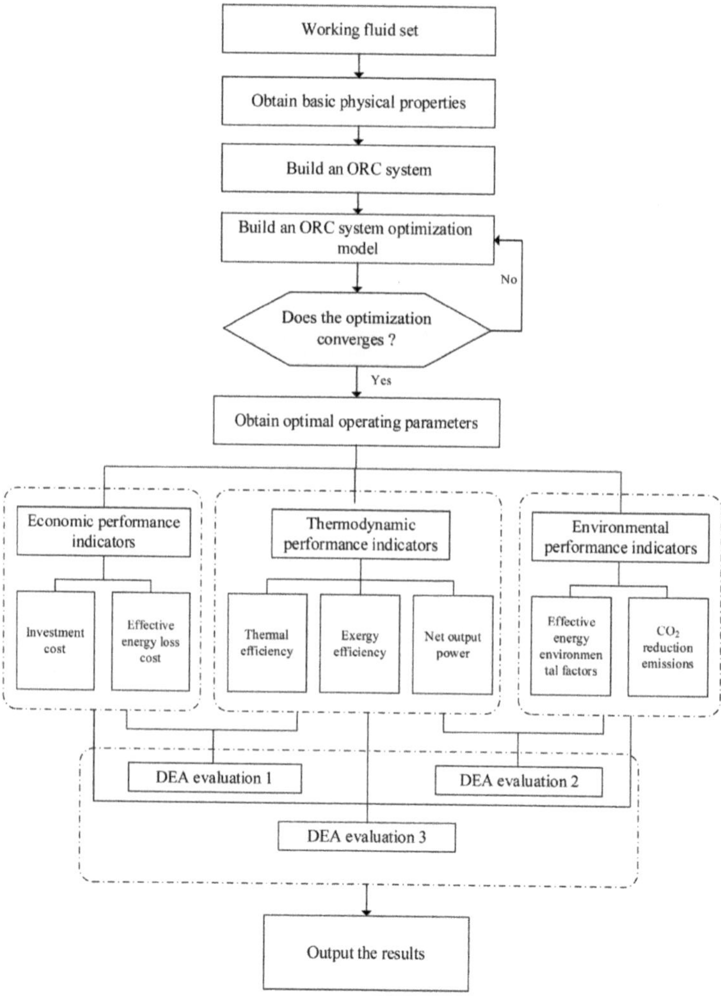

Figure 4. Multi-objective evaluation strategy of ORC working fluids based on DEA.

Table 2. Working fluid to be evaluated.

CAS	Alias or English Name	CAS	Alias or English Name	CAS	Alias or English Name
75-69-4	R11	71-43-2	Benzene	460-73-1	R245fa
75-71-8	R12	108-88-3	Toluene	677-56-5	R236cb
76-13-1	R113	67-64-1	Acetone	690-39-1	R236fa
76-14-2	R114	756-13-8	Novec649	431-63-0	R236ea
76-15-3	R115	115-10-6	E170	406-58-6	R365mfc
74-98-6	R290	75-45-6	R22	156-59-2	R1132(Z)
75-19-4	R270	75-10-5	R32	156-60-5	R1132(E)
106-97-8	R600	75-43-4	R21	102687-65-0	R1233zd(E)
75-28-5	R600a	306-83-2	R123	29118-24-9	R1234ze(e)
78-78-4	R601a	63938-10-3	R124	754-12-1	R1234yf
109-66-0	R601	75-68-3	R142b	29118-25-0	R1234ze(Z)
107-83-5	NSC66496	76-19-7	R218	690-27-7	R1225zc
463-82-1	2,2-Dime	355-25-9	R3110	5528-43-8	R1225ye(Z)
110-82-7	CHX	115-25-3	RC318	116-15-4	R1216
110-54-3	Hexane	678-26-2	DDFP	677-21-4	R1243zf
142-82-5	Heptane	353-36-6	R161	359-11-5	R1123
111-65-9	1-Octane	811-97-2	R134a	355-42-0	FC-72
115-07-1	R1270	420-46-2	R143a	56-23-5	R10
106-98-9	Butene	75-37-6	R152a	616-38-6	DMC
115-11-7	Isobutene	431-89-0	R227ea	1885-48-9	RE245fa2
74-99-7	Propyne	679-86-7	R245ca		

5. Results

5.1. Single-Objective Evaluation Results

Leveraging the optimization and computational outcomes (normalization) of the ORC system, a univariate evaluation of the working fluid is undertaken, employing thermodynamic performance metrics such as thermal efficiency, net output work, and exergic efficiency. Furthermore, economic performance is gauged using investment cost and exergic loss cost, while environmental considerations encompass CO_2 emission reduction and exergic environmental factors. These parameters collectively serve as the evaluation indices for this comprehensive assessment of the working fluid.

5.1.1. Thermodynamic Index Evaluation Results

The results of the single-objective evaluation, utilizing thermodynamic performance as the primary index, are presented in Figure 5. Thermodynamic performance serves as a metric reflecting the extent of the ORC system's recovery and utilization of waste heat. Taking thermal efficiency as an illustrative example, a higher thermal efficiency signifies an improved waste heat recovery effect, consequently elevating the priority of the corresponding working fluid. The working fluids are categorized into four grades—A, B, C, and D—based on the thermal efficiency value, ranging from high to low, as depicted in Figure 5 (a), (b), (c), and (d), respectively. Grade A denotes the superior working fluid, followed by grades B and C, with Grade D representing the least favorable. It is evident that substantial disparities exist in the single-objective evaluation outcomes across the three indices. For instance, working fluid R1132(E) holds the 55th and 53rd positions in the thermal efficiency and net work output evaluations, designating it as a Grade D working fluid. However, it secures the fifth position in the exergic efficiency evaluation, classifying it as a Grade A working fluid. Notably, the figure illustrates a counteractive trend between net output work and thermal efficiency for certain working fluids like Heptane, R10, DDFP, Novec649, etc. This phenomenon is linked to the working fluid's capacity to absorb heat in the evaporator, where the rate of heat absorption surpasses the increase in net output work, resulting in a scenario where thermal efficiency decreases with the rise in net power generation.

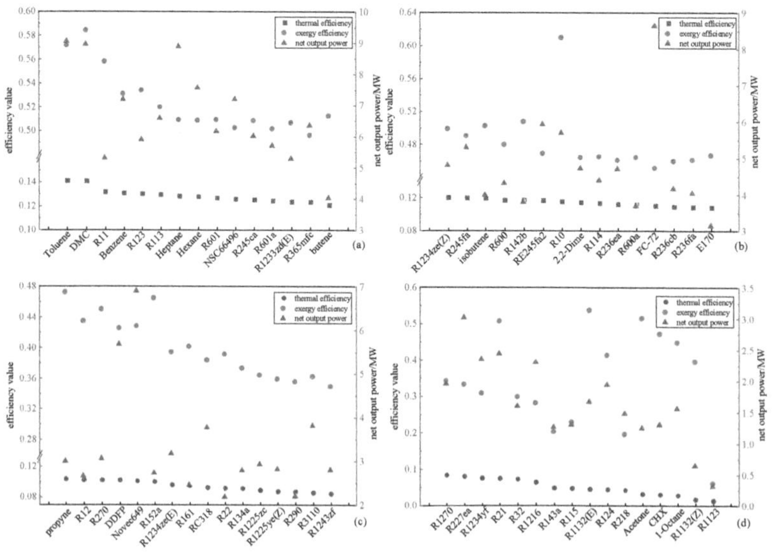

Figure 5. Results of single-objective evaluation of ORC working fluids using thermal efficiency, net power generation, and efficiency (**a**–**d**).

5.1.2. Evaluation Results of Economic Indicators

As depicted in Figure 6, both exergic cost and exergic loss cost represent expenses incurred during enterprise operations. A lower cost is advantageous for enterprise functionality, signifying a higher priority for the corresponding working fluid. Hence, the working fluids are categorized into four grades—A, B, C, and D—based on the order of investment cost, ranging from low to high. Reflecting Figure 6 (a), (b), (c), and (d), respectively, Grade A designates the superior working fluid, followed by Grades B and C, with Grade D representing the least favorable. Notably, substantial differences emerge in the ranking of working chemicals concerning investment cost and exergic loss cost, including instances of inverse rankings. For instance, R1123 is classified as Grade A and optimal in the assessment of investment cost, while earning a Grade C in the evaluation of exergic loss cost.

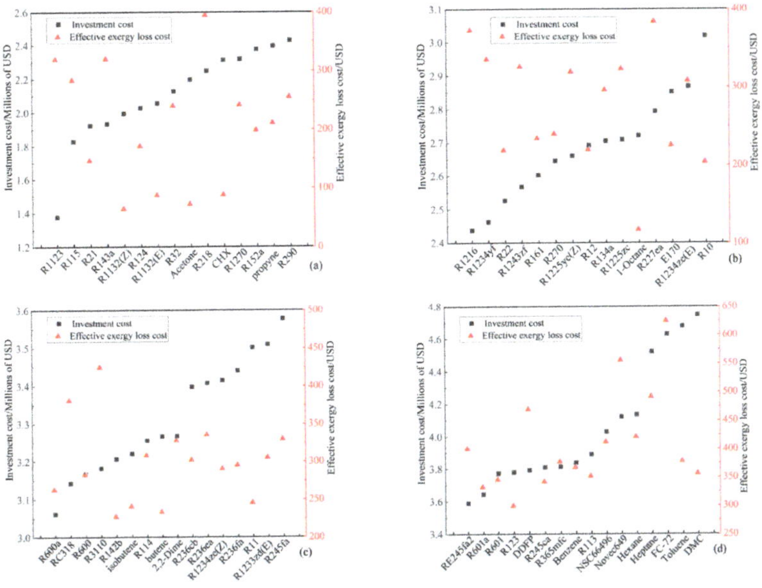

Figure 6. Results of single-target evaluation of ORC working fluids by investment cost and loss cost (**a**–**d**).

5.1.3. Environmental Index Evaluation Results

Exergic environmental factors and CO_2 reduction serve as measures to assess whether an enterprise's operations align with the principles of green and low-carbon development, as illustrated in Figure 7. A lower exergic environmental factor implies reduced environmental impact, elevating the priority of the corresponding working fluid. Exergic environmental factors associated with working fluids are categorized into four grades—A, B, C, and D—aligning with Figure 7 (a), (b), (c), and (d), respectively. Grade A denotes the superior working fluid, followed by Grades B and C, with Grade D representing the least favorable. Notably, working fluids such as R1132(Z), Acetone, 1-Octane, CHX, and R1132(E) exhibit minimal exergic factors, indicating optimal performance. Conversely, when CO_2 emission reduction is utilized as an evaluation index, Toluene, DMC, Heptane, and FC-72 emerge with the highest CO_2 emission reduction and superior performance. The two evaluation outcomes display inconsistencies, with some working fluids even securing opposing rankings. For instance, working fluid R1132(Z) claims the top position in the former evaluation, marking it as the best, while it ranks second to last in the latter assessment.

In conclusion, it becomes evident that the single-objective evaluation method, when applied to assess working fluids, tends to be one-sided and lacks objectivity. The adoption of multiple single-objective evaluation combinations may result in disparities or even contradictions in the categorization of working quality across different evaluation indexes, preventing the effective screening of working fluids. Consequently, we propose the subsequent multi-objective evaluation method based on the DEA method.

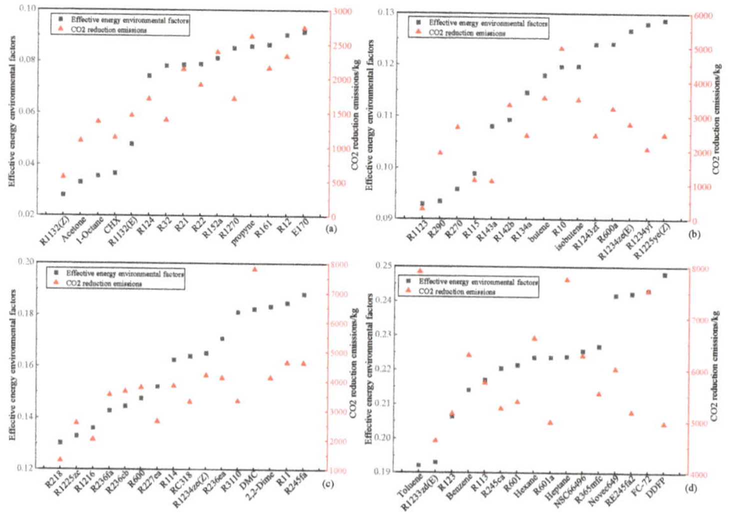

Figure 7. Results of single-objective evaluation of ORC working fluids using environmental factors and CO_2 reduction as indicators (**a**–**d**).

5.2. Results of Multi-Objective Evaluation

Exergic loss cost, investment cost, and exergic environmental factors are designated as input indices, whereas exergic efficiency, net output work, exergic efficiency, and CO_2 emission reduction are designated as output indices. This consideration is based on the thermodynamic, economic, and environmental performance of the ORC system during operation, stemming from the optimization and computation results of various indices. The evaluation and screening of the working fluid for the ORC system are undertaken using the DEA multi-objective evaluation method.

5.2.1. Evaluation of Thermodynamic and Economic Performance

Figure 8 illustrates the outcomes of the DEA multi-objective evaluation of organic working fluids in the ORC system, utilizing thermodynamic and economic performance parameters as indices at an evaporation temperature of 120 °C. The working fluids are categorized into four grades—A, B, C, and D—based on the effective values of the DEA evaluation, corresponding to Figure 8 (a), (b), (c), and (d), respectively, with Grade D representing the least favorable category. Working substances such as Heptane, Propyne, Toluene, Acetone, R21, R1132(E), and R10 achieved Grade A status, each with an evaluation effective value of 1. This denotes that these working substances exhibit the most comprehensive performance among all evaluated options and should be prioritized when selecting working substances for the ORC system. Conversely, working fluids like R1123, R218, R143a, and others falling under Grade D demonstrate a poor overall performance, making them unsuitable for the ORC system.

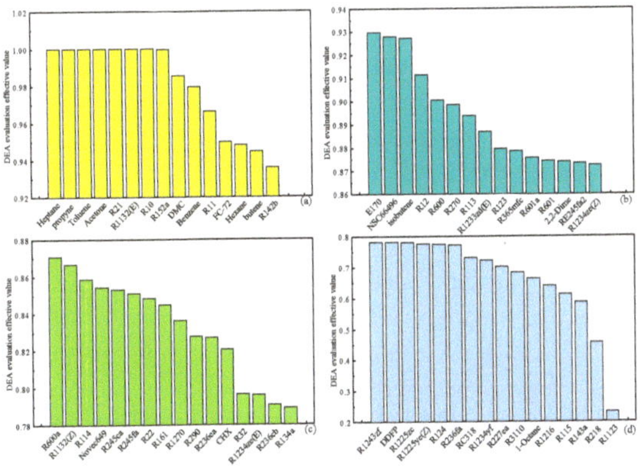

Figure 8. Effective values of DEA of organic working fluids from thermodynamic and economic perspectives at 120 °C (a–d).

5.2.2. Evaluation Based on Thermodynamic and Environmental Performance

Figure 9 depicts the outcomes of the multi-objective DEA evaluation for organic working fluids in the ORC system, utilizing thermodynamic and environmental performance parameters as indices at an evaporation temperature of 120 °C. The working fluids are classified into four grades—A, B, C, and D—based on the effective values of the DEA evaluation, corresponding to Figure 9 (a), (b), (c), and (d), respectively, with Grade D being the least favorable. Notably, working fluids such as 1-Octane, Acetone, R152a, R1132(E), R10, and DMC achieved an effective value of 1, showcasing the most optimal comprehensive performance when applied in the ORC system. On the contrary, working fluids like R1123, R218, R143a, and others falling under Grade D exhibited lower effective values and an inferior comprehensive performance compared to other options, rendering them unsuitable for use as circulating working fluids in the ORC system.

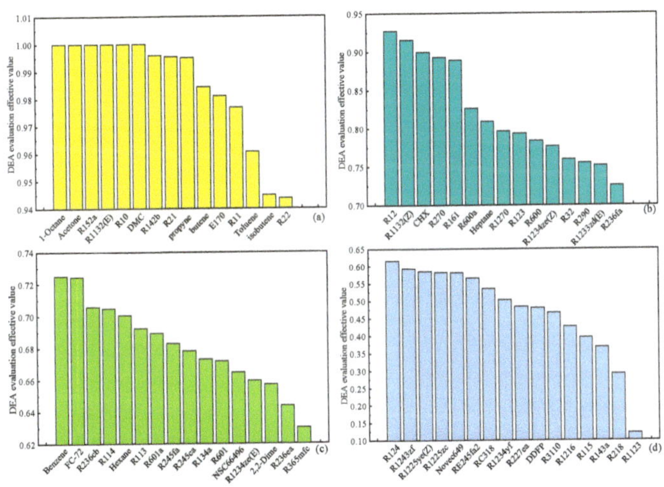

Figure 9. Effective values of DEA of organic working fluids from thermodynamic and environmental perspectives at 120 °C (a–d).

5.2.3. Evaluation Based on Thermodynamic, Economic, and Environmental Performance

The evaluation outcomes of ORC system working fluids utilizing the DEA multi-objective evaluation method are illustrated in Figure 10. The effective value derived from the DEA evaluation serves as an indicator of the comprehensive performance of a given working fluid under the current evaluation criteria. A higher effective value correlates with a more superior comprehensive performance of the working fluid. As denoted in Figure 10 (a), (b), (c), and (d), respectively, Grade A represents the optimal working fluid, Grade B follows, then Grade C, and finally Grade D is deemed the least favorable. Notably, Heptane, 1-Octane, Propyne, Toluene, Acetone, R21, R152a, R1132(E), R10, and DMC are classified as Grade A working substances, each achieving an effective value of 1 in the DEA evaluation. This signifies that the comprehensive performance of these working substances surpasses that of others, recommending them as the preferred circulating working fluids for the ORC system. Conversely, Class D working fluids exhibit an inferior performance compared to others, warranting their exclusion from the recommended circulating working fluids for the ORC system.

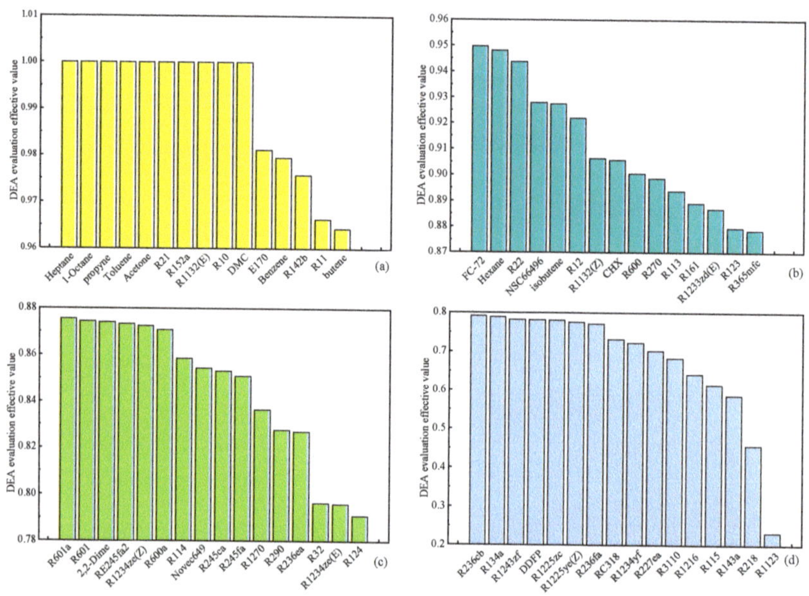

Figure 10. Effective values of DEA of organic working fluids from thermodynamic, economic, and environmental perspectives at 120 °C (**a**–**d**).

The findings reveal that the count of optimal working fluid rises as the evaporation temperature declines, aligning with the characteristic of system performance fluctuations in response to evaporation temperature variations. Discrepancies in the optimal working fluid across the three scenarios signify that the DEA method comprehensively assesses and ranks the performance of organic working fluids, taking into account variations in evaluation indices. The overall trend observed in the majority of working fluids remains consistent, indicating that the assessment ranking of working fluids adheres to the performance patterns of these fluids under each evaluation index.

In conclusion, upon contrasting the outcomes of multi-objective evaluation with those of single-objective evaluation, it becomes apparent that the evaluation and ranking outcomes of working fluid quality by the DEA method generally exhibit greater conformity with the single-objective evaluation criteria for working fluid. This alignment indicates consistency with the corresponding system performance characteristics of the working fluid. Nevertheless, within a limited scope, the ranking of working fluid deviates from the single-objective result, suggesting a comprehensive evaluation outcome achieved through the reconciliation of multiple evaluation indicators.

5.3. Application Cases of Multi-Objective Evaluation

To validate the precision of the multi-objective evaluation outcomes, we conducted applied research on the working fluid Heptane, chosen due to its commendable performance in the multi-objective evaluation results. A comparative analysis of the application effects between Heptane and the working fluid R245fa commonly used in the existing literature was undertaken [24,25]. The results, depicted in Figure 11, reveal notable advantages for Heptane. Specifically, in contrast to R245fa, Heptane demonstrates a 5.02% reduction in exergy damage, a 0.24% increase in thermal efficiency, and a substantial 2.04% augmentation in net work output. These findings underscore the superior performance of Heptane, affirming the efficacy of the multi-objective evaluation approach in selecting the optimal working fluid.

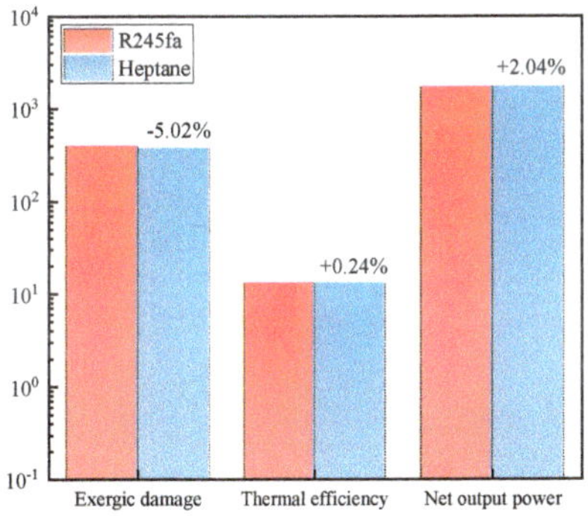

Figure 11. Comparison of the application effect of Heptane and R245fa.

6. Conclusions

In this study, single-objective and multi-objective evaluations were conducted to study the screening of working fluids in the ORC system. The limitations inherent in the single-objective evaluation approach were duly addressed. A multi-objective evaluation method, employing data envelopment analysis (DEA), was introduced to comprehensively assess and rank organic working fluids within the ORC system, facilitating the selection of the optimal working fluid.

Although some traditional refrigerants have acceptable performance in terms of CO_2 emission reduction, some may have high global warming potential (GWP) values that could offset the benefits brought about by the emission reduction. Some are also accompanied by potential hazards due to their toxicity. Therefore, they need to be selected with caution. We suggest giving priority to working fluids with a low GWP and complying with safety standards to meet international environmental protection regulations and safety requirements, and to enhance the practical guiding value of the research.

The key conclusions derived from this study are as follows:

(1) Single-objective evaluations were performed for 62 working fluids, resulting in disparate screening outcomes due to the singular nature of the evaluations and subjective influences. The results indicated that Toluene and DMC have a better advantage in thermo-

dynamic performance. The investment cost of R1123 is the lowest, while the exergy loss of R1132(Z) is the lowest.

(2) The DEA multi-objective evaluation method emerged as a robust approach capable of harmonizing diverse evaluation indicators. It swiftly and accurately gauged the comprehensive performance of various working fluids within the ORC system, delineating their prioritized applications. This method effectively overcame the limitations associated with single-objective evaluations.

(3) The working fluid Heptane obtained by DEA multi-objective evaluation showed better advantages than R245fa, the working fluid commonly used in the literature. The application results of multi-objective evaluation indicated that Heptane demonstrates a 5.02% reduction in exergy damage, a 0.24% increase in thermal efficiency, and a substantial 2.04% augmentation in net work output.

The proposed multi-objective evaluation method, grounded in DEA, offers an innovative approach for assessing and ranking ORC working fluids under distinct evaporation temperatures and varied combination indexes. This method circumvents issues related to a limited working fluid dataset and reliance on singular evaluation indices. The strategy provides precise, concise data, offering practical guidance for the intricate process of selecting working fluids, especially when confronted with variations in low-temperature heat source temperatures and diverse operational requisites in industrial production. Hence, this screening strategy holds substantial practical application value.

Author Contributions: L.Z.: Writing—original draft, Visualization, Software, Investigation, Formal analysis. L.W.: Writing—review and editing, Software, Conceptualization. X.S.: Writing—review and editing, Supervision, Conceptualization. L.X.: Supervision, Funding acquisition, Conceptualization. S.T.: Writing—review and editing, Supervision, Funding acquisition, Conceptualization. S.X.: Supervision, Funding acquisition, Conceptualization. S.J.: Supervision, Conceptualization. All authors have read and agreed to the published version of the manuscript.

Funding: This work is supported by the National Natural Science Foundation of China (No. 22178190), the National Youth Natural Science Foundation of China (No. 22108138, 22408195), the Key Research and Development Program of Shandong Province (Major Scientific and Technological Innovation Project) (No. 2024CXGC010405), and Shandong Provincial Key Laboratory of Chemical Process Simulation and Optimization Industrial Software (PKL2024F23).

Data Availability Statement: The original contributions presented in this study are included in the article. Further inquiries can be directed to the corresponding author.

Conflicts of Interest: The authors declare no conflicts of interest.

Nomenclature

$\dot{m}$	mass flow rate
$\dot{W}$	power
$\dot{Q}$	heat duty
I_{sys}	overall exergic loss of all components in the system
E_{sys}	exergic loss of the system
I_{Eva}	exergy loss of the evaporator
I_{Exp}	exergy loss of the turbine
I_{Con}	exergy loss of the condenser
T_m	temperature of the reservoir
E_{XQ}	exergy of constant-temperature heat source Q
Y	cost of each component
h	specific enthalpy, $(J \cdot kg^{-1})$

T_0	ambient temperature, (25 °C)
M	unit component cost index
Z_B	base cost of the unit component
Q_B	basic capacity
V	annual cost of electricity consumption
C_{ele}	unit price of electricity
t	operating time of the system device in a year
U	annual water cost
f_m	correction factor for materials of construction
f_p	correction factor for design pressure
f_t	correction factor for design temperature
$C_{D_{sys}}$	overall exergic loss cost of the system
I_k	exergic cost of a unit part k
C_{F_k}	average cost of fuel per exergic unit
β_{CO_2}	CO_2 emission coefficient
ε	non-Archimedean infinitesimally small quantity
θ	effective value of DMUk
λ	combination proportion of the jth decision-making unit
f_{ei}	exergic environmental factor
θ_{ei}	environmental damage effect factor
C_{ei}	coefficient of environmental impact
f_{es}	exergy stability coefficient
Abbreviations	
ORC	organic Rankine cycle
CCR	Charnes–Cooper–Rhodes
BCC	Banker–Charnes–Cooper
DEA	data envelopment analysis
CON	condenser
EVA	evaporator
TUR	turbine

References

1. BP. *BP Statistical Review of World Energy*; BP: London, UK, 2021.
2. Xu, Z.Y.; Wang, R.Z.; Yang, C. Perspectives for Low-Temperature Waste Heat Recovery. *Energy* **2019**, *176*, 1037–1043. [CrossRef]
3. Liu, Z.; Preston, D.J. Enhanced Condensation for Improved Energy Efficiency. *Joule* **2019**, *3*, 1182–1184. [CrossRef]
4. Yang, S.; Yang, S.; Wang, Y.; Qian, Y. Low Grade Waste Heat Recovery with a Novel Cascade Absorption Heat Transformer. *Energy* **2017**, *130*, 461–472. [CrossRef]
5. Emadi, M.A.; Chitgar, N.; Oyewunmi, O.A.; Markides, C.N. Working-Fluid Selection and Thermoeconomic Optimisation of a Combined Cycle Cogeneration Dual-Loop Organic Rankine Cycle (ORC) System for Solid Oxide Fuel Cell (SOFC) Waste-Heat Recovery. *Appl. Energy* **2020**, *261*, 114384. [CrossRef]
6. Roumpedakis, T.C.; Loumpardis, G.; Monokrousou, E.; Braimakis, K.; Charalampidis, A.; Karellas, S. Exergetic and Economic Analysis of a Solar Driven Small Scale ORC. *Renew. Energy* **2020**, *157*, 1008–1024. [CrossRef]
7. Desai, N.B.; Bandyopadhyay, S. Thermo-Economic Comparisons between Solar Steam Rankine and Organic Rankine Cycles. *Appl. Therm. Eng.* **2016**, *105*, 862–875. [CrossRef]
8. Li, J.; Ge, Z.; Liu, Q.; Duan, Y.; Yang, Z. Thermo-Economic Performance Analyses and Comparison of Two Turbine Layouts for Organic Rankine Cycles with Dual-Pressure Evaporation. *Energy Convers. Manag.* **2018**, *164*, 603–614. [CrossRef]
9. Heberle, F.; Preißinger, M.; Brüggemann, D. Zeotropic Mixtures as Working Fluids in Organic Rankine Cycles for Low-Enthalpy Geothermal Resources. *Renew. Energy* **2012**, *37*, 364–370. [CrossRef]
10. Fathi, B.; Ashena, M.; Bahari, A.R. Energy, Environmental, and Economic Efficiency in Fossil Fuel Exporting Countries: A Modified Data Envelopment Analysis Approach. *Sustain. Prod. Consum.* **2021**, *26*, 588–596. [CrossRef]
11. Mohd Chachuli, F.S.; Ahmad Ludin, N.; Md Jedi, M.A.; Hamid, N.H. Transition of Renewable Energy Policies in Malaysia: Benchmarking with Data Envelopment Analysis. *Renew. Sustain. Energy Rev.* **2021**, *150*, 111456. [CrossRef]

12. Wang, S.; Xing, L.; Chen, X.; Song, M. Evaluating and Enhancing Natural Resource Asset Management Efficiency in China: A Data Envelopment Analysis Study. *Resour. Policy* **2024**, *92*, 105000. [CrossRef]
13. Xiang, S.; Wang, L.; Wang, Y.; Bi, R.; Xia, L.; Sun, X. Exploration of Gradient Energy-Saving Separation Processes for Ethylene Glycol Mixtures Based on Energy, Exergy, Environment, and Economic Analyses. *Sep. Purif. Technol.* **2021**, *279*, 119787. [CrossRef]
14. Wang, L.; Niu, B.; Xia, L.; Zhao, W.; Sun, X.; Xiang, S. Towards on Energy-Saving Separation of Ethylene Glycol Mixture from Production Plants Based on Energy, Exergy, Economic, Environmental, Exergoeconomic, and Exergoenvironmental Analyses. *J. Clean. Prod.* **2022**, *377*, 134388. [CrossRef]
15. Yang, J.; Wu, J.; Li, X.; Zhu, Q. Sustainability Performance Analysis of Environment Innovation Systems Using a Two-Stage Network DEA Model with Shared Resources. *Front. Eng. Manag.* **2022**, *9*, 425–438. [CrossRef]
16. Henriques, C.O.; Gouveia, C.M.; Tenente, M.; Da Silva, P.P. Employing Value-Based DEA in the Eco-Efficiency Assessment of the Electricity Sector. *Econ. Anal. Policy* **2022**, *73*, 826–844. [CrossRef]
17. Peng, C.; Feng, D.; Guo, S. Material Selection in Green Design: A Method Combining DEA and TOPSIS. *Sustainability* **2021**, *13*, 5497. [CrossRef]
18. Bi, G.; Shao, Y.; Song, W.; Yang, F.; Luo, Y. A Performance Evaluation of China's Coal-Fired Power Generation with Pollutant Mitigation Options. *J. Clean. Prod.* **2018**, *171*, 867–876. [CrossRef]
19. Song, M.-L.; Zhang, L.-L.; Liu, W.; Fisher, R. Bootstrap-DEA Analysis of BRICS' Energy Efficiency Based on Small Sample Data. *Appl. Energy* **2013**, *112*, 1049–1055. [CrossRef]
20. Jradi, S.; Chameeva, T.B.; Delhomme, B.; Jaegler, A. Tracking Carbon Footprint in French Vineyards: A DEA Performance Assessment. *J. Clean. Prod.* **2018**, *192*, 43–54. [CrossRef]
21. Banker, R.D.; Charnes, A.; Cooper, W.W. Some Models for Estimating Technical and Scale Inefficiencies in Data Envelopment Analysis. *Manag. Sci.* **1984**, *30*, 1078–1092. [CrossRef]
22. Lukawski, M.Z.; Tester, J.W.; DiPippo, R. Impact of Molecular Structure of Working Fluids on Performance of Organic Rankine Cycles (ORCs). *Sustain. Energy Fuels* **2017**, *1*, 1098–1111. [CrossRef]
23. Wang, L.; Zhao, J.; Teng, J.; Dong, S.; Wang, Y.; Xiang, S.; Sun, X. Study on an Energy-Saving Process for Separation Ethylene Elycol Mixture through Heat-Pump, Heat-Integration and ORC Driven by Waste-Heat. *Energy* **2022**, *243*, 122985. [CrossRef]
24. Wang, E.; Zhang, H.; Fan, B.; Wu, Y. Optimized Performances Comparison of Organic Rankine Cycles for Low Grade Waste Heat Recovery. *J. Mech. Sci. Technol.* **2012**, *26*, 2301–2312. [CrossRef]
25. Hu, B.; Guo, J.; Yang, Y.; Shao, Y. Selection of Working Fluid for Organic Rankine Cycle Used in Low Temperature Geothermal Power Plant. *Energy Rep.* **2022**, *8*, 179–186. [CrossRef]

processes

MDPI

Article

Optimizing Solar PV Deployment in Manufacturing: A Morphological Matrix and Fuzzy TOPSIS Approach

Citlaly Pérez Briceño [1], Pedro Ponce [1,*], Aminah Robinson Fayek [2], Brian Anthony [3], Russel Bradley [3], Therese Peffer [4], Alan Meier [5] and Qipei Mei [6]

1 Institute of Advanced Materials for Sustainable Manufacturing, Tecnologico de Monterrey, Monterrey 64849, N.L., Mexico; a01336766@tec.mx
2 Hole School of Construction Engineering, Department of Civil and Environmental Engineering, University of Alberta, 7-203 Donadeo Innovation Centre for Engineering, 9211 116 St NW, Edmonton, AB T6G 1H9, Canada; aminah.robinson@ualberta.ca
3 Department of Mechanical Engineering, Massachusetts Institute of Technology, 77 Massachusetts Avenue, Room 3-173, Cambridge, MA 02139, USA; banthony@mit.edu (B.A.); russelb@mit.edu (R.B.)
4 California Institute for Energy and Environment, University of California, Berkeley, CA 94720, USA; tpeffer@berkeley.edu
5 Energy and Efficiency Institute, University of California, Davis, CA 95616, USA; akmeier@ucdavis.edu
6 Civil and Environmental Engineering Department, University of Alberta, 6-263 Donadeo Innovation Centre for Engineering, 9211 116 St, Edmonton, AB T6G 1H9, Canada; qipei.mei@ualberta.ca
* Correspondence: pedro.ponce@tec.mx

Abstract: The growing energy demand of the industrial sector and the need for sustainable solutions highlight the importance of efficient decision making in solar photovoltaic (PV) implementation. Selecting optimal PV configuration is complex due to the interdependent technical, economic, environmental, and social factors involved. This study introduces an integrated decision-making method combining a morphological matrix and fuzzy TOPSIS to systematically select and rank optimal PV system configurations for manufacturing firms. While the morphological matrix exhaustively examines possible design solutions based on sensing, smart, sustainable, and social (S4) attributes, the fuzzy TOPSIS method ranks the alternatives by handling uncertainty in decision making. A case study conducted in a Mexican manufacturing company validates the methodology's effectiveness. The optimal PV configuration identified comprehensively addresses operational and sustainability criteria, covering all lifecycle stages. This approach demonstrates quantitative superiority and greater robustness compared to existing fuzzy TOPSIS-based methods for solar PV applications. The findings highlight the practical value of data-driven, multi-criteria decision making for industrial solar energy adoption, enhancing project feasibility, cost efficiency, and environmental compliance. Future research will incorporate discrete event simulation (DES) to further refine energy consumption strategies in manufacturing.

Keywords: solar energy; decision making; manufacturing; morphological matrix; fuzzy TOPSIS; sustainability

1. Introduction

In recent decades, the rapid rise in energy demand has positioned energy consumption as a critical issue, primarily because most of the energy is produced by fossil fuels [1]. These fuels are significant contributors to environmental problems such as climate change and global warming, leading to health problems for humans [2]. These challenges have been instrumental in accelerating the adoption of alternative energies, including solar radiation, wind power, biomass, water, and geothermal resources. These resources are characterized by their unlimited availability, minimal to zero greenhouse gas (GHG) emissions, and lower costs than traditional energy sources [3]. Solar energy, with the potential to generate 6500 TW (sufficient to surpass global energy demands), is regarded as one of the cleanest forms of energy available [4]. Solar energy offers numerous environmental benefits over conventional energy sources, including the reduction in CO_2 emissions and other harmful greenhouse gases, and contributes to energy independence and security [5]. Solar energy is vitally important in the global transition toward sustainable energy solutions. Solar energy is one of the critical strategic choices to meet future local and international energy needs. Solar energy is pivotal in integrating renewable energy sources to mitigate adverse climate change effects [6]. However, integrating solar energy and other alternative sources requires comprehensive evaluation methods to maximize energy generation benefits while minimizing environmental impacts. Additionally, assessing their technical, environmental, economic, and social impacts is crucial for the effective utilization of solar energy and other renewables.

Despite its many advantages, in 2020, solar energy accounted just for 3.8% of the electricity produced worldwide, just behind hydropower (16.7%) and nuclear energy (9.8%). Meanwhile, fossil fuels produce more than 61% of the electricity [7]. Moreover, China remains by far the main producer of electricity by solar energy [8].

Although industrial production is crucial for economic growth, it also leads to an increase in energy consumption [4]. The significance of energy in the advancement of industries is highly critical, as a large portion of energy is consumed during industrial processes. The energy supplied to the industrial sector is employed in four main areas, which are construction, agriculture, mining, and manufacturing [9]. Furthermore, in 2019, the industrial sector consumed more than 40% of the electricity produced worldwide [10], thus, the use of alternative and renewable energy sources in industries lead to sustainable economic development and low-carbon production [11].

Solar energy is attracting attention as an excellent promising option to be used in industry sector, mainly for two applications: the solar thermal and the photovoltaic (PV). The most common applications of solar energy in industry are: hot water, steam, drying and dehydration processes, preheating, concentration, pasteurization, sterilization, washing, cleaning, chemical reaction, industrial space heating, food, plastic, building, textile and even business concerns [9].

The selection and configuration of PV systems have become complex as industries aim to transition towards more sustainable practices [12]. Traditional decision-making methods may not sufficiently encompass the diverse technical, social, economic, and environmental considerations pertinent to PV systems throughout their lifecycle. Consequently, this research is driven by the necessity to equip industries with a systematic and data-driven approach for assessing and selecting optimal PV system configurations connected with their specific technical demands and social and environmental obligations. By integrating decision-making techniques, such as the morphological matrix and fuzzy TOPSIS (Technique for Order of Preference by Similarity to Ideal Solution), this research enhances the efficiency and effectiveness of PV system selection within industrial contexts, thereby contributing to the advancement of sustainable energy solutions.

Existing MCDM approaches, such as AHP (Analytic Hierarchy Process), ANP (Analytic Network Process), and traditional TOPSIS, have been applied to solar PV selection. However, these methods have limitations:

- Rigid decision structures: AHP and ANP rely on strict hierarchical structures, making them less adaptable to dynamic industrial needs.

- Lack of comprehensive lifecycle assessment: Most studies focus on site selection or economic feasibility, neglecting the complete PV system lifecycle (analysis, installation, operation, and disposal).
- Limited handling of uncertainty: Traditional TOPSIS does not effectively incorporate fuzzy logic, which is essential for dealing with uncertain or subjective decision-making criteria.

This research aims to formulate a comprehensive decision-making framework for selecting and configuring PV solar energy systems in industrial environments. This framework will incorporate technical, economic, environmental, and social aspects throughout the entire lifecycle of the systems, including phases of analysis, installation, operation, and disposal. This work proposes the utilization of a morphological matrix to identify potential solution trajectories for solar energy projects by considering various attributes, including sensing, smart, sustainability, and social factors. Subsequently, a fuzzy TOPSIS approach is employed to evaluate and determine the most appropriate configuration for a specific solar energy project. Compared to traditional methods, such as AHP, ANP, and traditional TOPSIS, the proposed hybrid approach of this work demonstrates enhanced capability to manage decision uncertainty, greater flexibility in incorporating lifecycle-based criteria, and improved adaptability to dynamic industrial scenarios. This methodology ensures that decisions are holistic, balancing all relevant factors and leading to the most advantageous outcomes for industrial solar energy utilization.

The structure of this work is as follows: Section 2 shows a literature review of multi-criteria decision making methods and the definition of the morphological matrix; Section 3 presents basic definitions of the fuzzy TOPSIS method, and the methodology proposed; a case study is presented in Section 4; results and discussion are in Section 5; and finally, Section 6 exposes the conclusions and policy implications.

2. Literature Review

2.1. Multi-Criteria Decision Making

According to the University of Massachusetts Dartmouth [13], decision making is the process of making choices by identifying a decision, gathering information, and assessing alternatives resolution. An adequate decision-making process, together with the use of an effective method, can help to analyze all the relevant information, define alternatives, and increase the chances of choosing the most suitable option possible that meets the needs and objectives of the problem.

Multi-criteria decision making (MCDM) is a research area that involves different decision-making methods that compare and analyze various available choices for one particular problem considering multiple and often conflicting qualitative or quantitative criteria with the objective of selecting the most satisfying alternative in certain, uncertain, fuzzy or risky environments [14,15]. MCDM improves the quality of decision making to become more rational and efficient and it can be applied in different disciplines and areas including economics and finance, engineering, health, energy, construction, among others [16].

According to [17,18], key steps in MCDM problems are: (1) problem identification, (2) goal definition, (3) criteria selection, (4) alternative identification, (5) criteria weighting, (6) alternative ranking using a suitable method, and (7) evaluation of outcomes.

There are many methods for solving MCDM problems and the choice is made based on the nature of the problem and the level of complexity of the decision-making process [19]. In [20], the authors divided MCDM problems into two general sub-categories:

- Multi-attribute decision making (MADM): problems with a finite (or discrete) number of alternatives. Goals, attributes (criteria), and options are clear; limitations are not.
- Multi-objective decision making (MODM): problems with a infinite (or continuous) number of alternatives. Goals are not clear, but limitations are.

In [21], Ilbahar et al. conducted a state-of-the-art review about MADM methods used throughout the renewable energy literature for the following purposes: evaluation of renewable energy sources, evaluation of renewable energy technologies, evaluation of renewable energy facility location, evaluation of renewable energy projects or investments, and design of renewable energy systems. The review concluded that the MADM methods most commonly used, including their various combinations, for solving the renewable energy problems mentioned before are Analytic Hierarchy Process (AHP) [22,23], Analytic Network Process (ANP) [22,23], Elimination Et Choix Traduisant la Realité (ELECTRE) [22,24] and Technique for Order of Preference by Similarity to Ideal Solution (TOPSIS) [22,25].

While MCDM techniques have been widely applied to solar energy projects, they often fail to comprehensively address the complexity of industrial solar PV deployment, particularly in manufacturing settings. One of the identified limitations is that most studies focus on the site selection or financial evaluation phase, without considering the installation, operation, and disposal phases critical to industrial settings. In addition, methods such as AHP and ANP assume statistic hierarchical structures, which do not adapt well to complex and dynamic industrial needs. Traditional TOPSIS assigns fixed numerical values, failing to incorporate expert judgement, uncertainty, and qualitative assessments, which are critical in solar PV investment decisions.

TOPSIS is one of the fundamental MCDM methods and is a highly regarded, applied, and adopted method because of its simplicity and ease of applicability. For these reasons, this method was chosen for the proposal presented in this work.

TOPSIS is based on the premise that the best solution has the shortest Euclidean distance from the positive ideal solution and the longest distance from the negative ideal one [26]. In a general view, TOPSIS assigns and measures weights for each parameter, normalizes scores, and determines the numerical difference for each alternative with respect to the optimal alternative and ranks them to determine the optimal one [27].

Traditional MCDM methods struggle with handling subjectivity and uncertainty in decision making, which is where fuzzy TOPSIS provides an advantage.

2.2. Fuzzy Sets

Humans predominantly use natural language to describe their everyday experiences. Nevertheless, using crips numerical values like 0 and 1 to represent truth and falsity is not always entirely accurate. For instance, when evaluating energy efficiency, traditional approaches often categorize appliances or systems as either 'energy efficient' or 'not energy efficient', which, in mathematical terms, equate to the discrete numbers 1 and 0, respectively. By employing fuzzy sets, individuals can provide more nuanced descriptions. Individuals can characterize the system as 'highly energy efficient', 'moderately energy efficient', 'marginally energy efficient', 'energy neutral', 'slightly energy inefficient', 'highly energy inefficient', etc. Fuzzy sets enable to convey the subtleties in a more natural and descriptive way, reducing the need for binary, all-or-nothing categorizations [28].

According to [29], the presence of imperfect information in the decision-making process poses challenges in establishing precise preferences among decision makers. This imperfection in decision-related information arises from various factors, including the limitations of decision makers, the complexity of available options, and the influence of psychological biases. Fuzzy numbers allow for a more flexible and realistic way to model this uncertainty in decision-making problems.

Decision making often involves imprecision and vagueness which can be effectively managed by fuzzy sets and fuzzy decision-making techniques [30]. Fuzzy numbers help to express linguistic variables and describe the subjective judgement of a decision maker (DM) in a quantitative manner, thus, fuzzy sets can deal with incomplete and uncertain knowledge and information [31]. Triangular, trapezoidal, and Gaussian are the fuzzy numbers most often used.

In TOPSIS, fuzzy numbers make it simple for evaluation and a realistic form of a modelling and countervailing method which include or exclude alternative solutions based on ruling-out process [32].

2.3. Applications of Fuzzy TOPSIS in Solar PV Energy

Table 1 shows a literature review about the fuzzy TOPSIS method applied in PV solar energy. As can be observed, this approach has primarily been used in the Asian context. This review also reveals that most studies (five) employed this technique to determine suitable locations for solar PV system installations [33–37]. Two studies focused on identifying risk and barriers associated with solar PV energy deployment [38,39], two explored the best alternative energy sources for investment [40,41], and one study centered on the selection of solar panels [42].

Table 1. Literature review fuzzy TOPSIS applied in solar photovoltaic energy.

Authors, (Year)	Title	Location/Country	Main Problem Addressed
Hooshangi et al. (2023) [33]	Evaluation of potential sites in Iran to localize solar farms using a GIS based Fermatean Fuzzy TOPSIS	Iran	PV plant location
Ranganath et al. (2022) [38]	Application of fuzzy TOPSIS method for risk evaluation in development and implementation of solar park in India	India	Risks identification and analysis
Qian et al. (2021) [34]	Fuzzy Technique Application in Selecting Photovoltaic Energy and Solar Thermal Energy Production in Belt and Road Countries	Case study in China	PV plant location
Ali Sadat et al. (2021) [39]	Barrier analysis of solar PV energy development in the context of Iran using fuzzy AHP-TOPSIS method	Iran	PV energy production barriers
Anser et al. (2020) [35]	Assessing the integration of solar power projects: SWOT-based AHP–F-TOPSIS case study of Turkey	Turkey	PV plant location
Taylan et al. (2020) [40]	Assessment of Energy Systems Using Extended Fuzzy AHP, Fuzzy VIKOR, and TOPSIS Approaches to Manage Non-Cooperative Opinions	Saudi Arabia	Energy source evaluation and selection
Sasikumar and Ayyappan (2019) [42]	Multi-criteria Decision Making for Solar Panel Selection Using Fuzzy Analytical Hierarchy Process and Technique for Order Preference by Similarity to ideal Solution (TOPSIS): An Empirical Study	Not specified	Solar panel selection
Ligus and Peternek (2018) [41]	Determination of most suitable low-emission energy technologies development in Poland using integrated fuzzy AHP-TOPSIS method	Poland	Energy source evaluation and selection
Sindhu et al. (2017) [36]	Investigation of feasibility study of solar farms deployment using hybrid AHP-TOPSIS analysis: Case study of India	India	PV plant location
Kengpol et al. (2013) [37]	A Decision Support System for Selection of Solar Power Plant Locations by Applying Fuzzy AHP and TOPSIS: An Empirical Study	Thailand	PV plant location

The limited number of research works involving fuzzy TOPSIS applied to PV solar energy within the industrial sector, especially within the context of Latin America countries, underscores the importance of this proposal. Furthermore, the assessment of solar technologies presents an opportunity for the application of MCDM methods due to the numerous criteria that come into play during the decision-making process.

This proposal takes advantage of the S4 Framework introduced by Molina et al. [43] and Ponce et al. [44], which has arisen as a response to the rethinking and reshaping of business processes initiated by the adoption of innovative practices and operational strategies, facilitated by the integration of new technologies. This approach has enhanced the development of products and services. The incorporation of distinct features and functionalities into S4 products, processes, manufacturing systems, and enterprises provide additional value, thereby serving as a strategy to maintain competitiveness in the market by providing solutions to current societal challenges and addressing evolving consumer needs [43].

While previous studies have successfully applied fuzzy TOPSIS for site selection, risk assessment, and technology comparison, none have combined it with a morphological matrix for industrial PV system design. This research fills that gap by integrating both methodologies to optimize PV configurations based on sensing, smart, sustainable, and social (S4) factors.

2.4. The Morphological Matrix

General Morphological Analysis (GMA) is a method for identifying and analyzing the total set of possible relationships or configurations that compose a given multidimensional, non-quantifiable, and complex problem. Its principle is decomposing a complex (multivariate) concept into a number of ("simple") one dimensional concepts and thus, to establish which of the configurations are possible, viable, and practical, and which are not [45,46].

Morphological matrices are a very powerful, simple, and systematic combinatorial tool which use the principle of morphological analysis to explore the complete set of possible relationships within a system, it helps to reduce the total problem space to a relevant solution space [47].

In the morphological matrix, the functions are listed along the column and the possible solutions for each function are listed along the corresponding rows. Figure 1 shows the morphological matrix proposed in this work, considering the S4 Framework proposed by Pérez et al. [48], thus, the sensing, smart, sustainable, and social needs are considering as functions, while the solutions will be the sensing, smart, sustainable, and social features.

The morphological matrix helps to visualize all the possible combinations existing for each function. As observed in Figure 1, the output for the matrix will be the "solution paths" created from each combination.

Since in a solar energy project it may exist a huge number of solution paths, it is important to first identify the technical and social needs of the project, in this case, the needs are divided in sensing, smart, sustainable, and social, and thus, reduce the number of solutions contained in the matrix, so the solution paths resulted will be those that best meet the needs. In this case, the fuzzy TOPSIS method will help to choose the best solution path created from a morphological matrix.

Based on the literature review, several key gaps in existing research have been identified. First, most studies focus on PV site selection or economic analysis, but few address installation, operation, and disposal. Furthermore, while morphological matrices have been used in engineering design, their application in MCDM for solar PV selection remains underexplored. Finally, previous applications of fuzzy TOPSIS focus on site selection and risk assessment, not industrial PV system configuration.

This research addresses these gaps by proposing a hybrid framework that integrates a morphological matrix with fuzzy TOPSIS to systematically generate PV system configurations tailored to industrial manufacturing firms and accounts for uncertainty in decision making using fuzzy logic.

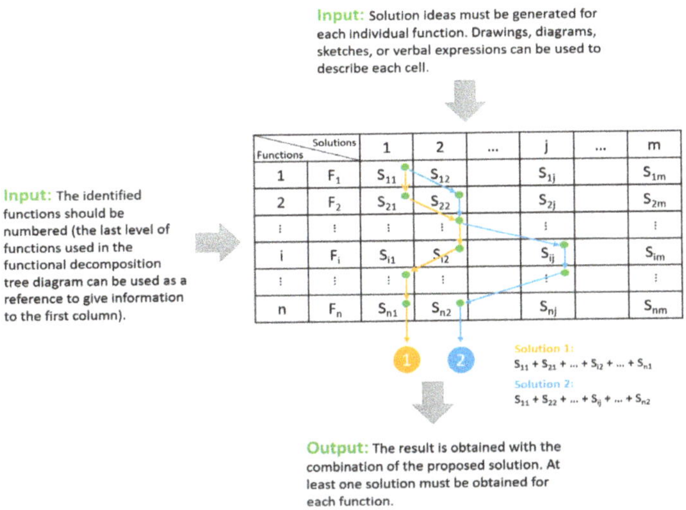

Figure 1. Example of a morphological matrix.

3. Methodology

Below, some basic definitions of fuzzy sets, fuzzy numbers, and linguistic variables are presented [49]. Then, the morphological matrix and the fuzzy TOPSIS approach proposed are presented.

3.1. Fuzzy TOPSIS Definitions

Definition 1. *A fuzzy set, denoted as $\tilde{a}$, ascertained within a designated universe of discourse, X, is characterized by a membership function symbolized as $\mu\tilde{a}(x)$. This function allocates a real number within the range of 0 to 1, inclusive, to each constituent element, x, within X. The computed numerical value, $\mu\tilde{a}(x)$, elucidates the degree to which element x is associated with the fuzzy set $\tilde{a}$.*

Definition 2. *A fuzzy number is defined as a specific subset within the universe of discourse, X, that possesses the dual characteristics of normality and convexity. As depicted in Figure 2, a representative fuzzy number, denoted as $\tilde{a}$, duly satisfies these prerequisites within the predefined universe of discourse, X.*

A triangular fuzzy number, denoted as $\tilde{a}$, can be encapsulated through a triadic representation composed of elements (a_1, a_2, a_3). Its intellectual framework and the corresponding mathematical elucidation are demonstratively represented in Equation (1).

$$\mu_{\tilde{a}}(x) = \begin{cases} 0, & x \leq a_1, \\ (x - a_1)/(a_2 - a_1), & a_1 < x \leq a_2 \\ (a_3 - x)/(a_3 - a_2), & a_2 < x \leq a_3 \\ 0, & x > a_3 \end{cases} \tag{1}$$

Definition 3. *The foundational operations pertinent to fuzzy triangular numbers are hereby established, presuming that $\tilde{a} = (a_1, a_2, a_3)$ and $\tilde{b} = (b_1, b_2, b_3)$ are both real number configurations:*

$\tilde{a}$ multiplied by $\tilde{b}$ results in $(a_1 \times b_1, a_2 \times b_2, a_3 \times b_3)$ representing multiplication, whereas the sum of $\tilde{a}$ and $\tilde{b}$ produces $(a_1 + b_1, a_2 + b_2, a_3 + b_3)$ representing addition.

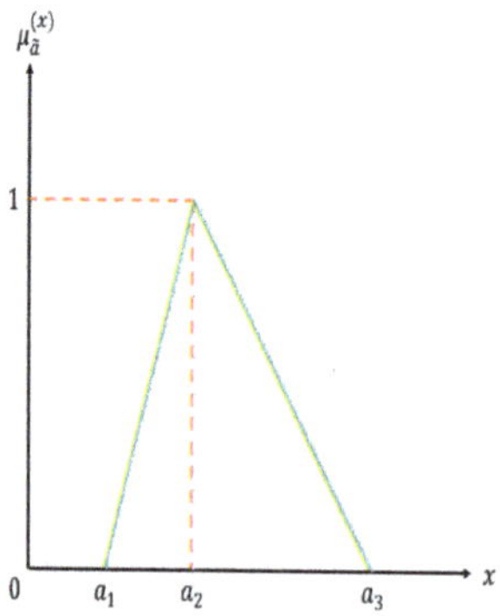

Figure 2. A triangular fuzzy number $\tilde{a}$.

Definition 4. *A matrix, symbolized as $\tilde{D}$, is acknowledged as a fuzzy matrix on the condition that it incorporates at least one constituent element identifiable as a fuzzy number.*

Definition 5. *A linguistic variable constitutes a variable expressed in linguistic terminologies. This concept is advantageous in describing complex situations that elude traditional quantitative definitions. For instance, "weight" serves as a linguistic variable capable of accommodating values such as very low, low, medium, high, and very high.*

Definition 6. *The fuzzy MADM construct can be transcribed using matrices as observed in Equations (2) and (3):*

$$\begin{array}{c} \begin{matrix} C_1 & C_2 & C_3 & \cdots & C_n \end{matrix} \\ \tilde{D} = \begin{matrix} A_1 \\ A_2 \\ A_3 \\ \vdots \\ A_m \end{matrix} \begin{bmatrix} \tilde{x}_{11} & \tilde{x}_{12} & \tilde{x}_{13} & \cdots & \tilde{x}_{1n} \\ \tilde{x}_{21} & \tilde{x}_{22} & \tilde{x}_{23} & \cdots & \tilde{x}_{2n} \\ \tilde{x}_{31} & \tilde{x}_{32} & \tilde{x}_{33} & \cdots & \tilde{x}_{3n} \\ \vdots & \vdots & \vdots & \ddots & \vdots \\ \tilde{x}_{m1} & \tilde{x}_{m1} & \tilde{x}_{m3} & \cdots & \tilde{x}_{mn} \end{bmatrix}, \end{array} \quad (2)$$

$$\tilde{W} = \left[\tilde{w}_1, \tilde{w}_2, \cdots, \tilde{w}_n\right], \quad (3)$$

where $\tilde{x}_{ij}$ ($i = 1, 2, \ldots, m$; $j = 1, 2, \ldots, n$) and $\tilde{w}_j$ ($j = 1, 2, \ldots, n$) denote the linguistic triangular fuzzy numbers such that $\tilde{x}_{ij} = (a_{ij}, b_{ij}, c_{ij})$ and $\tilde{w}_j = (w_{j1}, w_{j2}, w_{j3})$. Here, $\tilde{w}_j$ characterizes the weight of the jth criteria, C_j, and $\tilde{x}_{ij}$ denotes the performance rating of the ith alternative, Ai, concerning the jth criteria, C_j, evaluated by k evaluators. In their study, Mahdavi et al. (2008) [49] *deployed the method of average value to amalgamate the fuzzy performance score $\tilde{x}_{ij}$ for k evaluators concerning the same evaluation criteria of Equation (4):*

$$\tilde{x}_{ij} = (1/k)\left(\tilde{x}_{ij}^{1} + \tilde{x}_{ij}^{2} + \cdots + \tilde{x}_{ij}^{k}\right), \quad (4)$$

where $\tilde{x}_{ij}^{k}$ is the rating of alternative A_i concerning criterion C_j evaluated by kth evaluators, and $\tilde{x}_{ij}^{k} = \left(\tilde{a}_{ij}^{k}, \tilde{b}_{ij}^{k}, \tilde{c}_{ij}^{k}\right)$.

Definition 7. *The normalized fuzzy decision matrix denoted by* $\widetilde{R}$ *is shown as Equation (5):*

$$\widetilde{R} = \left[\widetilde{r}_{ij}\right]_{m\times n}, \tag{5}$$

In the case of possessing triangular fuzzy numbers denoted by $\widetilde{x}_{ij}$, *with i spanning from 1 to m and j from 1 to n, the procedure for conducting normalization can be articulated as follows in Equations (6) and (7):*

$$\widetilde{r}_{ij} = \left(a_{ij}/c_j^*,\ b_{ij}/c_j^*,\ c_{ij}/c_j^*\right) \qquad i = 1,2,\ldots m,\ j \in B \tag{6}$$

$$\widetilde{r}_{ij} = \left(a_j^-/c_{ij},\ a_j^-/b_{ij},\ a_j^-/a_{ij}\right) \qquad i = 1,2,\ldots m,\ j \in C \tag{7}$$

where B and C are the set of benefit criteria and cost criteria, respectively, and

$$c_j^* = \max c_{ij}\ j \in B$$

$$a_j^- = \min a_{ij}\ j \in C$$

The normalized entities, denoted as $\widetilde{r}_{ij}$, *remain as triangular fuzzy numbers. The process of normalization can be implemented in a similar manner for trapezoidal fuzzy numbers. Equation (8) illustrates the weighted fuzzy normalized decision matrix:*

$$\widetilde{V} = \begin{bmatrix} \widetilde{v}_{11} & \widetilde{v}_{12} & \widetilde{v}_{13} & \cdots & \widetilde{v}_{1n} \\ \widetilde{v}_{21} & \widetilde{v}_{22} & \widetilde{v}_{23} & \cdots & \widetilde{v}_{2n} \\ \widetilde{v}_{31} & \widetilde{v}_{32} & \widetilde{v}_{33} & \cdots & \widetilde{v}_{3n} \\ \vdots & \vdots & \vdots & \ddots & \vdots \\ \widetilde{v}_{m1} & \widetilde{v}_{m1} & \widetilde{v}_{m3} & \cdots & \widetilde{v}_{mn} \end{bmatrix} = \begin{bmatrix} \widetilde{w}_1\widetilde{r}_{11} & \widetilde{w}_2\widetilde{r}_{12} & \cdots & \widetilde{w}_j\widetilde{r}_{1j} & \cdots & \widetilde{w}_n\widetilde{r}_{1n} \\ \widetilde{w}_1\widetilde{r}_{21} & \widetilde{w}_2\widetilde{r}_{22} & \cdots & \widetilde{w}_j\widetilde{r}_{2j} & \cdots & \widetilde{w}_n\widetilde{r}_{2j} \\ \vdots & \vdots & \ddots & \vdots & \ddots & \vdots \\ \widetilde{w}_1\widetilde{r}_{i1} & \widetilde{w}_2\widetilde{r}_{i2} & \cdots & \widetilde{w}_j\widetilde{r}_{ij} & \cdots & \widetilde{w}_n\widetilde{r}_{in} \\ \vdots & \vdots & \ddots & \vdots & \ddots & \vdots \\ \widetilde{w}_1\widetilde{r}_{m1} & \widetilde{w}_2\widetilde{r}_{m2} & \cdots & \widetilde{w}_j\widetilde{r}_{mj} & \cdots & \widetilde{w}_n\widetilde{r}_{mn} \end{bmatrix} \tag{8}$$

3.2. Proposed Methodology

Bulleted lists look like this: This section delineates the methodology proposed, providing an initial overview of the utilization of the morphological matrix and the fuzzy TOPSIS method, followed by a more comprehensive exploration of each method separately.

The methodology employed herein draws upon the S4 Framework proposed by Pérez et al. [48]. The framework is bifurcated in its application, firstly in the derivation of solutions via the morphological matrix, and secondly in the selection of criteria used for the assessment of alternatives within the fuzzy TOPSIS method. The S4 Framework is subdivided into four principal categories:

- Sensing (s1): Sensors presented in the analysis, installation, operation, and disposal of the PV system for measuring meteorological and power variables, as well as for maintenance and monitoring purposes.
- Smart (s2): Features that collect data/information through automated reasoning for decision-making purposes.
- Sustainable (s3): Environmental impact analysis of the PV system component through their lifespan.
- Social (s4): Legal and regulatory framework for installing, operating, and disposing a PV system; economic viability, and analysis of population acceptance.

Figure 3 provides an overview of the methodological steps, highlighting the integration of the morphological matrix and fuzzy TOPSIS for industrial solar PV decision making. Initially, the needs identified under the sensing, smart, sustainable, and social categories, along with their potential solutions for the PV project, are allocated within the morphological matrix. Consequently, the matrix generates solution paths tailored to each functional requirement. This matrix output subsequently functions as the input for the fuzzy TOPSIS method. The fuzzy TOPSIS method is then executed, taking into consideration the S4 Framework's criteria. The ultimate output from this methodology will be the optimal solution path that adeptly satisfies all the technical and social requirements intrinsic to the PV solar project.

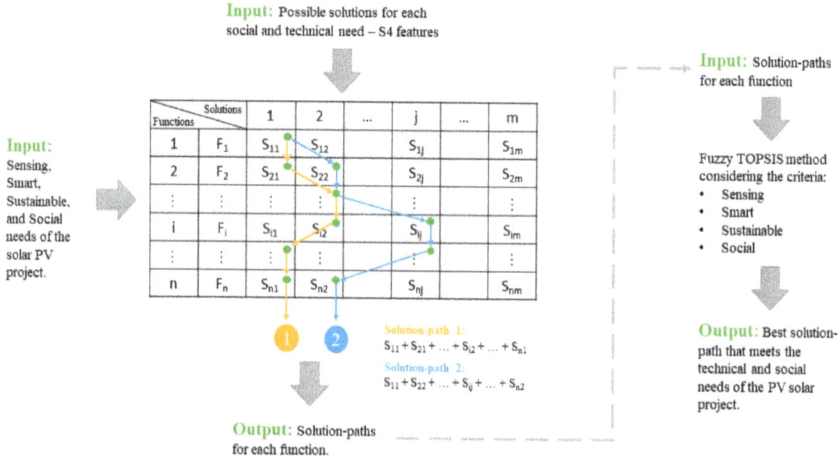

Figure 3. Proposed method: morphological matrix + fuzzy TOPSIS for solar PV projects.

Figure 4 summarizes the proposed methodology and Figure 5 shows the flow diagram for the use of (a) the morphological matrix and (b) the fuzzy TOPSIS method in PV solar energy projects.

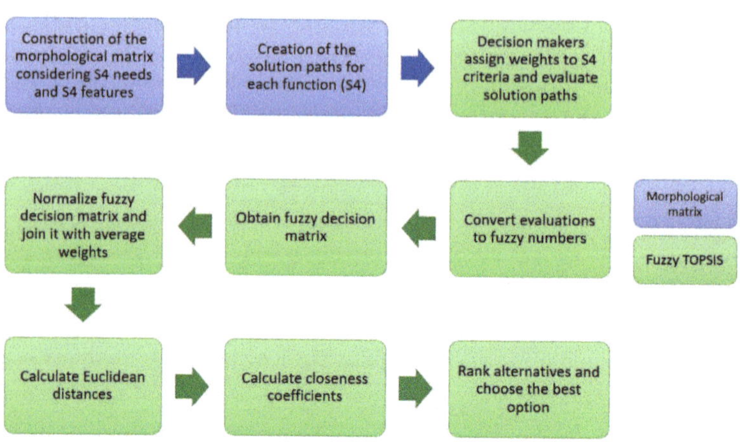

Figure 4. Summarized proposed methodology.

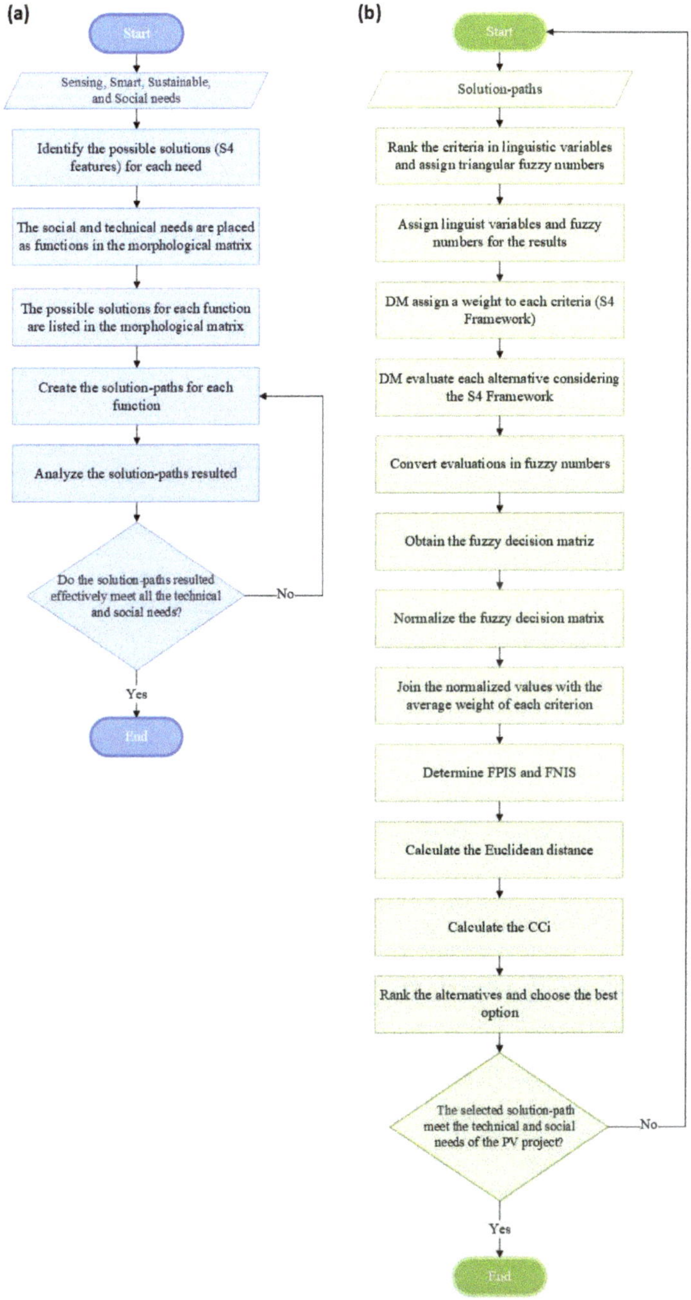

Figure 5. Flowchart for the use of the morphological matrix and fuzzy TOPSIS method in PV solar energy projects. (**a**) Flowchart for the use of the morphological matrix; (**b**) flowchart for the use of the fuzzy TOPSIS method.

3.2.1. The Morphological Matrix Method

The proposed method for the use of morphological matrices in solar PV projects is as follows:

Step 1. Determine the sensing, smart, sustainable, and social needs of the project. According to [48], the needs of a PV project can be divided in technical and social needs:

- Technical needs: electricity consumption, available area to install the PV system, and solar radiation received.
- Social needs: investment considering governmental incentives, and governmental restriction (current regulations).

In addition, depending on the PV project, there may be more needs that must be considered for each stage of the system's lifecycle (analysis, installation, operation, and disposal), such as

- Measurement of meteorological variables (radiation, air, temperature, etc.).
- Measurement of power variables (voltage and current).
- Data storage and analysis (store main variables and adjust sample time autonomously).
- Monitoring system.
- Battery management system (if the system is stand-alone).
- Environmental impact analysis.
- Legal and regulatory framework applicable.
- Population acceptance.

The needs mentioned above are just some of the many that may exist depending on the PV solar project, so the user can add or remove needs depending on their context. Moreover, according to the definition of sensing, smart, sustainable, and social mentioned above, technical and social needs can be further divided into S4 needs.

Step 2. Identify the possible solutions for each S4 need detected in the previous step. Depending on the need, the solutions will be sensing, smart, sustainable, or social features. Figure 6 presents the features of each "S" category and a more detailed description of each feature can be consulted at [48].

Step 3. Place the S4 needs as functions along the morphological matrix columns.

Step 4. Listed the possible solutions (S4 features) identified in Step 2 along the rows of the morphological matrix.

Step 5. Create the solution paths for each function. In this step, one solution for each function (need) must be selected.

Step 6. Analyze the solution paths resulted and verify which of them effectively meet all the technical and social needs. If there is not any solution path that satisfy all the needs, the user must return to Step 5.

3.2.2. The Fuzzy TOPSIS Method

In this work, the fuzzy TOPSIS approach proposed by Ponce et al. [50] is expanded to create a methodology for addressing a multi-attribute decision-making challenge within the solar energy environment. Considering the uncertainty present in the decision data, linguistic variables are utilized to evaluate the significance of each criterion and the performance rating of each alternative concerning these criteria. After pooling the fuzzy ratings provided by decision makers, the decision matrix is transformed into a fuzzy decision matrix and subsequently establishes a weighted normalized fuzzy decision matrix. Then, the degree of fuzzy similarity of each alternative concerning the fuzzy positive ideal solution (FPIS) and the fuzzy negative ideal solution (FNIS) is calculated. Ultimately, a closeness coefficient is defined for each alternative to ascertain their rankings. A higher closeness coefficient signifies that an alternative is closer to FPIS and, simultaneously, farther from FNIS.

Once the solution paths were created (it is necessary to have more than one to run this process), the fuzzy TOPSIS method is as follows.

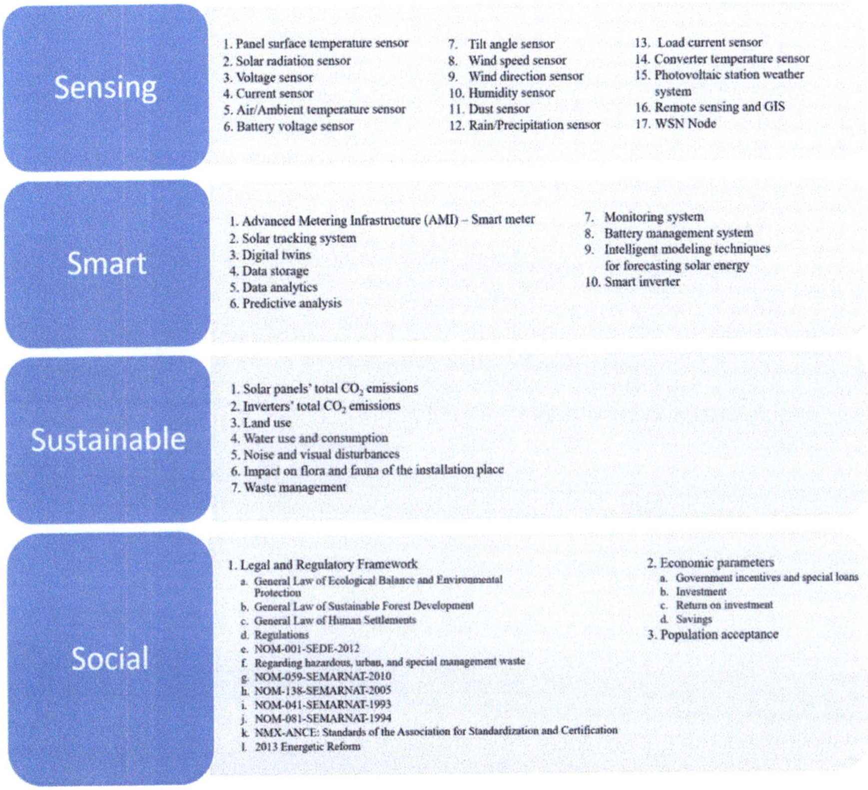

Figure 6. S4 Framework features.

Step 1. Rank each criterion in linguistic variables and assign a triangular fuzzy number from 0–1 ($\tilde{x}_{ij}, i = 1, 2, \ldots, m, j = 1, 2, \ldots, n$) (Table 2). For this instance, the $\tilde{D}$ defined by Equation (2) es equivalent to the $\tilde{R}$ defined by Equation (5).

Table 2. Linguistic variables for the importance weight of each criterion.

Linguistic Variables	Fuzzy Numbers
Very low (VL)	(0; 0; 0.1)
Low (L)	(0; 0.1; 0.3)
Medium–low (ML)	(0.1; 0.3; 0.5)
Medium (M)	(0.3; 0.5; 0.7)
Medium–high (MH)	(0.5; 0.7; 0.9)
High (H)	(0.7; 0.9; 1.0)
Very high (VH)	(0.9; 1.0; 1.0)

Step 2. Assign linguistic variables ($\tilde{w}_j$, $j = 1, 2, \ldots, n$) to rank the results of each alternative, in this case, the alternatives are the solution paths created from the morphological matrix (Table 3).

Step 3. DM assign a weight to each criterion (sensing, smart, sustainable, and social) using the linguistic variables from Table 2.

Step 4. DM evaluate each solution path according to the criteria and the linguistic variables from Table 3.

Step 5. The evaluations are converted to fuzzy triangular numbers defined in step 1.

Table 3. Linguistic variables for the ratings.

Category/Benefit Criteria	Linguistic Variable	Fuzzy Numbers
s1, s2, s3, s4	Very poor (VP)	(0; 0; 1)
	Poor (P)	(0; 1; 3)
	Medium poor (MP)	(1; 3; 5)
	Fair (F)	(3; 5; 7)
	Medium good (MG)	(5; 7; 9)
	Good (G)	(7; 9; 10)
	Very good (VG)	(9; 10; 10)

Step 6. Obtain the fuzzy decision matrix by determining the average score of each criterion for each solution path and calculating each criterion's average weight.

Step 7. Normalize the fuzzy decision matrix by finding the fuzzy number with which each criterion is maximized (the maximum number) and dividing all the fuzzy numbers of that criterion by that maximum value. This ensures that all the criteria are given appropriate importance in the decision-making process.

Step 8. The normalized values are joined with the average of the weight of each criterion by multiplying the normalized values with the weight averages obtained in step 6.

Step 9. Determine the FPIS and the FNIS. In this method, the FPIS and the FNIS for all the criteria is as follows, considering that 1 is the best possible score (very high) and 0 is the worst possible (very low):

$$A^* = (1.00; 1.00; 1.00)$$

$$A^- = (0.00; 0.00; 0.00)$$

Step 10. Calculate the Euclidean distance for each solution path concerning the FPIS and FNIS using Equation (9) and Equation (10).

$$D = \sqrt{\frac{1}{3}\left[\sum_{i=1}^{3}(b_i - a_i)^2\right]} \tag{9}$$

$$D_{A_m} = \sum_{i=1}^{4} D_i \tag{10}$$

where for triangular fuzzy numbers $\tilde{a} = (a_1, a_2, a_3)$ and $\tilde{b} = (b_1, b_2, b_3)$, in this case, $\tilde{b}$ represents values for FPIS and FNIS.

Step 11. Calculate each solution paths' closeness coefficient (CC_i) using Equation (11).

$$CC_i = \frac{D^-_{A_m}}{D^*_{A_m} + D^-_{A_m}} \quad (11)$$

Step 12. Rank the solution paths according to CC_i and choose the best option, which will be the one closer to 1.

Step 13. Analyze if the selected solution path meets the company's preferences and objectives, including the S4 needs; if it does not, start the process again; if it does, end the process.

4. Case Study with a Medium-Sized Mexican Manufacturing Company

To validate the proposed methodology, a case study was conducted in a medium-sized Mexican manufacturing company. This section presents the main characteristics of the company and the evaluations made by the decision makers.

A medium-sized Mexican manufacturing company is interested in installing a grid-connected PV system in its production plant. The plant is located in Monterrey, Mexico, the sun's radiance is an abundant and reliable source of energy that makes the installation of solar panels a viable and attractive option for companies looking to reduce their carbon footprint and energy costs. With an average of over 300 days of sunshine per year, Monterrey is ideally situated for solar energy production. By harnessing the sun's energy, the company can generate electricity that is both cost-effective and environmentally friendly. The installation of solar panels will not only reduce the company's reliance on traditional energy sources but also contribute to a more sustainable future. The sun's radiance in Monterrey provides an excellent opportunity for companies to invest in clean energy and demonstrate their commitment to sustainability. As can be observed in Figure 7, although Monterrey does not receive as much solar radiation as other states, it is still a state with a lot of potential to install PV solar energy.

Figure 7. Photovoltaic power potential for Monterrey, Mexico (own elaboration with data from [51]).

The manufacturing company is specialized in developing car parts, so it has a variety of production processes and energy consumption patterns. In terms of energy consumption, the company's main energy usage comes from its production processes. Depending on the specific car parts being manufactured, the company uses a variety of manufacturing methods such as casting, stamping, or injection molding. These processes typically require significant amounts of electricity and may involve the use of large machinery or equipment. Additionally, the company uses other forms of energy such as natural gas, propane, or diesel fuel to power its equipment and heating systems. The amount of energy consumed depends on the production required. Given the increasing focus on sustainability and environmental responsibility, the company must explore novel ways to reduce its energy consumption and carbon footprint. They are looking into renewable energy sources such as solar or wind power, or implementing energy-efficient practices such as LED lighting and energy-efficient equipment.

Overall, a manufacturing company that develops car parts will likely have a significant focus on research and development and will consume energy primarily through its production processes. As the industry continues to evolve, there may be an increasing emphasis on sustainability and the use of renewable energy sources. In the case of PV panels, the company's needs have been divided in sensing, smart, sustainable, and social, and are shown in Table 4.

Five DM from the manufacturing company have analyzed these S4 needs, and they have chosen the S4 features that best meet them according to their judgement. Then, each DM has analyzed all the possible solutions and created solutions-paths by choosing the combinations of features that best meet the company's needs. Figure 8 shows the morphological matrix created with the S4 features chosen for the S4 needs by each DM, as well as the solutions-paths created by them; the number of the features corresponds to the number assigned in Figure 6.

Solutions / Functions	DM1	DM2	DM3	DM4	DM5
1. Sensing (S1)	1,2,3,4,6,13	1,2,3,4,6,11,12, 13,14,15	1,2,3	1,2,3,4,6,11,12, 13,14,15,16	1,2,3,4,5,6,7,8, 9,10,15
2. Smart (S2)	4,5,6,7,8,9,10	4,5,6,7,8	1,2,4,5,6,7,8,9, 10	4,5,6,7,8,9,10	4,5,6,7,8,9,10
3. Sustainable (S3)	1,2,3,4,5,6	1,2,3,4,5	1,2,3	1,2,3	1,2,3
4. Social (S4)	1,2	2,3	1,2	2	1,2

Solution-path 1 (SP1):
$S_1(1,2,3,4,6,13) + S_2(4,5,6,7,8) + S_3(1,2,3) + S_4(2)$

Solution-path 2 (SP2):
$S_1(1,2,3,4,6,11,12,13,14,15) + S_2(4,5,6,7,8,9,10) + S_3(1,2,3,4,5) + S_4(1,2)$

Solution-path 3 (SP3):
$S_1(1,2,3) + S_2(4,5,6,7,8,9,10) + S_3(1,2,3,4,5,6) + S_4(2,3)$

Solution-path 4 (SP4):
$S_1(1,2,3,4,6,11,12,13,14,15,16) + S_2(4,5,6,7,8,9,10) + S_3(1,2,3) + S_4(1,2)$

Solution-path 5 (SP5):
$S_1(1,2,3,4,5,6,7,8,9,10,15) + S_2(1,2,4,5,6,7,8,9,10) + S_3(1,2,3) + S_4(1,2)$

Figure 8. The morphological matrix with the solution paths created by the decision makers.

Table 4. Sensing, smart, sustainable, and social needs from the manufacturing company.

Sensing	Smart	Sustainable	Social
Measurement of meteorological and power variables.	Data management and analysis for decision making and monitoring.	Environmental impact analysis applicable for a photovoltaic system installed on the ground	Legal framework applicable.
Battery system for backup.			Analysis of investment, return of investment, and savings considering governmental incentives.
			Adaptability for the user.

Table 5 presents the weight assignation for each criterion given by the DM using the linguistic variables from Table 2. Then, DM analyzed the solution paths created in Figure 8 and evaluated them considering the linguistic variables from Table 3 and the S4 criteria; the results are in Table 6.

Table 5. Weight of each criterion by the decision makers of the manufacturing company.

Criteria	DM1	DM2	DM3	DM4	DM5
s1	H	VH	L	VH	H
s2	MH	VH	VH	VH	H
s3	L	H	VH	M	L
s4	VH	VH	H	VH	VH

Table 6. Evaluation of each alternative by the decision makers from the manufacturing company.

Criteria	Alternative	DM1	DM2	DM3	DM4	DM5
s1	SP1	G	F	MG	F	MP
	SP2	VG	VG	VG	MG	G
	SP3	MP	P	P	MP	P
	SP4	VG	VG	VG	VG	VG
	SP5	VG	VG	VG	VG	VG
s2	SP1	MG	VG	MG	F	MP
	SP2	F	VG	G	MG	MG
	SP3	G	VG	VG	MG	MG
	SP4	G	VG	VG	G	G
	SP5	VG	VG	VG	VG	G
s3	SP1	P	MP	F	MP	MP
	SP2	G	F	G	MG	MG
	SP3	G	VG	G	G	VG
	SP4	F	MP	F	F	MP
	SP5	F	MP	F	F	MP
s4	SP1	VG	P	MP	MG	MP
	SP2	VG	VG	G	G	VG
	SP3	G	MG	VG	MG	F
	SP4	VG	VG	G	G	VG
	SP5	VG	VG	G	G	VG

5. Results and Discussion

In this section, the results of the fuzzy TOPSIS approach proposed for selecting the solution path that best meet the needs of the manufacturing company are presented.

Table 7 shows the transformation of the evaluations made by the DM (Table 6) to triangular fuzzy numbers considering the values of Table 3.

Table 7. Triangular fuzzy numbers of the evaluations of each alternative.

Criteria	Alternative	DM1	DM2	DM3	DM4	DM5
s1	SP1	(7; 9; 10)	(3; 5; 7)	(5; 7; 9)	(3; 5; 7)	(1; 3; 5)
	SP2	(9; 10; 10)	(9; 10; 10)	(9; 10; 10)	(5; 7; 9)	(7; 9; 10)
	SP3	(1; 3; 5)	(0; 1; 3)	(0; 1; 3)	(1; 3; 5)	(0; 1; 3)
	SP4	(9; 10; 10)	(9; 10; 10)	(9; 10; 10)	(9; 10; 10)	(9; 10; 10)
	SP5	(9; 10; 10)	(9; 10; 10)	(9; 10; 10)	(9; 10; 10)	(9; 10; 10)
s2	SP1	(5; 7; 9)	(9; 10; 10)	(5; 7; 9)	(3; 5; 7)	(1; 3; 5)
	SP2	(3; 5; 7)	(9; 10; 10)	(7; 9; 10)	(5; 7; 9)	(5; 7; 9)
	SP3	(7; 9; 10)	(9; 10; 10)	(9; 10; 10)	(5; 7; 9)	(5; 7; 9)
	SP4	(7; 9; 10)	(9; 10; 10)	(9; 10; 10)	(7; 9; 10)	(7; 9; 10)
	SP5	(9; 10; 10)	(9; 10; 10)	(9; 10; 10)	(9; 10; 10)	(7; 9; 10)
s3	SP1	(0; 1; 3)	(1; 3; 5)	(3; 5; 7)	(1; 3; 5)	(1; 3; 5)
	SP2	(7; 9; 10)	(3; 5; 7)	(7; 9; 10)	(5; 7; 9)	(5; 7; 9)
	SP3	(7; 9; 10)	(9; 10; 10)	(7; 9; 10)	(7; 9; 10)	(9; 10; 10)
	SP4	(3; 5; 7)	(1; 3; 5)	(3; 5; 7)	(3; 5; 7)	(1; 3; 5)
	SP5	(3; 5; 7)	(1; 3; 5)	(3; 5; 7)	(3; 5; 7)	(1; 3; 5)
s4	SP1	(9; 10; 10)	(0; 1; 3)	(1; 3; 5)	(5; 7; 9)	(1; 3; 5)
	SP2	(9; 10; 10)	(9; 10; 10)	(7; 9; 10)	(7; 9; 10)	(9; 10; 10)
	SP3	(7; 9; 10)	(5; 7; 9)	(9; 10; 10)	(5; 7; 9)	(3; 5; 7)
	SP4	(9; 10; 10)	(9; 10; 10)	(7; 9; 10)	(7; 9; 10)	(9; 10; 10)
	SP5	(9; 10; 10)	(9; 10; 10)	(7; 9; 10)	(7; 9; 10)	(9; 10; 10)

Table 8 presents the fuzzy decision matrix and the average weight of each criterion calculated from Table 5.

Table 8. The fuzzy decision matrix.

Alternative	s1	s2	s3	s4
SP1	(3.80; 5.80; 7.60)	(4.60; 6.40; 8.00)	(1.20; 3.00; 5.00)	(3.20; 4.80; 6.40)
SP2	(7.80; 9.20; 9.80)	(5.80; 7.60; 9.00)	(5.40; 7.40; 9.00)	(8.20; 9.60; 10.0)
SP3	(0.40; 1.80; 3.80)	(7.00; 8.60; 9.60)	(7.80; 9.40; 10.0)	(5.80; 7.60; 9.00)
SP4	(9.00; 10.0; 10.0)	(7.80; 9.40; 10.0)	(2.20; 4.20; 6.20)	(8.20; 9.60; 10.0)
SP5	(9.00; 10.0; 10.0)	(8.60; 9.80; 10.0)	(2.20; 4.20; 6.20)	(8.20; 9.60; 10.0)
Weight	(0.64; 0.78; 0.86)	(0.78; 0.92; 0.98)	(0.38; 0.52; 0.66)	(0.86; 0.98; 1.00)

Table 9 shows the fuzzy normalized decision matrix.

Table 9. The fuzzy normalized decision matrix.

Alternative	s1	s2	s3	s4
SP1	(0.38; 0.58; 0.76)	(0.46; 0.64; 0.80)	(0.12; 0.30; 0.50)	(0.32; 0.48; 0.64)
SP2	(0.78; 0.92; 0.98)	(0.58; 0.76; 0.90)	(0.54; 0.74; 0.90)	(0.82; 0.96; 1.00)
SP3	(0.04; 0.18; 0.38)	(0.70; 0.86; 0.96)	(0.78; 0.94; 1.00)	(0.58; 0.76; 0.90)
SP4	(0.90; 1.00; 1.00)	(0.78; 0.94; 1.00)	(0.22; 0.42; 0.62)	(0.82; 0.96; 1.00)
SP5	(0.90; 1.00; 1.00)	(0.86; 0.98; 1.00)	(0.22; 0.42; 0.62)	(0.82; 0.96; 1.00)

Table 10 presents the fuzzy weighted normalized decision matrix.

Table 10. The fuzzy weighted normalized decision matrix.

Alternative	s1	s2	s3	s4
SP1	(0.24; 0.45; 0.65)	(0.36; 0.59; 0.78)	(0.05; 0.16; 0.33)	(0.28; 0.47; 0.64)
SP2	(0.50; 0.72; 0.84)	(0.45; 0.70; 0.88)	(0.21; 0.38; 0.59)	(0.71; 0.94; 1.00)
SP3	(0.03; 0.14; 0.33)	(0.55; 0.79; 0.94)	(0.30; 0.49; 0.66)	(0.50; 0.74; 0.90)
SP4	(0.58; 0.78; 0.86)	(0.61; 0.86; 0.98)	(0.08; 0.22; 0.41)	(0.71; 0.94; 1.00)
SP5	(0.58; 0.78; 0.86)	(0.67; 0.90; 0.98)	(0.08; 0.22; 0.41)	(0.71; 0.94; 1.00)

In Table 11, the Euclidean distance of each solution path with respect to the FPIS and FNIS, calculated with Equations (9) and (10), is found.

Table 11. The Euclidean distance of each solution path.

Alternative	A^*	A^-
SP1	2.42	1.78
SP2	1.51	2.72
SP3	2.00	2.22
SP4	1.47	2.74
SP5	1.43	2.77

With Equation (11), CC_i was calculated for each solution path and then, the ranking was made. The results can be observed in Table 12.

Table 12. CC_i of each solution path.

Rank	Alternative	CC_i	Recommendation
1	SP5	0.66	Best choice
2	SP4	0.65	Feasible alternative
3	SP2	0.64	Acceptable option
4	SP3	0.53	Less optimal
5	SP1	0.42	Not recommended

As observed in Table 12, the best solution path for the manufacturing company is SP5 since it obtained the CC_i closest to the FPIS, which is (1.00, 1.00, 1.00). Thus, according to Figure 6, the S4 features that best meet the company's needs can be found in Table 13.

Table 13. S4 features for each solution path.

Sensing	Smart	Sustainable	Social
1. Panel surface temperature sensor 2. Solar radiation sensor 3. Voltage sensor 4. Current sensor 5. Air/Ambient temperature sensor 6. Battery voltage sensor 7. Tilt angle sensor 8. Wind speed sensor 9. Wind direction sensor 10. Humidity sensor 11. PV station weather sensor	1. Smart meter 2. Solar tracking system 3. Data storage 4. Data analytics 5. Predictive analysis 6. Monitoring system 7. Battery management system 8. Intelligent modelling techniques for forecasting solar energy 9. Smart inverter	1. Solar panels' total CO2 emissions 2. Inverters' total CO2 emissions 3. Land use	1. Legal and regulatory framework 2. Economic parameters

In alignment with the firm's sensing and smart requirements, it is unequivocally evident that the Sensing and Smart characteristics of SP5 optimally fulfil its requirements, principally pertaining to the measurement of meteorological and power variables, in addition to data analysis for informed decision making. Despite the PV system's grid-connection, the consideration of backup batteries has necessitated the inclusion of additional features for a battery system by the DM.

However, in terms of sustainability and social factors, SP5 lacks crucial features necessary to address the company's needs. For instance, considering the ground installation of the system, it is essential to account for the potential impacts of noise and visual disturbances, as well as effects on local flora and fauna. Furthermore, water usage and consumption are paramount factors, particularly during the operational phase, chiefly for maintenance and cleaning purposes. Finally, waste management, an aspect with substantial environmental impact, should have been considered given the significance of correct waste disposal in PV systems. It is noteworthy that none of the decision makers opted for waste management analysis (as shown in Figure 8), thus, hypothetically, none of the solution paths would adequately address the company's sustainability needs.

In relation to the Social category, SP5 does not account for population acceptance, a factor integral to ensuring a harmonious relationship between the user and the PV system. While some regulations may be applicable depending on the system, all three Social features are indispensable for PV systems. In this case, as illustrated in Figure 8, no decision maker selected all three Social features. Therefore, similar to the Sustainable category, no solution path adequately covers the social needs of the company.

5.1. *Sensitivity Analysis*

Sensitivity analysis is essential to evaluate the robustness of the methodology proposed and to determine how changes in decision criteria weights and fuzzy number assignments affect the final ranking of solar PV configurations.

The sensitivity analysis consists of two tests:

1. Weight variation analysis: Decision criteria weights are systematically adjusted to examine their impact on rankings and weights are increased/decreased for one criterion at a time while keeping the others constant.
2. Alternative fuzzy number assignments: Slight modifications to linguistic variable values are made (e.g., changing "Good" from (0.7, 0.9, 1.0) to (0.6, 0.8, 1.0)). The ranking stability is then evaluated under these variations.

5.1.1. Test 1: Weight Variation Analysis

Weights assigned to S4 criteria—sensing, smart, sustainable, and social—were varied in increments of ±10% while keeping the total sum equal to 1. A total of five weight variations were tested (Table 14).

Table 14. Weight scenarios.

Scenario	s1	s2	s3	s4
Base case	0.30	0.30	0.20	0.20
Scenario 1	0.40	0.25	0.20	0.15
Scenario 2	0.25	0.40	0.20	0.15
Scenario 3	0.25	0.25	0.30	0.20
Scenario 4	0.30	0.25	0.15	0.30

After applying fuzzy TOPSIS for each scenario, the CC_i of each solution path was recalculated (Table 15).

Table 15. Ranking results under different weight scenarios.

Scenario	SP1	SP2	SP3	SP4	SP5	Best Alternative
Base case	0.42	0.64	0.53	0.65	0.66	SP5
Scenario 1 (higher technical weight)	0.42	0.62	0.55	0.67	0.68	SP5
Scenario 2 (higher economic weight)	0.41	0.68	0.52	0.63	0.64	SP2
Scenario 3 (higher environmental weight)	0.40	0.60	0.55	0.66	0.67	SP5
Scenario 4 (higher social weight)	0.43	0.61	0.54	0.67	0.66	SP4

As observed in Table 14, SP5 remained the top-ranked solution in most scenarios, providing the robustness of the proposed methodology. SP2 moved up in Scenario 2, showing that when economic criteria were prioritized, SP2 was slightly more attractive. SP4 became the best choice in Scenario 4, indicating that if social aspects were weighted more heavily, SP4 would be the preferred alternative.

5.1.2. Test 2: Alternative Fuzzy Number Assignments

In this test, the linguistic variables assignments for fuzzy numbers were slightly modified to test their impact on rankings:

- "Good" changed from (0.7, 0.9, 1.0) to (0.6, 0.8, 1.0),
- "Fair" changed from (0.3, 0.5, 0.7) to (0.4, 0.6, 0.8), and
- "Poor" changed from (0.0, 0.1, 0.3) to (0.0, 0.2, 0.4).

Fuzzy TOPSIS was recalculated with the modified fuzzy number assignments (Table 16).

Table 16. Ranking results under different fuzzy number assignments.

Scenario	SP1	SP2	SP3	SP4	SP5	Best Alternative
Original fuzzy numbers	0.42	0.64	0.53	0.65	0.66	SP5
Modified fuzzy numbers	0.44	0.63	0.55	0.66	0.67	SP5

Table 16 shows that minor changes in fuzzy number assignments did not significantly impact rankings, confirming the stability of the decision-making process. SP5 remained the top choice of the decision-making process.

From the sensitivity analysis, it can be concluded that the proposed methodology is highly stable, as demonstrated by minimal ranking fluctuations under different conditions. SP5 remained the top-ranked alternative in almost all scenarios, showing that the decision is robust to minor variations in inputs. The methodology adapts to changes in decision criteria weights, allowing customization based on industry priorities. Future studies could extend sensitivity analysis by incorporating Monte Carlo simulations to further validate the robustness of the framework.

5.2. Comparison with Other Proposals

In Table 17, a comparison between this proposal and the proposals mentioned in the literature review (Table 1) that employ fuzzy TOPSIS to address various solar PV energy problems is presented. This comparison is intended to illustrate how the fuzzy TOPSIS method is applied, at which stage of the PV system's lifecycle the proposals are centered, and the type of criteria they focus on.

Table 17. Comparison between this proposal and others using the fuzzy TOPSIS method.

Reference	Method	Lifecycle Stage				Criteria			
		Analysis	**Installation**	**Operation**	**Disposal**	**Technical**	**Economic**	**Environment**	**Social**
This research	Morphological matrix and fuzzy TOPSIS	x	x	x	x	x	x	x	x
[33]	GIS-based Fermatean Fuzzy TOPSIS	x	x			x	x		x
[38]	Fuzzy TOPSIS	x	x	x		x			x
[34]	Fuzzy AHP-TOPSIS	x		x		x	x	x	x
[39]	Fuzzy AHP-TOPSIS	x	x	x		x	x		x
[35]	SWOT-based AHP, fuzzy VIKOR and fuzzy TOPSIS	x	x	x		x	x	x	x
[40]	Delphi method, fuzzy AHP, fuzzy VIKOR and TOPSIS	x		x		x	x	x	x
[42]	Fuzzy AHP-TOPSIS	x		x		x	x		x
[41]	Fuzzy AHP-TOPSIS	x		x	x		x	x	x
[36]	AHP and fuzzy TOPSIS	x	x	x		x	x	x	x
[37]	Fuzzy AHP-TOPSIS	x	x			x		x	

Regarding the fuzzy TOPSIS method, except for Ranganath et al. (2022) [38], all the studies combined the method with other MCDM methods, mainly with AHP, since the combination of AHP and fuzzy TOPSIS can provide a more comprehensive and robust decision-making framework and can effectively deal with the complexity of MCDM problems [52].

This proposal considers the complete lifecycle of PV systems, since each stage is essential because it contributes to the overall success, efficiency, and sustainability of the system. Furthermore, the analysis of each stage ensures that the system is well located, complies with regulations, is efficient in its operation and safe for personnel, and it is environmentally responsible throughout its lifecycle. Regarding the other proposals, as observed in Table 17, all the studies focus on the analysis stage, where the technical and economic viability of the project can be assessed. The second most considered stage is operation phase. During this stage, the system generates electricity, which involves monitoring to ensure it continues to function optimally, meeting economic, environmental, and social goals. The installation stage is addressed by six proposals, taking into account the structure, safety precautions, component compatibility, and proper documentation. Among all the studies reviewed, only Ligus and Peternek (2018) [41] address the disposal stage for waste management purposes. The fact that only one study considers disposal stage is critical because this stage is essential for environmental responsibility and regulatory compliance.

On the other hand, the two most considered criteria in the revised proposals were technical and social. The former primarily ensures the system's effectiveness, while the latter is essential for compliance with legal requirements. The economic criterion, focused on analyzing the economic viability of the project, was the third most considered. Finally, environmental criteria were addressed in only six studies, covering subcriteria such as waste management, GHG emissions, and impact on land and on flora and fauna.

While methods like AHP or Delphi could provide structured ways of determining weights, fuzzy TOPSIS was selected due to its ability to directly incorporate uncertainty and linguistic judgements, making it particularly suited to the complex, multi-criteria environment of industrial PV selection.

Decision making constitutes a critical process in the deployment and operation of solar energy systems, as corroborated by the findings of this study. The employment of morphological matrices enables a broader understanding of the system requirements and potential solutions, as well as the possible solution combinations. The application of a multi-criteria methodology, specifically fuzzy TOPSIS, in consideration of the S4 features and criteria, provides a holistic evaluation of solar energy systems, tailored to the distinct needs of manufacturing enterprises. This evaluation covers the complete lifecycle and takes into account essential technical, economic, environmental, and social factors at each stage.

However, it is essential for decision makers to meticulously analyze the needs and judiciously select potential solutions, thereby ensuring the effective achievement of the objectives of the decision-making tools proposed in this study. Conclusively, a strategy predicated on the previous results is proposed for the current and future operations of the company, as illustrated in Table 18.

Table 18. Strategy for present and future actions for the manufacturing company.

Present Actions	Future Actions
1. Install appropriate sensors (e.g., panel surface temperature, solar radiation, voltage, current, air/ambient temperature, battery voltage, tilt angle, wind speed, wind direction, and humidity) to collect accurate and reliable data for efficient solar panel operation and maintenance. 2. Implement a smart meter system to monitor energy production and consumption, enabling better energy management and facilitating grid integration. 3. Implement a solar tracking system to maximize solar energy capture by adjusting the panels' orientation according to the sun's position. 4. Develop a monitoring system to track and manage the performance of solar panels and related components in real time. 5. Implement a battery management system to optimize battery charging/discharging cycles, prolonging battery life and ensuring efficient energy storage.	1. Expand data storage and analytics capabilities to process and analyze sensor data, enabling the identification of trends and optimization opportunities. 2. Implement predictive analysis techniques to forecast solar energy production, allowing for better grid management and integration. 3. Utilize intelligent modelling techniques for forecasting solar energy to improve the accuracy of solar energy predictions, thus enhancing overall system efficiency. 4. Implement a smart inverter system to manage energy production, consumption, and storage more effectively and to provide grid support when needed. 5. Continuously monitor and evaluate solar panels' and inverters' total CO_2 emissions to assess and improve their environmental sustainability. 6. Implement land use strategies that minimize the environmental impact of solar panel installations, such as using previously disturbed lands or integrating solar panels into building designs. 7. Advocate for and adhere to a comprehensive legal and regulatory framework that promotes solar energy adoption and ensures responsible practices. 8. Monitor and consider economic parameters such as costs, incentives, and market trends to ensure the financial viability of solar panels installation and operation.

5.3. Practical Implications

The proposed morphological matrix + fuzzy TOPSIS framework provides a structured, data-driven approach for decision making in industrial solar PV deployment, offering manufacturing firms a systematic way to explore, evaluate, and select optimal system configurations based on multiple lifecycle-based criteria. By integrating technical, economic, environmental, and social factors, the methodology ensures that PV system adoption aligns with both operational efficiency and sustainability goals. The framework is highly adaptable, allowing companies to customize criteria weights based on strategic priorities, regulatory constraints, or financial considerations; thus, the proposed methodology can be customized for diverse industrial sectors beyond manufacturing, including automotive, textiles, and food processing industries, as well as firms of varying sizes and geographical locations.

While this proposed decision-making framework facilitates the optimal selection of solar PV configuration, manufacturing companies may face several other implementation challenges beyond the decision-making stage. Regulatory barriers often present significant hurdles, including unclear permitting procedures, grid access limitations, and complex compliance requirements, which can delay or deter solar PV adoption. Additionally, financial risks stemming from high initial investment costs, limited financing options, or uncertain payback periods could restrict companies' willingness to pursue renewable energy projects. Technical constraints, such as integration difficulties with existing power infrastructure and the availability of trained personnel to manage PV systems, may also pose substantial barriers. Addressing these practical challenges through targeted policies, financial incentives, streamlined regulatory frameworks, and specialized training programs is crucial to successfully implementing solar PV projects in industrial settings.

Mexico has experienced significant fluctuations in renewable energy policies over recent years, impacting solar PV adoption. Notably, reforms and regulatory changes have introduced uncertainty into the market. Initiatives such as the Electric Industry Law (Ley de la Industria Eléctrica-LIE) and recent changes proposed by the federal government have prioritized state-owned utilities and fossil fuel-generated power, indirectly creating barriers to renewable energy expansion and investment.

Conversely, supportive policy frameworks like net-metering, distributed generation incentives, and renewable energy auctions have historically encouraged solar adoption among industrial and commercial sectors. For instance, distributed generation policies allow industrial companies to offset energy costs by selling surplus solar-generated power back to the grid. However, regulatory uncertainties surrounding grid access, unclear permitting procedures, and the lack of consistent financial incentives have created significant hurdles for manufacturing firms aiming to adopt solar technologies.

Thus, to bolster solar PV adoption in Mexico's industrial sector, stable and transparent policy frameworks, clear regulations for renewable integration, financial incentives, streamlined permitting processes, and comprehensive renewable energy standards are critical. Policymakers should address these barriers proactively to unlock Mexico's considerable solar energy potential fully.

5.4. Limitations of the Proposed Methodology

Despite its advantages, the proposed methodology has certain limitations. First, the selection of criteria weights and linguistic variable assignments in fuzzy TOPSIS relies on expert judgement, which may introduce subjectivity and bias into the decision-making process in real-world scenarios. While sensitivity analysis helps mitigate this concern, further validation through larger expert panels or machine learning-based optimization techniques could enhance objectivity. Second, the computational complexity increases as the number of decision criteria and solution paths grows, making it less scalable for highly complex industrial applications with hundreds of potential configurations. Finally, the methodology has been tested in a single case study in a Mexican manufacturing company, and additional case studies across different industries and geographic regions would be needed to generalize its applicability further.

6. Conclusions

This research proposes a hybrid methodology integrating a morphological matrix with fuzzy TOPSIS, supported by the S4 Framework (sensing, smart, sustainable, and social), to select optimal PV configuration for manufacturing firms. The morphological matrix visually represents all possible PV system configurations based on identified requirements, while fuzzy TOPSIS effectively ranks these alternatives, managing uncertainty and incorporating multiple decision criteria.

A case study involving a Mexican manufacturing firm validated the proposed methodology. Five decision makers analyzed company-specific needs and established the morphological matrix, generating multiple solution paths. Fuzzy TOPSIS ranked these solutions, identifying Solution Path 5 as optimal based on decision-maker inputs. However, upon evaluation against the S4 requirements, it was found that this selection failed to cater to the Sustainable and Social needs of the company, thus rendering it unfit as the best possible alternative. Sensitivity analysis confirmed the stability and robustness of the ranking results, even when criteria weights and fuzzy logic inputs were varied.

Although the proposed methodology proves effective in identifying the best combination of S4 features for manufacturing companies considering the implementation of photovoltaic systems, the role of decision makers is paramount in the evaluation of S4 requirements and the selection of the features that best satisfy them. This is crucial to ensure that the selected alternative is indeed the most suitable.

This study emphasizes the practical importance of informed, multi-criteria decision making for industrial solar PV adoption, advocating for comprehensive understanding and customization based on sustainability goals, cost-efficiency, and unique organizational needs. Future research work includes the integration of discrete event simulation to optimize solar energy systems and energy consumption in the manufacturing process, leading to cost reduction, extended system lifespan, and maintenance savings by simulating real-time performance and operational variability, providing dynamic feedback to refine system configurations and decision-making processes. Furthermore, future studies should include detailed lifecycle cost analyses and comprehensive environmental impact assessments to further strengthen the decision-making framework's completeness and robustness. Incorporating artificial intelligence and optimization algorithms could also significantly improve the scalability and automation of the decision-making process, since criteria weighting could be automated based on historical project performance data, while optimization algorithms could rapidly evaluate thousands of configuration alternatives.

Finally, the practical and policy implications suggest the necessity of creating an environment that encourages Mexican manufacturing companies and other industrial sectors to embrace solar PV systems. Effective policies should include financial incentives, educational initiatives, research and development funding, environmental standards, and adaptability and training, encouraging widespread solar PV integration and sustainable industrial growth.

Author Contributions: Conceptualization, C.P.B., B.A., R.B., A.R.F. and P.P.; methodology, C.P.B., Q.M., R.B., A.M., B.A., R.B, T.P, A.R, Q.M. and P.P.; software, C.P.B.; validation, P.P., B.A., R.B., A.R.F., T.P. and A.M.; formal analysis, C.P.B., P.P and A.R.F.; investigation, C.P.B., P.P. and A.R; resources, P.P.; data curation, C.P.B.; writing—original draft preparation, C.P.B., P.P., B.A., R.B., T.P., A.M. and Q.M, writing—review and editing, P.P., A.R.F., B.A., T.P., A.M., A.M. and Q.M.; visualization, C.P.B.; supervision, P.P., A.R.F. and Q.M.; project administration, P.P.; funding acquisition, P.P. All authors have read and agreed to the published version of the manuscript.

Funding: This research project is supported by the Institute of Advanced Materials for Sustainable Manufacturing, Tecnologico de Monterrey, Mexico, and the University of Alberta, Canada.

Data Availability Statement: Data will be made available on request.

Acknowledgments: The authors would like to acknowledge the financial and technical support of the Institute of Advanced Materials for Sustainable Manufacturing, Tecnologico de Monterrey, in the production of this work, and the University of Alberta.

Conflicts of Interest: The authors declare no conflicts of interest.

Abbreviations

The following abbreviations are used in this manuscript:

AHP	Analytic Hierarchy Process
ANP	Analytic Network Process
CC_i	closeness coefficient
CO2	carbon dioxide
DM	decision maker
ELECTRE	Elimination Et Choix Traduisant la Realité
F	fair
FNIS	fuzzy negative ideal solution
FPIS	fuzzy positive ideal solution
G	good
GHG	greenhouse gas
GIS	geographic information system
GMA	General Morphological Analysis
H	high
KPIs	key performance indicators
L	low
M	medium
MADM	Multi-attribute decision making
MCDM	Multi-criteria decision making
MG	medium good
MH	medium—high
ML	medium—low
MODM	Multi-objective decision making
MP	medium poor
P	poor
PV	photovoltaic
ROI	return on investment
R&D	research and development
s1	sensing
s2	smart
s3	sustainable
s4	social
S4	sensing, smart, sustainable, and social
SP	solution path
SWOT	strengths, weaknesses, opportunities, and threats
TOPSIS	Technique for Order of Preference by Similarity to Ideal Solution
VG	very good
VH	very high
VL	very low
VP	very poor
WSN	wireless sensor network

References

1. Zou, C.; Zhao, Q.; Zhang, G.; Xiong, B. Energy Revolution: From a Fossil Energy Era to a New Energy Era. *Nat. Gas Ind. B* **2016**, *3*, 1–11. [CrossRef]
2. Kåberger, T. Progress of Renewable Electricity Replacing Fossil Fuels. *Glob. Energy Interconnect.* **2018**, *1*, 48–52. [CrossRef]
3. Marks-Bielska, R.; Bielski, S.; Pik, K.; Kurowska, K. The Importance of Renewable Energy Sources in Poland's Energy Mix. *Energies* **2020**, *13*, 4624. [CrossRef]

4. Kuşkaya, S.; Bilgili, F.; Muğaloğlu, E.; Khan, K.; Hoque, M.E.; Toguç, N. The Role of Solar Energy Usage in Environmental Sustainability: Fresh Evidence through Time-Frequency Analyses. *Renew. Energy* **2023**, *206*, 858–871. [CrossRef]
5. Aman, M.M.; Solangi, K.H.; Hossain, M.S.; Badarudin, A.; Jasmon, G.B.; Mokhlis, H.; Bakar, A.H.A.; Kazi, S.N. A Review of Safety, Health and Environmental (SHE) Issues of Solar Energy System. *Renew. Sustain. Energy Rev.* **2015**, *41*, 1190–1204. [CrossRef]
6. Abdel-Basset, M.; Gamal, A.; ELkomy, O.M. Hybrid Multi-Criteria Decision Making Approach for the Evaluation of Sustainable Photovoltaic Farms Locations. *J. Clean. Prod.* **2021**, *328*, 129526. [CrossRef]
7. U.S. Energy Information Administration (EIA). International—Electricity. Available online: https://www.eia.gov/international/data/world/electricity/electricity-generation?pd=2&p=0000000000000000000000000000fvu&u=0&f=A&v=mapbubble&a=-&i=none&vo=value&t=C&g=00000000000000000000000000000000000000000000000001&l=249-ruvvvvvfvtvnvv1vrvvvvfvvvvvvfvvvou20evvvvvvvvvvnvvvs0008&s=315532800000&e=1609459200000& (accessed on 21 March 2023).
8. IRENA. Country Rankings. Available online: https://www.irena.org/Data/View-data-by-topic/Capacity-and-Generation/Country-Rankings (accessed on 20 March 2023).
9. Mekhilef, S.; Saidur, R.; Safari, A. A Review on Solar Energy Use in Industries. *Renew. Sustain. Energy Rev.* **2011**, *15*, 1777–1790. [CrossRef]
10. International Energy Agency. World Electricity Final Consumption by Sector, 1974–2019. Available online: https://www.iea.org/data-and-statistics/charts/world-electricity-final-consumption-by-sector-1974-2019 (accessed on 11 July 2022).
11. Paramati, S.R.; Apergis, N.; Ummalla, M. Dynamics of Renewable Energy Consumption and Economic Activities across the Agriculture, Industry, and Service Sectors: Evidence in the Perspective of Sustainable Development. *Environ. Sci. Pollut. Res.* **2018**, *25*, 1375–1387. [CrossRef]
12. Reiter, S.; Schulze, P.; Somers, K. Reimagining Industrial Operations. *McKinsey Quarterly*. 19 May 2020. Available online: https://www.mckinsey.com/capabilities/sustainability/our-insights/reimagining-industrial-operations (accessed on 16 October 2023).
13. University of Massachusetts Dartmouth. Decision-Making Process. Available online: https://www.umassd.edu/fycm/decision-making/process/ (accessed on 20 February 2023).
14. Uzun Ozsahin, D.; Denker, A.; Kibarer, A.G.; Kaba, S. Chapter 4—Evaluation of Stage IV Brain Cancer Treatment Techniques. In *Applications of Multi-Criteria Decision-Making Theories in Healthcare and Biomedical Engineering*; Ozsahin, I., Ozsahin, D.U., Uzun, B., Eds.; Academic Press: Cambridge, MA, USA, 2021; pp. 59–69. ISBN 978-0-12-824086-1.
15. Haddad, M.; Sanders, D. Selection of Discrete Multiple Criteria Decision Making Methods in the Presence of Risk and Uncertainty. *Oper. Res. Perspect.* **2018**, *5*, 357–370. [CrossRef]
16. Eltarabishi, F.; Omar, O.H.; Alsyouf, I.; Bettayeb, M. Multi-Criteria Decision Making Methods and Their Applications—A Literature Review. In Proceedings of the International Conference on Industrial Engineering and Operations Management, Dubai, United Arab Emirates, 10–12 March 2020.
17. Taherdoost, H.; Madanchian, M. Multi-Criteria Decision Making (MCDM) Methods and Concepts. *Encyclopedia* **2023**, *3*, 77–87. [CrossRef]
18. Haddad, M.J.M. A Framework That Uses Sensitivity Analysis to Select Multi Criteria Decision Making Methods. *J. Comput. Syst. Eng.* **2017**, 413–419. Available online: http://www.jcise.electronic-development.co.uk/ (accessed on 26 March 2025).
19. Karunathilake, H.; Bakhtavar, E.; Chhipi-Shrestha, G.; Mian, H.R.; Hewage, K.; Sadiq, R. Chapter Seven—Decision Making for Risk Management: A Multi-Criteria Perspective. In *Methods in Chemical Process Safety*; Khan, F.I., Amyotte, P.R., Eds.; Advanced Methods of Risk Assessment and Management; Elsevier: Amsterdam, The Netherlands, 2020; Volume 4, pp. 239–287.
20. Hwang, C.-L.; Yoon, K. *Multiple Attribute Decision Making*; Lecture Notes in Economics and Mathematical Systems; Springer: Berlin/Heidelberg, Germany, 1981; Volume 186, ISBN 978-3-540-10558-9.
21. Ilbahar, E.; Cebi, S.; Kahraman, C. A State-of-the-Art Review on Multi-Attribute Renewable Energy Decision Making. *Energy Strategy Rev.* **2019**, *25*, 18–33. [CrossRef]
22. Velasquez, M.; Hester, P. An Analysis of Multi-Criteria Decision Making Methods. *Int. J. Oper. Res.* **2013**, *10*, 56–66.
23. Alsalem, M.A.; Zaidan, A.; Bahaa, B.; Hashim, M.; Albahri, O.S.; Albahri, A.S.; Mohsin, A.; Mohammed, K. Systematic Review of an Automated Multiclass Detection and Classification System for Acute Leukaemia in Terms of Evaluation and Benchmarking, Open Challenges, Issues and Methodological Aspects. *J. Med. Syst.* **2018**, *42*, 204. [CrossRef] [PubMed]
24. Sabaei, D.; Erkoyuncu, J.; Roy, R. A Review of Multi-Criteria Decision Making Methods for Enhanced Maintenance Delivery. *Procedia CIRP* **2015**, *37*, 30–35. [CrossRef]
25. Widianta, M.M.D.; Rizaldi, T.; Setyohadi, D.P.S.; Riskiawan, H.Y. Comparison of Multi-Criteria Decision Support Methods (AHP, TOPSIS, SAW & PROMENTHEE) for Employee Placement. *J. Phys. Conf. Ser.* **2018**, *953*, 012116. [CrossRef]
26. Chakraborty, S. TOPSIS and Modified TOPSIS: A Comparative Analysis. *Decis. Anal. J.* **2022**, *2*, 100021. [CrossRef]

27. Ozsahin, I.; Uzun Ozsahin, D.; Uzun, B.; Mustapha, M.T. Chapter 1—Introduction. In *Applications of Multi-Criteria Decision-Making Theories in Healthcare and Biomedical Engineering*; Ozsahin, I., Ozsahin, D.U., Uzun, B., Eds.; Academic Press: Cambridge, MA, USA, 2021; pp. 1–2. ISBN 978-0-12-824086-1.
28. Alam, N.M.F.H.N.B.; Ku Khalif, K.M.N.; Jaini, N.I.; Gegov, A. The Application of Z-Numbers in Fuzzy Decision Making: The State of the Art. *Information* **2023**, *14*, 400. [CrossRef]
29. Aliev, R.; Guirimov, B.; Huseynov, O.; Aliyev, R. A Consistency-Driven Approach to Construction of Z-Number-Valued Pairwise Comparison Matrices. *Iran. J. Fuzzy Syst.* **2021**, *18*, 37–49. [CrossRef]
30. Sarkar, B. 14—Fuzzy Decision Making and Its Applications in Cotton Fibre Grading. In *Soft Computing in Textile Engineering*; Majumdar, A., Ed.; Woodhead Publishing Series in Textiles; Woodhead Publishing: Sawston, UK, 2011; pp. 353–383. ISBN 978-1-84569-663-4.
31. Nădăban, S.; Dzitac, S.; Dzitac, I. Fuzzy TOPSIS: A General View. *Procedia Comput. Sci.* **2016**, *91*, 823–831. [CrossRef]
32. Kore, M.N.B.; Ravi, K.; Patil, B. A Simplified Description of FUZZY TOPSIS Method for Multi Criteria Decision Making. *Int. Res. J. Eng. Technol.* **2017**, *4*, 2047–2050.
33. Hooshangi, N.; Mahdizadeh Gharakhanlou, N.; Ghaffari Razin, S.R. Evaluation of Potential Sites in Iran to Localize Solar Farms Using a GIS-Based Fermatean Fuzzy TOPSIS. *J. Clean. Prod.* **2023**, *384*, 135481. [CrossRef]
34. Qian, X.; Bai, Y.; Huang, W.; Dai, J.; Li, X.; Wang, Y. Fuzzy Technique Application in Selecting Photovoltaic Energy and Solar Thermal Energy Production in Belt and Road Countries. *J. Energy Storage* **2021**, *41*, 102865. [CrossRef]
35. Anser, M.K.; Mohsin, M.; Abbas, Q.; Chaudhry, I.S. Assessing the Integration of Solar Power Projects: SWOT-Based AHP–F-TOPSIS Case Study of Turkey. *Environ. Sci. Pollut. Res.* **2020**, *27*, 31737–31749. [CrossRef] [PubMed]
36. Sindhu, S.; Nehra, V.; Luthra, S. Investigation of Feasibility Study of Solar Farms Deployment Using Hybrid AHP-TOPSIS Analysis: Case Study of India. *Renew. Sustain. Energy Rev.* **2017**, *73*, 496–511. [CrossRef]
37. Kengpol, A.; Rontlaong, P.; Tuominen, M. A Decision Support System for Selection of Solar Power Plant Locations by Applying Fuzzy AHP and TOPSIS: An Empirical Study. *J. Softw. Eng. Appl.* **2013**, *6*, 470–481. [CrossRef]
38. Ranganath, N.; Sarkar, D.; Patel, P.; Patel, S. Application of Fuzzy TOPSIS Method for Risk Evaluation in Development and Implementation of Solar Park in India. *Int. J. Constr. Manag.* **2022**, *22*, 2764–2774. [CrossRef]
39. Ali Sadat, S.; Vakilalroaya Fini, M.; Hashemi-Dezaki, H.; Nazififard, M. Barrier Analysis of Solar PV Energy Development in the Context of Iran Using Fuzzy AHP-TOPSIS Method. *Sustain. Energy Technol. Assess.* **2021**, *47*, 101549. [CrossRef]
40. Taylan, O.; Alamoudi, R.; Kabli, M.; AlJifri, A.; Ramzi, F.; Herrera-Viedma, E. Assessment of Energy Systems Using Extended Fuzzy AHP, Fuzzy VIKOR, and TOPSIS Approaches to Manage Non-Cooperative Opinions. *Sustainability* **2020**, *12*, 2745. [CrossRef]
41. Ligus, M.; Peternek, P. Determination of Most Suitable Low-Emission Energy Technologies Development in Poland Using Integrated Fuzzy AHP-TOPSIS Method. *Energy Procedia* **2018**, *153*, 101–106. [CrossRef]
42. Sasikumar, G.; Ayyappan, S. Multi-Criteria Decision Making for Solar Panel Selection Using Fuzzy Analytical Hierarchy Process and Technique for Order Preference by Similarity to Ideal Solution (TOPSIS): An Empirical Study. *J. Inst. Eng. India Ser. C* **2019**, *100*, 707–715. [CrossRef]
43. Molina, A.; Ponce, P.; Miranda, J.; Cortés, D. *Enabling Systems for Intelligent Manufacturing in Industry 4.0: Sensing, Smart and Sustainable Systems for the Design of S3 Products, Processes, Manufacturing Systems, and Enterprises*; Springer International Publishing: Cham, Switzerland, 2021; ISBN 978-3-030-65546-4.
44. Ponce, P.; Meier, A.; Miranda, J.; Molina, A.; Peffer, T. The Next Generation of Social Products Based on Sensing, Smart and Sustainable (S3) Features: A Smart Thermostat as Case Study. *IFAC-PapersOnLine* **2019**, *52*, 2390–2395. [CrossRef]
45. Ritchey, T. General Morphological Analysis: An Overview. *Acad. Lett.* **2022**. [CrossRef]
46. Barbu, B.; Purton, S. *Morphological Analysis: Big Words, Simple Idea*; NATO: Brussels, Belgium, 2017.
47. George, D. Concept Generation Using Morphological and Options Matrices. Master's Thesis, Clemson University, Clemson, SC, USA, 2012.
48. Pérez, C.; Ponce, P.; Meier, A.; Dorantes, L.; Sandoval, J.O.; Palma, J.; Molina, A. S4 Framework for the Integration of Solar Energy Systems in Small and Medium-Sized Manufacturing Companies in Mexico. *Energies* **2022**, *15*, 6882. [CrossRef]
49. Mahdavi, I.; Mahdavi-Amiri, N.; Heidarzade, A.; Nourifar, R. Designing a Model of Fuzzy TOPSIS in Multiple Criteria Decision Making. *Appl. Math. Comput.* **2008**, *206*, 607–617. [CrossRef]
50. Ponce, P.; Pérez, C.; Fayek, A.R.; Molina, A. Solar Energy Implementation in Manufacturing Industry Using Multi-Criteria Decision-Making Fuzzy TOPSIS and S4 Framework. *Energies* **2022**, *15*, 8838. [CrossRef]

1. SOLARGIS. World Bank Group Photovoltaic Power Potential—Mexico. Available online: https://globalsolaratlas.info/map?c=23.926013,-102.590332,5&r=MEX (accessed on 28 August 2022).
2. Rane, N.; Choudhary, S. Fuzzy AHP and Fuzzy TOPSIS as an Effective and Powerful Multi-Criteria Decision-Making (MCDM) Method for Subjective Judgements in Selection Process. *Int. Res. J. Mod. Eng. Technol. Sci.* **2023**, *5*, 3786–3799. [CrossRef]

processes

MDPI

Article

Research on the Process for Removing Heat-Stable Salts from Organic Amine Absorbents via Electrodialysis

Guang Chen [1], Qian Liu [2], Li Liu [3], Shilong Zhang [4], Guixian Li [2], Hongwei Li [2] and Dongliang Wang [2,*]

1 Gansu Yinguang Juyin Chemical Co., Ltd., Baiyin 730900, China; gsygjyhgyxgscg@163.com
2 School of Petrochemical Engineering, Lanzhou University of Technology, Lanzhou 730050, China; liuq.68@outlook.com (Q.L.); lgxwyf@163.com (G.L.); lhwzqy@163.com (H.L.)
3 Hefei Sanyuan Engineering & Technology Service Co., Ltd., Hefei 230041, China; liuli@chinaecec.com
4 Gansu Odewangs Material Technology Co., Ltd., Lanzhou 730050, China; xhqllb2411@163.com
* Correspondence: wangdl@lut.edu.cn

Abstract

The use of organic amine absorbents in CO_2 capture technologies is highly significant. The widespread application of this technique is limited by the heat-stable salts (HSSs) produced during the cyclic absorption–desorption process. This research focused on the HSS removal process using electrodialysis technology and systematically examined the effects of operating voltage, initial concentration, pH, current density, the ratio of liquid volume in the enriched chamber to that in the diluting chamber, and the type of ion-exchange membrane on desalination efficiency, energy consumption, and amine loss. An increase in both voltage and initial concentration significantly enhances the rate of water migration. The rate of ion migration is observed to follow the order of $Cl^- > SO_4^{2-} > F^-$ in a homogeneous membrane, while in a heterogeneous membrane, the order is $SO_4^{2-} > Cl^- > F^-$. The optimal operating voltage is 10 V, with a pH level of 8 resulting in the highest SO_4^{2-} removal efficiency. An industrial scenario validated the optimized process conditions, which balanced energy consumption with desalination efficiency. This methodology is essential not only for providing a viable solution for the industrial purification of organic amines but also for promoting the environmentally sustainable development of carbon capture technologies.

Keywords: electrodialysis technology; HSS removal; separation performance

1. Introduction

One of the hottest topics in carbon emission reduction technology is CO_2 capture using organic amine absorbents. Among the challenges faced are the formation of heat-stable salts (HSSs) and the high energy consumption associated with the capture process, both of which hinder the widespread use of organic amine CO_2 capture technology. The organic amine loses its active sites for absorbing CO_2 as a direct result of HSS production, which reduces the effectiveness of the decarburization process. The cost of CO_2 capture is further increased by issues like increased equipment corrosion and material foaming in the absorption and regeneration towers [1]. Therefore, as the main factor affecting the absorption effect of the organic amine solution [2], its removal is a key link in the carbon capture process using the organic amine method. Several HSS purification technologies for the organic amine CO_2 capture process have been created both domestically and internationally, such as distillation recovery [3], adsorption [4], ion exchange, and electrodialysis [5].

The distillation recovery method regenerates the active amine in the lean amine solution by heating and distillation. Tavan Y et al. conducted an investigation into the elimination of HSS from the amine solution via vacuum distillation, and they juxtaposed the empirical industrial data with simulation outcomes derived from Aspen Hysys [6]. The results were basically consistent, indicating the effectiveness and applicability of the vacuum distillation technology in amine purification. Cai P. et al. investigated the application of vacuum distillation technology within the context of membrane distillation, successfully regenerating a non-functional methyl diethanolamine (MDEA) solution under conditions of 10.0 kPa and 65 °C [7]. The study reported a recovery rate of 97% for the MDEA solution. However, the energy consumption of distillation technology is generally high. Especially when recovering amine liquids under vacuum conditions, the high temperature and the corrosiveness of the distillation effluent will damage the equipment. Even if corrosion-resistant materials such as stainless steel are used, it is difficult to completely avoid it. Moreover, the process impurities generated are often harmful to the environment, which will increase the difficulty and cost of subsequent treatment.

The adsorption method is based on the selective adsorption of impurity molecules by adsorption materials such as activated carbon, thereby realizing the purification and separation of amine solutions. Edathil et al. successfully synthesized a magnetic alginate gel material (Ca-Fe_2O_3) and applied it to remove organic acid anions in the MDEA solution, showing applicability in specific application scenarios [8]. Yan et al. developed a synthetic material for the fixed-bed adsorption process [9]. This synthetic material can deeply purify and remove impurities in various foaming pollutants, especially the sulfate impurities abundantly present in the amine solution [10]. Compared with traditional activated carbon, these new adsorbents with unique surface structures and chemical properties have shown significant advantages in key indicators such as purification efficiency and economic cost. However, there is still a lack of detailed research and reports on the regeneration ability of adsorption materials. Furthermore, the strength of the adsorbent's adsorption of H_2O can also cause a multiple-fold change in energy consumption. Therefore, its high energy consumption has become a significant challenge hindering its commercialization [11].

Ion-exchange technology is a technology developed based on the purification of HSSs, which can perform reversible ion exchange between the solid phase and the liquid phase. Research on the utilization of anion-exchange resins as adsorbents for the removal of HSSs from lean amine solvents remains limited. Morgan et al. described the first commercial application of ion-exchange resins in amine recovery [12]. The recycled amine solution contaminated with sodium chloride flows downward through the bed of strong basic anion-exchange resins, where chloride ions are replaced by hydroxide ions, and the resin is regenerated with the solution of 1 mol·L^{-1}. Jameh used an anion-exchange resin to remove HSSs from the solution, but did not mention the properties or composition of the anion-exchange resin used [13]. It is important to note that during the ion-exchange process, the resin may become contaminated over prolonged usage, leading to a reduction in its operational efficacy. Consequently, regular performance assessments and timely regeneration treatments are essential [14]. Nevertheless, the financial investment and scale required for the ion-regeneration process and the treatment of high salt concentrations are relatively substantial [15]. Furthermore, the diverse origins and types of HSS encountered in the CO_2 capture process significantly impair the absorption efficiency of organic amines, indicating that the existing advanced purification technologies for HSSs in organic amine solutions remain inadequate. Therefore, it is necessary to develop and propose in-depth purification technologies for HSSs to better cope with the current environmental protection and resource conservation challenges in China and promote green and sustainable development.

Electrodialysis is a continuous and efficient separation technique that leverages the osmotic selectivity of ion-exchange membranes, specifically anion-exchange membranes (AEMs) and cation-exchange membranes (CEMs) to facilitate the migration of charged ions from the raw diluting chamber to the enriched chamber when subjected to a direct current electric field. This process effectively retains uncharged components within the freshwater chamber, thereby achieving the separation of charged and uncharged species. The application of this technology is extensive, encompassing the desalination of brackish and alkaline waters, the desalination of organic products, and the treatment of saline waste liquids. Furthermore, it is currently recognized as a prominent method within the field of purification. Hong M et al. used a small electrodialysis device to remove HSSs from the aqueous alkanolamine solution when exploring the purification method of the aqueous alkanolamine solution, which not only improved the purity of the aqueous alkanolamine solution but also helped to optimize the entire production process [16]. Zhou Z et al. proposed a novel bipolar membrane electrodialysis (BMED) process to tackle the issues of low efficiency in removing heat-stable salts (HSSs) from lean amine solutions via the traditional resin exchange method, as well as the requirement for substantial amounts of alkali for regeneration. This offers guidance for the industrial application of HSS separation from desulfurization wastewater and other relevant sources [17].

Grushevenko et al. further advanced the utilization of electrodialysis technology by implementing a two-step electrodialysis procedure aimed at selectively removing sulfates from MEA solutions post CO_2 absorption [18]. This method is characterized by its systematic approach and has yielded significant results. Volkov et al. conducted a more comprehensive study, utilizing electrodialysis to eliminate HSSs from MEA with varying CO_2 loadings [19]. Their findings indicated that up to 70% of HSSs could be removed within a 30-min timeframe. This outcome not only underscored the efficacy of electrodialysis in HSS removal but also offered innovative strategies for addressing the HSS challenges associated with CO_2 absorption in alkanolamine solutions. Table 1 also summarizes the characteristics of various HSS purification technologies.

Table 1. Characteristics of various heat-stable salt purification technologies.

Project	Electrodialysis	Ion Exchange	Distillation Recovery	Adsorption Method
Technical principle	Removal of charged ions driven by voltage difference	Ion-exchange resin adsorbs ionic impurities	Thermal recovery of alkanolamine and water from amine solution degradation products and heat—stable salts	Adsorbent materials purify insoluble impurities
Feed requirements	Cold lean amine solution	Cold lean amine solution with solid particles and hydrocarbon impurities removed	Neutralized HSS	Pretreated cold lean amine solution
Required chemicals	An equivalent amount of R_3NH^+ and NaOH	NaOH and H_2SO_4 for resin regeneration	An equivalent amount of alkali to R_3NH^+	Agents for adsorbent regeneration
Types of impurities that can be purified	HSS	HSS impurities and some amides, amino acids, etc.	All non—volatile impurities	Insoluble impurities; HSS and surfactants

Table 1. *Cont.*

Project	Electrodialysis	Ion Exchange	Distillation Recovery	Adsorption Method
Amine solution recovery rate	Nearly 98%	Nearly 99%	85~95%	Nearly 99%
Cost	Moderate	Moderate	High	Relatively low
Technical advantages	No limit on the salt concentration in the raw material; low chemical consumption	Suitable for the treatment of low-concentration HSSs; low energy consumption	Capable of purifying a wide variety of pollutants with a small total amount after pollutant concentration	Simple operation, low cost, and a wide range of pollution purification
References	[5]	[12]	[3]	[4]

Distillation technology is characterized by high energy consumption and the adsorption method is constrained by the limited regeneration capacity of adsorbent materials. In ion-exchange processes, the regeneration of ion-exchange resins, coupled with the associated equipment costs and challenges related to handling high salt concentrations, remains relatively expensive. Consequently, these purification techniques are currently inadequate for achieving thorough removal of heat-stable salts (HSSs) in organic amine solutions. In contrast, electrodialysis, a continuous and efficient separation technique, selectively separates ionic species in aqueous solutions by applying a voltage across semi-permeable ion-exchange membranes. This paper primarily investigates the separation of HSSs (anions) from a mixed solution utilizing electrodialysis. The research explores the impact of various operating conditions, facilitated by a specially designed membrane stack, on the performance of electrodialysis separation and energy consumption. Key factors examined include operating voltage, initial concentration, initial value, current density, the volume ratio of the enriched to diluting chamber, and the characteristics of the membranes employed.

2. Materials and Methods

2.1. Electrodialysis Device

The electrodialyzer is the focal point of the electrodialysis apparatus, and Figure 1 illustrates the electrodialysis principle. Spacers divide the essential part, the ion-exchange membrane, which is alternately positioned between the positive and negative electrodes. Each membrane pair consists of one spacer, one AEM, and one CEM. The voltage for each pair of membranes is between 0.5 and 1 V. Zhejiang Lanran Technology Co., Ltd. (Hangzhou, China). supplied the desktop electrodialyzer utilized in this experiment, which was put together using the one-pole two-stage assembly method. The three pumps have a maximum flow rate of 1.8 min^{-1} and operate under constant voltage (maximum 18 V) or constant current (maximum 3 A). The three solution tanks (diluting chamber, enriched chamber, and electrode water chamber) have a combined maximum storage capacity of 1 L. These tanks are utilized for the storage of raw solution, concentrated solution, and electrode solution, respectively. The experimental setup employed a membrane stack consisting of ten cation-exchange membranes and ten anion-exchange membranes, with electrodes measuring 47 mm × 107 mm, constructed from titanium-coated ruthenium. The homogeneous and heterogeneous membranes utilized in this study were produced by Zhejiang Lanran Technology Co., Ltd. (Hangzhou, China). Upon the application of voltage, anions migrate from the freshwater chamber to the anode through the anion-exchange membrane, where they are subsequently captured by the cation-exchange membrane located in the enriched chamber. The experiment was conducted at ambient temperature, and samples were systematically collected from both the diluting chamber and the enriched chamber for analytical testing.

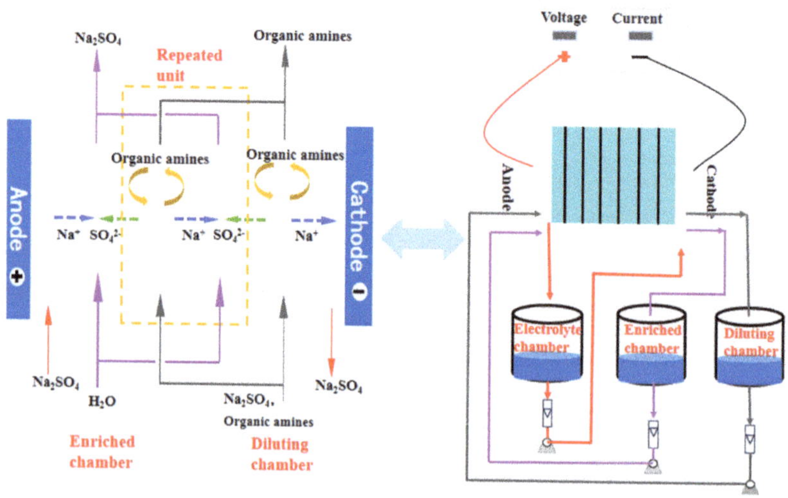

Figure 1. Schematic diagram of electrodialysis purification.

2.2. Ion-Exchange Membrane

The core component of electrodialysis technology is the ion-exchange membrane, which is fabricated from a specialized polymer material containing active ion-exchange groups. The membrane is divided by specially engineered partitions and is alternately positioned between the anode and cathode end plates. Anion-exchange membranes generally contain quaternary ammonium functional groups, with R representing the base membrane structure. In contrast, cation-exchange membranes typically incorporate sulfonic acid functional groups (R-SO_3H), where R denotes the base membrane, -SO^{3-} signifies the fixed ion, and H^+ corresponds to the dissociable ion. The characteristic functional group configurations of ion-exchange membranes are depicted in Figure 2. When an electric field is applied, only ions with specific charges are permitted to traverse the ion-exchange membrane, while other ions are effectively obstructed. This selective permeability enables the membrane to efficiently separate ions by accurately screening and allowing the passage of target ions.

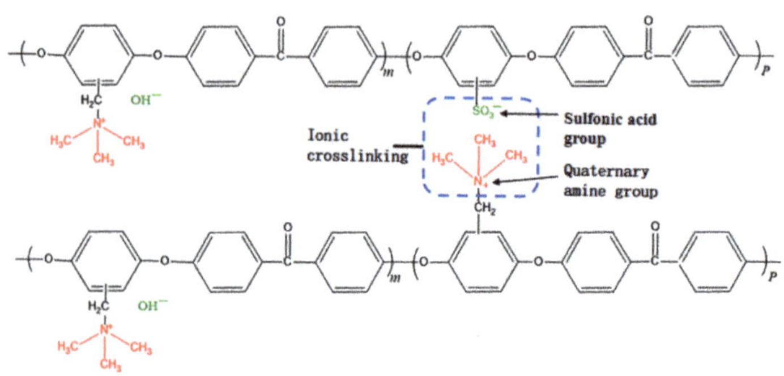

Figure 2. Functional group structure diagram of ion-exchange membrane.

Based on the macroscopic architecture of ion-exchange membranes, they can be categorized into homogeneous and heterogeneous types. Homogeneous membranes are fabricated by incorporating active functional groups into an inert matrix, resulting in a chemical structure characterized by small pores. In contrast, heterogeneous membranes are produced by blending powdered ion-exchange resin with a binder, leading to a chemically non-uniform structure. The ion-exchange channels of both homogeneous and heterogeneous membranes employed in the electrodialysis process are illustrated in Figure 3. To ensure sustained performance stability over extended periods, the ion-exchange membrane must exhibit high total diffusion resistance and robust chemical stability, thereby preventing degradation or damage from chemical exposure. The primary performance parameters of the ion-exchange membrane utilized in this study are summarized in Table 2.

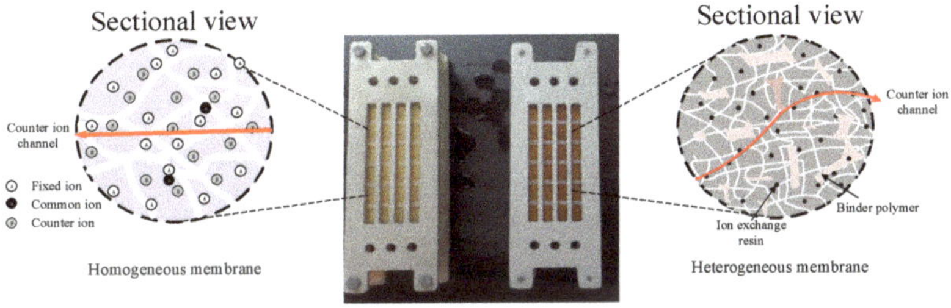

Figure 3. Ion-exchange channels.

Table 2. Ion-exchange membrane performance parameters.

Index Name	Homogeneous Membrane		Heterogeneous Membrane	
	Cation-Exchange Membrane	Anion-Exchange Membrane	Cation-Exchange Membrane	Anion-Exchange Membrane
Resistance $(\Omega \cdot cm^{-2})$	3.0	2.4	$2 \sim 3$	$2 \sim 3$
Bursting Strength (MPa)	≥0.40	≥0.25	0.6	0.6
Temperature Resistance(°C)	5~40	5~40	15~50	15~50
Resistance to pH Value	1~11	1~11	1~14	1~14
Diaphragm Thickness (mm)	0.17	0.14	0.2 ± 0.05	0.2 ± 0.05
Diaphragm Size (mm)	550 × 1100	550 × 1100	550 × 1100	550 × 1100

The performance parameters of the ion-exchange membrane were provided by Zhejiang Lanran Technology Co., Ltd.

Given that the performance of the ion-exchange membrane in the electrodialysis process is intricately linked to its structural characteristics, Fourier transform infrared (FTIR) spectroscopy and microscopic morphological analyses of both the cross-section and surface were conducted. These analyses aimed to assess the conditions of the membrane matrix morphology and the distribution of ion-exchange sites, specifically the resin particles. The ion-exchange membrane utilized in this study is a polymerizable composition created by impregnating an ion-exchange resin into a substrate made of polyolefin woven fabric. The FITR-infrared spectrogram is illustrated in Figure 4: the upper red line indicates the infrared peak corresponding to the AEM, while the lower black line represents the infrared peak associated with the CEM. The presence of residual hygroscopic water molecules within the unmodified resin component is evidenced by the O-H stretching vibrations and O-H in-plane scissor bending vibrations observed at the peaks of 3400 cm^{-1} and 1600 cm^{-1}, respectively.

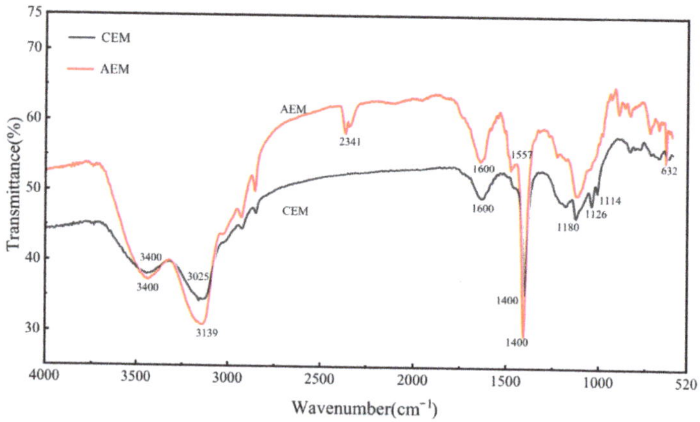

Figure 4. FITR-infrared spectrogram.

An SEM analysis of the ion-exchange membrane was conducted. The cross-sectional morphology of the CEM is illustrated in Figure 5, where resin particles that enhance the mechanical strength of the membrane are observed on the surface of the thick support layer, characterized by a loose porous structure, as well as within the membrane matrix. The functional groups present on the resin particles may provide conductive regions within the ion-exchange membrane, facilitating the formation of flow channels for ion transport during the electrodialysis process. The surface morphology of the AEM is depicted in Figure 6a,b. The rough surface structure of the AEM is advantageous as it can enhance water flux by increasing the likelihood of interaction between water molecules and various ions. Although Figure 6c,d do not reveal any prominent large pores, the discontinuity of the adhesive and the aggregation of resin particles contribute to the presence of numerous small pores and cracks. To provide a clearer representation of the distribution of functional groups on the membrane, a mapping test was conducted. As illustrated in Figure 7a,b, it was determined that the sulfur (S) element constituted 55% of the ion-exchange groups in the CEM, while the nitrogen (N) element accounted for 63% in the AEM.

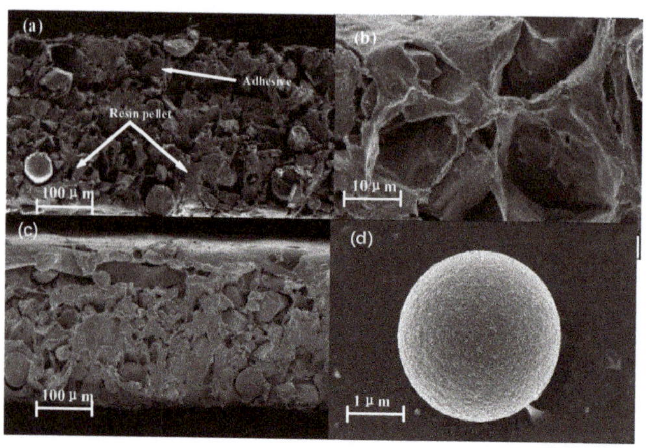

Figure 5. SEM images of cation exchange membranes at different scales (**a**) 100 μm, (**b**) 10 μm, (**c**) 100 μm, (**d**) 1 μm.

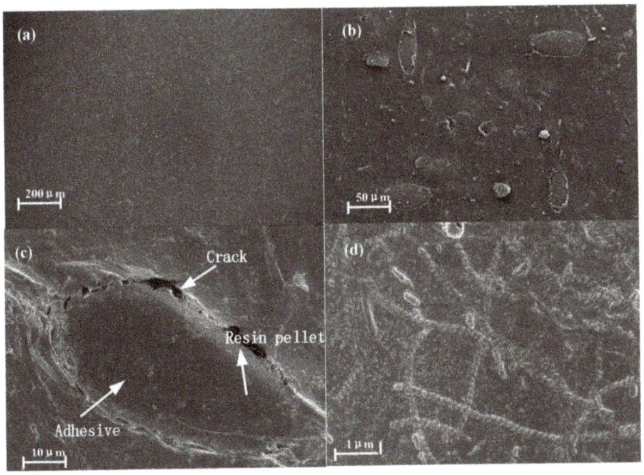

Figure 6. SEM images of anionic exchange membranes at different scales (**a**) 200 μm, (**b**) 50 μm, (**c**) 10 μm, (**d**) 1 μm.

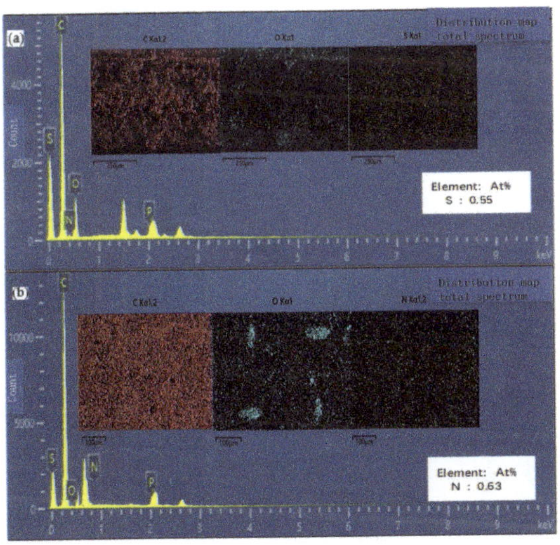

Figure 7. Mapping diagram (**a**) CEM; (**b**) AEM.

2.3. *Experimental Methods*

2.3.1. Operation Method

The electrodialysis experiment was conducted utilizing a constant-voltage mode, with applied voltages of 5 V and 15 V, respectively. Upon the application of an electric field, cations and anions migrated across the ion-exchange membrane into the enriched water chamber. The electrode water chamber was filled with a solution of Na_2SO_4 at a specified concentration (1~3%), which helped to create a stable electric field and successfully stopped the electrodes from corroding and becoming contaminated. Amine solutions and salt solutions of varying concentrations were introduced into the diluting chamber and the enriched chamber, respectively, with volumes ranging from 500 to 1500 mL. The circulation flow rate was maintained at approximately 70 L h^{-1}. Each solution was circulated for a duration of 30 to 80 min, during which the changes in conductivity of both two chambers were recorded synchronously. The experiment concluded when the conductivity fell below 1 $mS\cdot cm^{-1}$. All experimental procedures were conducted at ambient temperature [20].

2.3.2. Analysis and Calculation Methods

An ultraviolet spectrophotometer was employed to quantify various organic amines, specifically monoethanolamine (MEA) and methyldiethanolamine (MDEA). Anions were identified through ion chromatography. The concentration of sodium hydroxide was assessed using phenolphthalein as an indicator. Conductivity measurements were obtained with a conductivity meter. The pH of the solution was determined using a combined meter. Fluoride concentration was measured utilizing an ion-selective electrode (fluoride ion electrode) in conjunction with a standard reference electrode connected to a Metrohm ion meter. To adjust the total ionic strength, a total ionic strength adjustment buffer (TISAB) solution was prepared, consisting of 58.5 g of trisodium citrate and 85 g of sodium nitrate, with nitric acid added to achieve a pH range of 5 to 6. The chloride concentration was quantified through potentiometric titration using an automatic titrator, while the sulfate concentration was determined via gravimetric analysis.

The average flux J ($mmol\cdot(m^2\cdot s)^{-1}$) was used to represent the average quantity of salt and water transferred through the ion-exchange membrane per unit time to evaluate the performance of the ion-exchange membrane [21].

$$J_{Na_2SO_4} = \frac{C_t V_t - C_0 V_0}{NA_t} \tag{1}$$

$$J_{H_2O} = \frac{V_t - V_0}{18NA_t} \tag{2}$$

Among the terms, V_t and V_0 refer to the volume of the solution in the diluting chamber at time t and 0; t (s) refers to the operating time; C_t and C_0 ($mmol\cdot L^{-1}$) refer to the concentrations at time t and 0; A_t (0.055 m^2) refers to the effective membrane area; N (N = 10) refers to the number of repeating units.

The desalination rate is one of the key indicators for measuring the performance of electrodialysis and the separation effect of ion-exchange membranes. In the electrodialysis process, it refers to the ratio of the reduced salt amount in the freshwater chamber to the salt content of the raw material, which directly reflects the desalination capacity of the electrodialysis system [22]. Its calculation formula is

$$f = \frac{C_{d0} V_{d0} - C_{d2} V_{d2}}{C_{d0} V_{d0}} \times 100\% \tag{3}$$

Among the terms, V_{d0}, V_{d2} (L) refers to the initial and final volumes of diluting, f(%) refers to the desalination rate; C_{d0}, C_{d2} ($mol\cdot L^{-1}$) refers to the initial and final salt concentrations in the diluting chamber.

As an important parameter for evaluating the economic efficiency of the electrodialysis process, the value of power consumption directly reflects the energy consumption of the electrodialysis system during operation [23]. Its calculation formula is

$$W = \frac{\int_0^t UIdt}{n} \tag{4}$$

Among the terms, W ($Wh\cdot L^{-1}$) refers to the amount of energy consumed by an electrodialysis device in one hour of operation; U (V) refers to the voltage drop across the membrane in the electrodialyzer; t (h) refers to the time required for desalination; I (A) refers to the current; n (mol) refers to the molar amount of the regenerated amine solution.

The current efficiency (η, %) is calculated using the following formula [24]:

$$\eta = \frac{Z(C_t' - C_0')VF}{N\int_0^t Idt} \times 100\% \tag{5}$$

Among the terms, Z refers to the charge number of the ion; C_0' and C_t' (mol·L^{-1}) refer to the molar concentrations at time 0 and t; F (96,485 C·mol^{-1}) is the Faraday constant; V (L) refers to the solution volume; I (A) refers to the current; N (N = 10) refers to the number of repeating units in the membrane stack.

The loss rate is calculated by the following formula [25]:

$$\text{loss late} = \frac{\omega_0 - \omega_t}{\omega_0} \times 100\% \tag{6}$$

Among the terms, ω_0 (%) refers to the initial mass fraction of the amine solution in the diluting chamber; ω_t (%) refers to the mass fraction of the amine solution in the diluting chamber at time t.

3. Results

Following the regeneration process in the desorption tower, the amine solution is found to contain a significant concentration of HSSs, which are comprised of various anions. Each anion is capable of binding to a single amine molecule, resulting in a scenario where the salt is unable to release the associated organic amine molecules during the regeneration process that involves heating. Consequently, electrodialysis technology was employed to perform desalination experiments on a simulated HSS. The initial solutions in the diluting chamber included NaCl, NaF, and Na_2SO_4. The study aimed to investigate the migration behaviors of these HSS anions during the electrodialysis experiments. Additionally, the research examined the influence of several factors on the electrodialysis separation performance and energy consumption, including operating voltage, initial concentration, initial values, current density, the volume ratio of enriched to diluting chambers, and the types of membranes used. Furthermore, the migration behavior of anions in an actual amine solution sample was validated.

3.1. *Effect of Operating Voltage*

In this experiment, the electrodialysis process is conducted in a constant-voltage mode, with the operating voltage set within the range of 5 to 15 volts. To optimize energy efficiency in the electrode chamber, the number of membrane stacks is increased. Figure 8 illustrates the conductivity of the freshwater chamber at various operating voltages. The data presented in the figure indicate a temporal decrease in the conductivity of the diluting chamber across different operating voltages, signifying the effective removal of salt from the freshwater solution. As cations and anions migrate towards the enriched chamber via the CEM and AEM, respectively, the concentration of salt ions in the diluting chamber diminishes, leading to a corresponding reduction in conductivity. Notably, the slope of the conductivity curve exhibits a decreasing trend over time, with a rapid decline in conductivity observed during the initial phase of the experiment, followed by a deceleration in the ion migration process in the later stages. At the same time, it can be found from the figure that the time required for the desalination experiment decreases with the increase in the operating voltage because the driving force of electrodialysis is the current field. The greater the current, the greater the driving force.

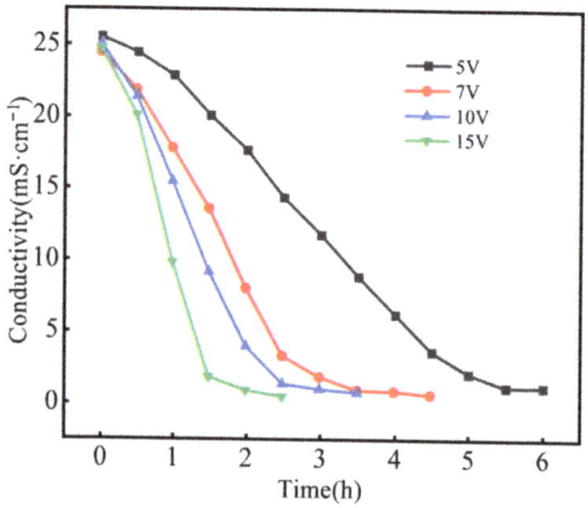

Figure 8. Effect of operating voltage on the change in conductivity of diluting chamber.

Different concentrations of SO_4^{2-} (16.9 g·L^{-1}), Cl^- (6.07 g·L^{-1}), and F^- (2.21 g·L^{-1}) were simulated and prepared, and the change laws under different voltages were discussed.

1. Effect of operating voltage on SO_4^{2-}.

Figure 9 illustrates the relationship between operating voltage and the reduction in salt concentration in the diluting chamber. As depicted in Figure 9a, an incremental increase in operating voltage correlates with a substantial decrease in salt concentration, which diminishes from an initial value of 16.9 g·L^{-1} to below 0.5 g·L^{-1}. Furthermore, the duration required for this desalination process decreases with increasing voltage, thereby demonstrating the beneficial impact of voltage on the desalination rate. An analysis of voltage and current variations during the initial 180 min allowed for the calculation of energy consumption and current efficiency at different voltages, as presented in Figure 9b. Notably, as the applied voltage escalates from 5 V to 15 V, the energy consumption necessary for the treatment of a unit volume of SO_4^{2-} solution consistently rises, reaching a significant level of 11.2 Wh·L^{-1} at 15 V. Additionally, the current efficiency exhibits a trend of initial increase followed by a decrease with rising operating voltage, achieving a peak current efficiency of 68.12% at 10 V.

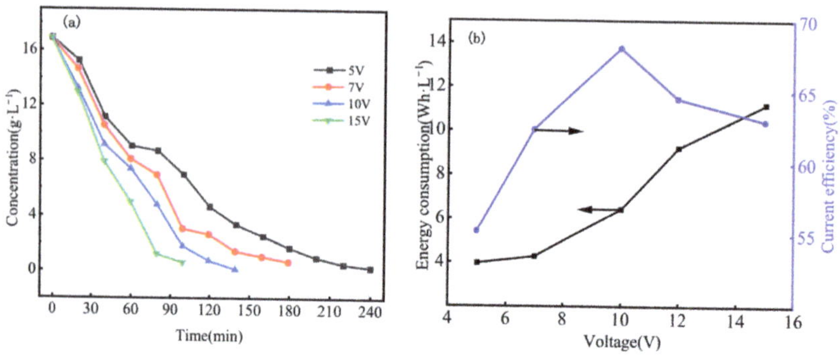

Figure 9. (**a**) Variation in SO_4^{2-} ion concentration in the diluting chamber and (**b**) relationship between energy consumption and current efficiency.

2. Influence of operating voltage on Cl^-.

Figure 10 shows that as the voltage gradually increases, the time required for the chloride salt content in the diluting chamber to decrease from 6.07 g·L^{-1} to below 0.5 g·L^{-1} gradually shortens. At a voltage of 15 V, the energy consumption significantly increases to 7.51 Wh·L^{-1}. The current efficiency first increases and then decreases as the operating voltage increases, reaching a maximum current efficiency of 65.2% at 10 V.

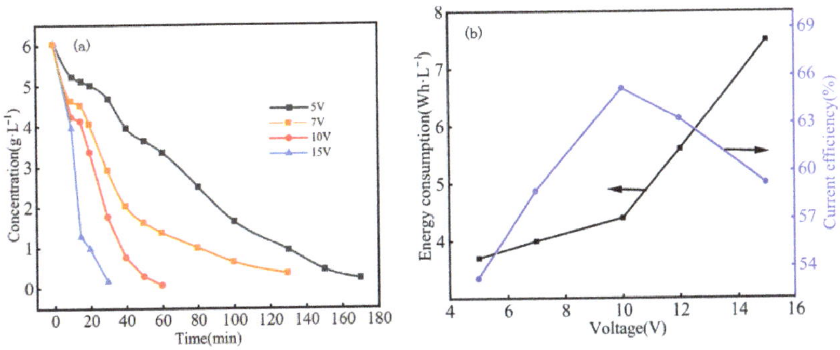

Figure 10. (**a**) Variation in Cl^- ion concentration in diluting chamber and (**b**) energy consumption and current efficiency relationship.

3. Influence of operating voltage on F^-.

Figure 11 shows that as the operating voltage gradually increases, the time required for the fluoride salt content in the diluting chamber decrease from 2.21 g·L^{-1} to below 0.5 g·L^{-1} gradually shortens. At a voltage of 15 V, the energy consumption significantly increases to 5.2 Wh·L^{-1}. The current efficiency first increases and then decreases as the operating voltage increases, reaching a maximum current efficiency of 55.4% at 10 V.

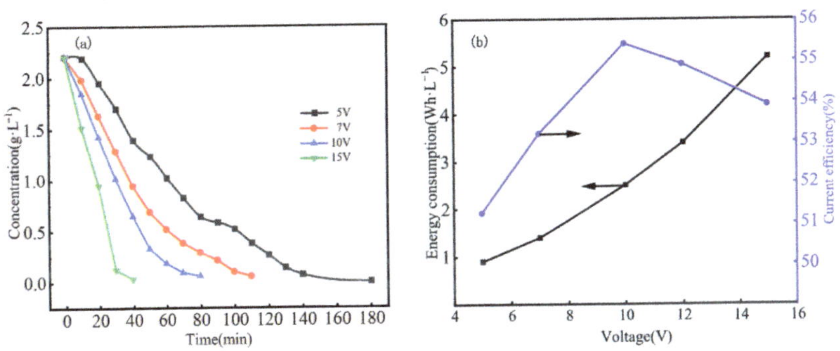

Figure 11. (**a**) Variation in F^- ion concentration in diluting chamber (**b**) and relationship between energy consumption and current efficiency.

The impact of electrodialysis on the removal of individual anions is significant. At lower voltages, the rate of desalination is comparatively slow, and energy consumption is minimal. However, as the operating voltage is elevated, the strength of the electric field increases, thereby enhancing the driving force for ion migration and mass transfer, which results in a greater number of ions traversing the ion-exchange membrane per unit time. Nonetheless, higher operating voltages necessitate increased energy to overcome resistance, leading to a corresponding rise in energy consumption. Additionally, with the increase in operating voltage, issues such as water decomposition, co-ion migration, and membrane surface scaling become more pronounced, ultimately diminishing current efficiency. Consequently, an operating voltage of 10 V has been identified as optimal.

3.2. Influence of Initial Concentration

The alteration in conductivity was investigated under an operating voltage of 10 V, with initial concentrations of SO_4^{2-}, Cl^-, and F^- in the diluting chamber solution set at 5 g·L^{-1}. The migration behaviors of the three ions within the combined solution were analyzed. In this phase of the experiment, heterogeneous ion-exchange membranes were utilized in the electrodialysis apparatus, which exhibit a pronounced steric hindrance effect on specific ion types due to their highly cross-linked structural characteristics. As illustrated in Figure 12a, the patterns of conductivity reduction during the migration of the three ions in the pure solution were largely consistent. However, when the ions coexisted in the same solution, Figure 12b reveals that SO_4^{2-} and Cl^- migrated to the enriched chambers more rapidly, while the migration of F^- significantly increased following the majority of SO_4^{2-} and Cl^- migrating to the enriched chamber. The migration of ions is driven by the electric field and concentration gradients, with divalent ions such as SO_4^{2-} occupying a greater number of ion-exchange sites compared to the monovalent Cl^-. Consequently, the migration of SO_4^{2-} surpasses that of Cl^-. The monovalent F^- ions demonstrate relatively lower removal efficiency during electrodialysis, attributed to their smaller ionic radius and higher solubility. Thus, the order of anion migration in electrodialysis utilizing heterogeneous membranes is established as $SO_4^{2-} > Cl^- > F^-$.

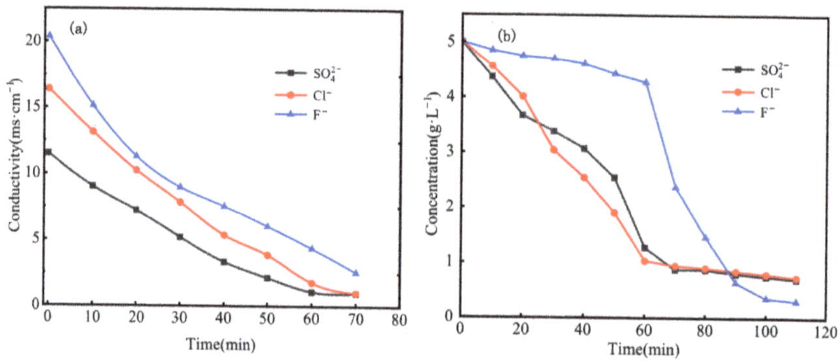

Figure 12. (**a**) Conductivity of pure solution. (**b**) Ion concentration of mixed solution V.

3.3. Influence of the Initial pH Value

The charge, degree of ionization, and the extent of bound water associated with ions are all affected by the pH of the solution. Figure 13 illustrates the influence of pH on the migration of SO_4^{2-}, Cl^-, and F^-. It is apparent that pH significantly impacts the migration of SO_4^{2-} and F^-, while it does not affect the migration of Cl^-. The removal rate of F^- under acidic conditions increases with rising pH, whereas the opposite trend is observed in alkaline conditions. In alkaline environments, the migration of F^- is markedly diminished due to competition from OH^- for mass transfer, which leads to the conversion of charged F^- into HF molecules. The maximum removal rate of SO_4^{2-} occurs at a pH of 8.

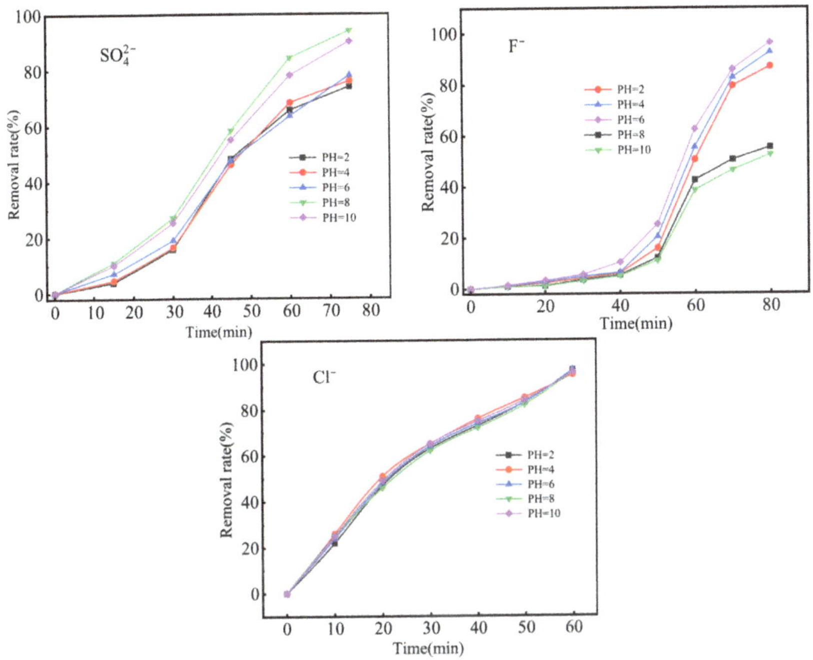

Figure 13. Effect of pH on the efficiency of SO_4^{2-}, Cl^-, and F^- removal.

3.4. Influence of Current Density

A clearance rate of 95.7% can be attained on an industrial scale by reducing the ion concentration to $1g \cdot L^{-1}$ under four current densities, as seen in Figure 14a, which is influenced by the current effect and concentration diffusion. At a fixed operating current density, the current efficiency falls as the current density drops in the same solution as seen in Figure 14b. The current efficiency will also drop because ion migration will transport some water molecules over the ion-exchange membrane, lowering the acidity of the solution in the concentrated water chamber. The current efficiency is in opposition to the energy consumption. Energy consumption increases as current efficiency decreases.

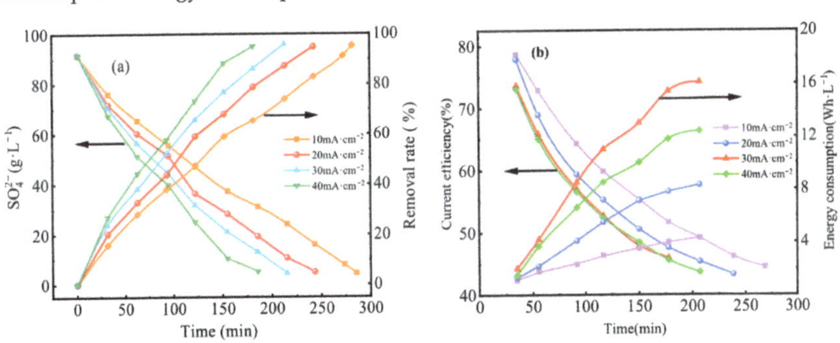

Figure 14. Different current density (**a**) removal rate and (**b**) current efficiency and energy consumption.

3.5. Influence of the Volume Ratio of the Enriched/Diluting Chambers

In the electrodialysis concentration experiment, the objective is to maximize the recovery of the original solution within the diluting chamber. This is achieved by varying the proportions of high-salt water introduced into both the diluting chamber and the enriched chamber, thereby facilitating a higher concentration of the solution. It is observed that, prior to reaching the maximum concentration limit of electrodialysis, an increased volume of solution in the diluting chamber correlates with a greater recovery of the original solution. By manipulating the volume of the diluting chamber, researchers can alter the volume ratio of the solutions in both the concentrated and diluting chambers, thereby investigating the concentration effects of electrodialysis across different volume ratios. Figure 15 shows the change in concentration with time under different volume ratios of the enriched/diluting chambers. In the circulation mode, the initial concentrations of the SO_4^{2-} solutions in the concentrated and dilute chambers are both 10.16 $g \cdot L^{-1}$. As the initial volume of the Na_2SO_4 solution in the dilute chamber increases continuously, the concentration change in the dilute chamber is relatively large, while that in the concentrated chamber is relatively small. The reason is that the concentration difference between the two sides of the membrane gradually increases, and the high concentration gradient leads to the reverse diffusion of ions from the enriched chamber to the diluting chamber.

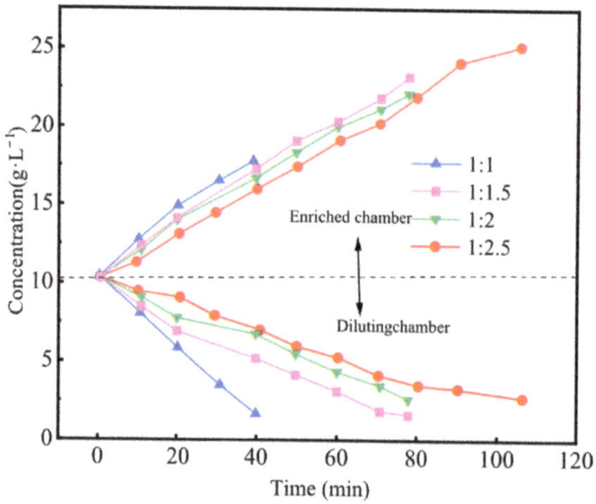

Figure 15. Effect of different volume ratios on the desalination performance of Na_2SO_4 solution.

3.6. *Influence of the Membrane*

Heterogeneous and homogeneous membranes were selected for the experiment. The migration of Cl^-, F^-, and SO_4^{2-} is shown in Figure 16. In the heterogeneous membrane, SO_4^{2-} has the fastest migration rate due to its high concentration, followed by Cl^-. In addition, because the fixed groups and active ions in the heterogeneous membrane have a strong affinity and selective permeability for SO_4^{2-}, it is more conducive to the migration of SO_4^{2-}.

Compared with the heterogeneous membrane, in the homogeneous membrane experiment, the migration rate of SO_4^{2-} slows down, while that of Cl^- speeds up, indicating that the homogeneous membrane has a higher selectivity for Cl^-. For F^-, regardless of the type of membrane, its migration rate is the lowest. The main reason is that F^- has the highest hydration free energy, which hinders the migration of F^-. Therefore, when using a heterogeneous membrane, the order of ion migration rates is $SO_4^{2-} > Cl^- > F^-$; when using a homogeneous membrane, the order is $Cl^- > SO_4^{2-} > F^-$.

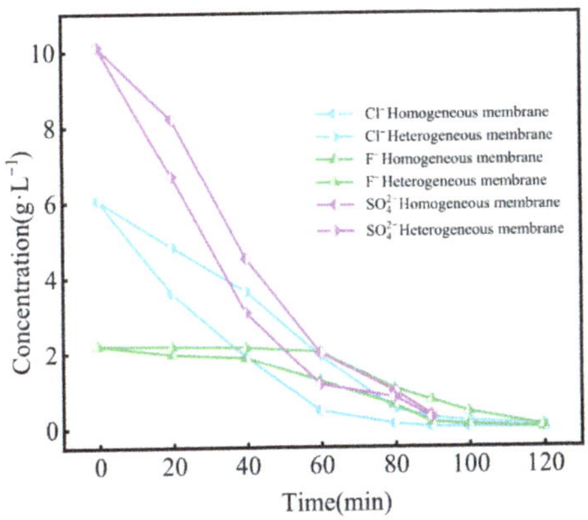

Figure 16. Migration efficiency of Cl^-, F^-, and SO_4^{2-} under different membranes.

3.7. Influence on Water Migration

Initial research on the water migration phenomenon during the electrodialysis desalination process has indicated that a portion of the water traverses the ion-exchange membrane [26–28]. This migration primarily occurs in the forms of osmotic water and electro-osmotic water. The predominant factor influencing the separation process in electrodialysis is electro-osmotic water [29]. This type of water movement involves the transport of certain water molecules in a hydrated state alongside ions, driven by the application of an electric field. Consequently, the effects of voltage and initial concentration on the water migration process were systematically examined. Solutions containing 25 g·L^{-1} SO_4^{2-} were prepared in the diluting chamber, and the apparatus was operated in a cyclical manner for a duration of 120 min.

1. Effect of voltage on water migration

As illustrated in Figure 17, the data presents the volume of water migration from the diluting chamber to the enriched chamber across various operating voltages. The findings indicate a progressive increase in water migration correlating with elevated voltage levels. Notably, a greater voltage differential results in a more pronounced increase in water migration. This phenomenon can be attributed to the fact that an increase in voltage enhances the driving force for ion migration, thereby accelerating the formation of bound water with ions and subsequently augmenting the overall volume of water migration.

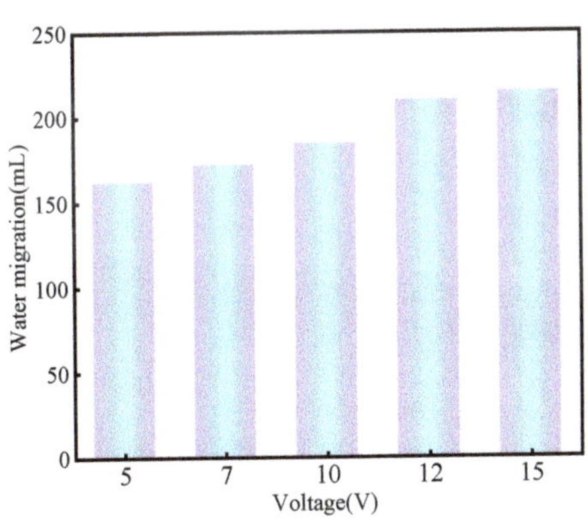

Figure 17. Water migration at different voltages.

2. The influence of initial concentration on water migration

As shown in Figure 18, the water migration amount increases with the increase in the initial concentration in the diluting chamber. When the initial concentration is above 10 g·L^{-1}, the increase in the water migration amount is obvious.

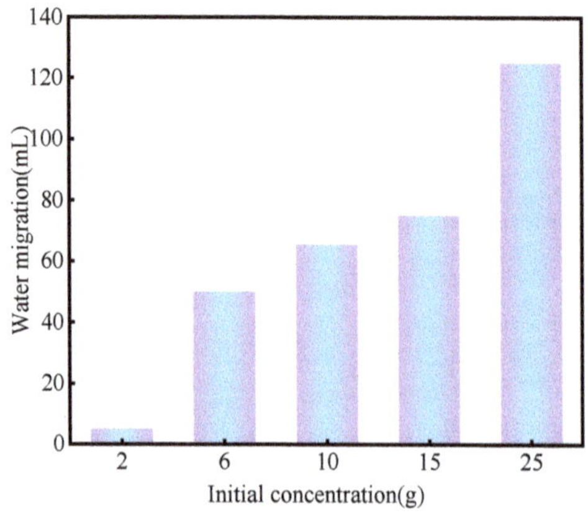

Figure 18. Water migration at different initial salt concentrations.

3.8. Effect on Amines

To investigate the impact of HSS concentration on the depletion of amine solution, MEA solutions with HSS mass fractions of 1%, 2%, and 3% were prepared. A 30 wt.% MEA solution was simulated for comparison. The process employed constant-pressure desalination, with a voltage maintained at 10 V. The solution was circulated through the electrodialysis apparatus at a flow rate of 1.16 L·min^{-1} for a duration of 180 min. Samples from both the chambers were collected at regular intervals for analysis.

The removal rates of anions and the concentration of amine were subsequently calculated, with amine loss determined based on the amine content in the diluting chamber. As indicated in Table 3, a significant increase in amine loss was observed with rising salt concentrations, suggesting that the HSS impede the purification process of the amine solution and adversely affect its recyclability.

Table 3. Mobility and MEA amine loss for different salt concentrations.

Salt Concentration (%)	Salt Concentration at the End (%)	Removal Rate (%)	Amine Concentration at the End (%)	Amine Loss (%)
1	0.19	81.0	30.25	16.70
2	0.43	78.5	30.47	17.13
3	0.77	74.3	31.04	19.25

The depletion of MDEA is also influenced by both electro-osmosis and concentration-gradient osmosis. Given that the initial amine originates from a lean amine solution, the removal of HSSs from the solution during the electrodialysis process inevitably leads to the diffusion of amine molecules from the diluting chamber to the enriched chamber. Consequently, this study primarily focused on the impact of the MDEA amine concentration gradient on the electrodialysis process. The findings indicate that the migration effect is optimized at a concentration of 30 wt.%, resulting in a reduced loss of amine. This observation is corroborated by the data presented in Table 4; it is consistent with the concentration of acidic gas removed by MDEA.

Table 4. Ion mobility and amine loss at different MDEA amine concentrations.

MDEA Concentration (%)	Cl^- Mobility (%)	SO_4^{2-} Mobility (%)	Amine Concentration at the End (%)	Amine Loss (%)
30	93.84	85.15	21.34	20.15
40	90.15	82.47	40.12	26.60
50	88.27	79.27	50.27	26.40

4. Industrial Case Analysis

The electrodialysis technique was employed to investigate and assess a genuine sample of the lean amine solution sourced from the sodium nitrite production facility of Jinchang Company. The results of the experiments are presented in Table 5. Prior to the purification process utilizing the electrodialysis apparatus, the effective loading of the amine solution was relatively low, while its HSS content was notably high. During the purification process, there was a significant reduction in the HSS content of the amine solution, indicating a considerable increase in effective loading. The purification effect likewise attained the level of a new absorbent and was readily apparent from appearance.

Table 5. Comparison of electrodialysis results of lean amine solution.

Inspection Items	Before Desalination	After Desalination	Removal Rate (%)	Inspection Standards
Amine concentration (%)	30	32.48	-	GB/T31589-2015 [30]
Amine loss (%)	-	18.7	-	-
Thermally stable salt content (%) (standard content < 1%)	6.24	0.65	89.6%	GB/T9725-2007 [31]
pH	3.68	3.63	-	GB/T6920-2019 [32]
Conductivity ($mS \cdot cm^{-1}$)	53.3	1.32	-	GB/T11007-2008 [33]
Fluoride ($mg \cdot L^{-1}$) (standard content < 10)	18.4	1.54	91.6%	SY/T7001-2014 [34]
Chloride ($mg \cdot L^{-1}$) (standard content < 500)	196.7	46.47	76.4%	SY/T7001-2014 [34]
Sulfate ($mg \cdot L^{-1}$) (standard content < 500)	92,680	75.82	99%	SY/T7001-2014 [34]
Sodium ion ($mg \cdot L^{-1}$) (empirical requirement < 500)	15,175	448	97%	SY/T7001-2014 [34]
Appearance	Yellowish-green	Pale yellow	Obvious	Visual inspection

5. Conclusions

An experimental investigation was carried out to examine the process of electrodialysis desalination in conjunction with lean amine solutions. The study involved the simulation of salt-laden solutions as well as the analysis of actual lean amine solution samples. Various parameters were assessed, including operating voltage, concentration of raw materials, value, current density, the volume ratio of concentrated to dilute chambers, and the characteristics of the membranes, all of which influence the electrodialysis regeneration process. The findings indicated the following:

1. An operating voltage of 10 V was identified as the optimal condition in terms of energy consumption and current efficiency. Under equivalent concentration conditions, Cl^- and SO_4^{2-} exhibited preferential migration. The migration of Cl^- was found to be

independent of the value, while the migration of F^- peaked at a value range of 4 to 6. The migration of SO_4^{2-} reached its maximum at a value of 8. Current efficiencies for SO_4^{2-}, Cl^-, and F^- under high current density conditions were recorded at 68.12%, 65.2%, and 55.4%, respectively. Correspondingly, the energy consumptions at the limiting voltage were measured at 11.2 $Wh \cdot L^{-1}$, 7.51 $Wh \cdot L^{-1}$, and 5.2 $Wh \cdot L^{-1}$.

2. Water permeation often leads to a decrease in current efficiency. The water migration amount gradually increases with the increase in voltage and the initial concentration of the freshwater chamber. The concentration of the diluting chamber varies greatly under different volume ratios of enriched/diluting chambers, and the ion removal rate in the freshwater chamber is greater than 98%. The order of ion migration rates for homogeneous membranes is $Cl^- > SO_4^{2-} > F^-$, and the order of ion migration rates for heterogeneous membranes is $SO_4^{2-} > Cl^- > F^-$.
3. Thermally stable salts play a role in deactivating and contributing to the losses of organic amine solutions during the carbon dioxide capture process. Research indicates that the electrodialysis method leads to an amine loss exceeding 15%. Among the various formulations, a 30 wt.% MEA lean amine solution exhibits the lowest amine loss at 16.7%, and its regeneration concentration is the most closely aligned with the concentration of the removed acid gas.

As a result, electrodialysis technology can efficiently extract sulfates from organic amines in addition to purifying and regenerating the lean amine solution. However, more tests are required to confirm the economic costs and operating impacts.

Author Contributions: Conceptualization, D.W. and Q.L.; methodology, D.W.; software, L.L., Q.L., and S.Z.; validation, G.C.; formal analysis, G.L. and H.L.; investigation, Q.L. and G.C.; resources, S.Z. and L.L.; writing—original draft preparation, Q.L.; writing—review and editing, D.W.; supervision, D.W.; project administration, D.W.; funding acquisition, G.C. All authors have read and agreed to the published version of the manuscript.

Funding: The authors would like to give thanks for the financial support received from the National Natural Science Foundation of China (No. 22268028), the Science and Technology Program of Gansu Province (No. 24ZDGF001), and the Program of Lanzhou Science and Technology Bureau (No. 2024-RC-6).

Data Availability Statement: The data presented in this study are available on request from the first author.

Conflicts of Interest: Author Guang Chen was employed by the Gansu Yinguang Juyin Chemical Co., Ltd. Author Li liu was employed by the Hefei Sanyuan Engineering & Technology Service Co., Ltd. Author Shilong Zhang was employed by the Gansu Odewangs Material Technology Co., Ltd. The remaining authors declare that the research was conducted in the absence of any commercial or financial relationships.

Nomenclature

HSS	Heat-stable salts
AEM	Anion-exchange membrane
CEM	Cation-exchange membrane
SEM	Scanning electron microscopy
MEA	Monoethanolamine
MEDA	Methyldiethanolamine
FTIR	Fourier transform infrared spectroscopy
TISAB	Total ionic strength adjustment buffer
J	Average flux

V_t	Volume of the solution in the diluting chamber at time t
V_0	Volume of the solution in the diluting chamber at time 0
C_t	Concentration at time t
C_0	Concentration at time 0
A_t	Effective membrane area (0.55 m^2)
N	Number of repeating units
f	Desalination rate
V_{d0}/V_{d2}	Initial and final volumes of diluting
C_{d0}/C_{d2}	Initial and final salt concentrations in the diluting chamber
W	Energy consumption of the electrodialysis process
U	Voltage drop across the membrane in the electrodialyzer
n	Molar amount of the regenerated amine solution
I	Current
η	Current efficiency
Z	Charge number of the ion
C_t'/C_0'	Molar concentrations at time 0 and t
F	Faraday constant
ω_0/ω_t	Initial mass fraction and that at time t of the amine solution in the diluting chamber

References

Wang, S. A brief analysis of the influence of heat-stable salts on amine liquid desulfurization units. *Petrochem. Technol. Econ.* **2016**, *32*, 46–49. [CrossRef]

Gao, J.; Wang, S.; Wang, J.; Cao, L.; Tang, S.; Xia, Y. Effect of SO_2 on the amine-based CO_2 capture solvent and improvement using ion exchange resins. *Int. J. Greenh. Gas Con.* **2015**, *37*, 38–45. [CrossRef]

Golubeva, I.; Dashkina, A.; Shulga, I. Demanding problems of amine treating of natural gas: Analysis and ways of solution. *Pet. Chem.* **2020**, *60*, 45–50. [CrossRef]

Bazhenov, S.; Rieder, A.; Schallert, B.; Vasilevsky, A.; Unterberger, S.; Grushevenko, E.; Volkov, V.; Volkov, A. Reclaiming of degraded MEA solutions by electrodialysis: Results of ED pilot campaign at post-combustion CO_2 capture pilot plant. *Int. J. Greenh. Gas Con.* **2015**, *42*, 593–601. [CrossRef]

Dumée, L.; Scholes, C.; Stevens, G.; Kentish, S. Purification of aqueous amine solvents used in post combustion CO_2 capture: A review. *Int. J. Greenh. Gas. Con.* **2012**, *10*, 443–455. [CrossRef]

Tavan, Y.; Moradi, M.; Rostami, A.; Azizpou, H. Theoretical and industrial aspects of amine reclaiming unit to separate heat stable salts. *Sep. Purif. Technol.* **2019**, *237*, 116314. [CrossRef]

Cai, P.; Wang, S.; Han, Y. An experimental study on MDEA regeneration by membrane distillation. *Nat. Gas. Chem. Ind.* **2009**, *34*, 25–29. [CrossRef]

Edathil, A.; Pal, P.; Banat, F. Alginate clay hybrid composite adsorbents for the reclamation of industrial lean methyldiethanolamine solutions. *Appl. Clay Sci.* **2018**, *156*, 213–223. [CrossRef]

Yan, X.; Wang, J.; Zhang, X. Research on the revival technology of CT-FSC alcohol amine desulfurization solution. *Chem. Eng. Oil Gas.* **2018**, *47*, 1–9. [CrossRef]

10. Garg, B.; Verheyen, T.; Pearson, P.; Feron, P.; Cousins, A. A technology review for regeneration of sulfur rich amine systems. *Int. J. Greenh. Gas. Con.* **2018**, *75*, 243–253. [CrossRef]

11. Zhao, J.; Zhou, A.; Chen, Y.; Zheng, J.; Ge, T. Current status of energy consumption of adsorption CO_2 direct air capture. *CIESC Journal* **2025**, *76*(4), 1375–1390. [CrossRef]

12. Morgan, C.; Klare, T. Chloride removal from DEA by ion exchange. In *Gas Conditioning Conference Proceedings*; University of Oklahoma: Norman, OK, USA, 1977; Available online: https://ac.huohua.me/ (accessed on 1 January 2025).

13. Jameh, A. Amine solution recovery package and controlling corrosion in regeneration tower. *World Acad. Sci. Eng. Technol.* **2010**, 45, 107–110. Available online: https://www.researchgate.net/publication/292746394_Amine_solution_recovery_package_and_controlling_corrosion_in_regeneration_tower (accessed on 1 January 2025).

14. Abdur, R.; Abdur, R.; Anthony, G. Design considerations in the development and application of microdisc electrode arrays (MDEAs) for implantable biosensors. *Biomed. Microdevices* **2009**, *4*, 36–39. [CrossRef]

15. Yan, N.; Wu, J.; Hua, L.; Gao, T.; Liu, R.; Lu, Z. Fabrication of highly dispersed aramid nanofibers based on cation-exchange resin as a proton donor: Low energy consumption, solvent closed-loop recovery. *Langmuir* **2025**, *41*, 20105–20115. [CrossRef] [PubMed]

16. Meng, H.; Zhang, S.; Li, C.; Li, L. Removal of heat stable salts from aqueous solutions of N-methyldiethanolamine using specially designed three-compartment configuration electrodialyzer. *J. Membr. Sci.* **2008**, *322*, 436–440. [CrossRef]
17. Zhou, Z.; Lin, Y.; Jin, Y.; Guan, K.; Yu, J. Removal of heat-stable salts from lean amine solution using bipolar membran electrodialysis. *J. Membr.* **2022**, *645*, 120213. [CrossRef]
18. Grushevenko, E.A.; Bazhenov, S.D.; Vasilebskii, V.P.; Novitskii, E.G.; Volkov, A.V. Two-step electrodialysis treatment of m noethanolamine to remove heat stable salts. *Russ. J. Appl. Chem.* **2018**, *91*, 602–610. [CrossRef]
19. Volkov, A.; Vasilevsky, V.; Bazhenov, S.; Volkov, V.; Riede, A.; Unterberger, S.; Schallert, B. Reclaiming of monoethanolamin (MEA) used in post-combustion CO_2-capture with electrodialysis. *Energy Procedia* **2014**, *51*, 148–153. [CrossRef]
20. Huang, C.; Xu, T.; Jacobs, M. Regenerating flue-gas desulfurizing agents by bipolar membrane electrodialysis. *AIChE J.* **2005**, 5 393–401. [CrossRef]
21. Wei, X.; Wang, Y.; Yan, H.; Wu, K.; Xu, T. Purification of methylsulfonylmethane from mixtures containing salt by convebtion electrodisalysis. *Membranes* **2020**, *10*, 23. [CrossRef]
22. Wang, X.; Wang, Y.; Zhang, X.; Feng, H.; Xu, T. In-situ combination of fermentation and electrodialysis with bipolar membrane for the production of lactic acid: Continuous operation. *Bioresour. Technol.* **2013**, *147*, 442–448. [CrossRef] [PubMed]
23. Li, J.; Morthensen, S.; Zhu, J.; Yuan, S.; Wang, J.; Volodine, A.; Lin, J.; Shen, J.; Bruggen, B. Exfoliated MoS_2 nanosheets loade on bipolar exchange membranes interfaces as advanced catalysts for water dissociation. *Sep. Purif. Technol.* **2018**, *194*, 416–42 [CrossRef]
24. Qiao, J.; Sun, Y.; Song, X.; Yu, J. Preparation of HBr and NaOH by Bipolar Membrane Electrolysis from Absorbed Complete Flui of Bromine Extraction. *J. East China Univ. Sci. Technol.* **2019**, *45*, 1–8. [CrossRef]
25. Kohl, A.L. *Gas Purification*; Gulf Pub. Co., Book Division: Houston, TX, USA, 1985.
26. Schmid, G.; Schwarz, H. Electrochemistry of capillary systems with narrow pores III. Electrical conductivityl. *Phys. Chem.* **195** *55*, 229–237. [CrossRef]
27. Oda, Y.; Yawataya, T. On the electro-osmotic water transport through cation-exchange resin membranes. *Bull. Chem. Soc. Jp* **1955**, *28*, 263–269. [CrossRef]
28. Strathmann, H. *Ion-Exchange Membrane Separation Processes*; Elsevier: Amsterdam, The Netherlands, 2004; ISBN 978008050940
29. Gurreri, L.; Tamburini, A.; Cipollina, A.; Micale, G. Electrodialysis applications in wastewater treatment for environment protection and resources recovery: A systematic review on progress and perspectives. *Membranes* **2020**, *10*, 146. [CrossRe [PubMed]
30. *GB/T31589-2015*; Analytical Method of Chemical Composition for Activated MDEA Decarbonization Desulfurization Agen National Library of Standards: Beijing, China, 2015.
31. *GB/T9725-2007*; Chemical Reagent-General Rule for Potentiometric Titration. National Library of Standards: Beijing, China, 200
32. *GB/T6920-2019*; Water Quality—Determination of pH Value—Glass Electrode Method. National Library of Standards: Beijin China, 2019.
33. *GB/T11007-2008*; Test Method of Electric Conductivity Analyzers. National Library of Standards: Beijing, China, 2008.
34. *SY/T7001-2014*; Analysis of Anion Composition of Thermally Stable Salts in Urea Desulfurization Solution by Ion Chromatograph National Petroleum Corporation: Beijing, China, 2014.

MDPI AG
Grosspeteranlage 5
4052 Basel
Switzerland
Tel.: +41 61 683 77 34

Processes Editorial Office
E-mail: processes@mdpi.com
www.mdpi.com/journal/processes

www.ingramcontent.com/pod-product-compliance
Lightning Source LLC
Chambersburg PA
CBHW061810210926
51850CB00006B/660
* 9 7 8 3 7 2 5 8 5 9 3 5 1 *